American Defense Preparedness Association
presents in cooperation with the
American Institute of Aeronautics &Astronics
American Society of Mechanical Engineers
Societyof Photo-Optical Insturmentation Engineers

An International Symposium & Exhibition on Active Materials & AdaptiveStructures

LIST OF ATTENDEES

November 4-8, 1991
at the Radisson Mark Plaza Hotel
Alexandria,Virginia

Lt. Gregory S. Agnes
USAF-Wright Laboratory
WL/FIBGC
Wright Patterson AFB OH 45433

Dr. Thomas E. Alberts
Old Dominion University
Mechanical Engineering Dept
Norfolk VA 23529-0247

Dr. Karen Albrecht
Martin Marietta
103 Chesapeake Park Plaza
MS E-420
Baltimore MD 21220

Dr. Stephen W. Allison
Oak Ridge National Lab
Applied Technology Div.
PO Box 2003
Oak Ridge TN 37831-7280

Dr. T. R. Alt
McDonnell Douglas Space Systems C
5301 Bolsa Avenue
Huntington Beach CA 92647

Mr. Ivan A. Amato
Science/AAAS
1333 H Street NW
Washington DC 20005

Mr. George G. Amir
NKF Engineering, Inc.
4200 Wilson Boulevard
Suite 900
Arlington VA 22203-1800

Mr. Eric H. Anderson
MIT
Bldg. 37-351, Rm. 37-331
Cambridge MA 02139

Mr. Jeffrey P. Andrews
Martin Marietta
103 Chesapeake Park Plaza
MS E-455
Baltimore MD 21220

Dr. Om Arora
David Taylor Research Center
Code 2812
Annapolis MD 21402

Dr. Farhang Aslani
Automated Analysis Corporation
2805 S. Industrial Highway
Suite K-100
Ann Arbor MI 48105-6791

Mr. Allan C. Aubert
Atlantic Applied Research Corpora
4 A Street
Burlington MA 01803

Mr. Fred Austin
Grumman Corporate Research Ctr
111 Stewart Avenue
Bethpage NY 11714

Mr. Johann Bals
German Aerospace Research Estab.
Institute of Flight Systems
Dynamics
Oberpfaffenhofen D8031 Federal Re

Dr. H. T. Banks
University of Southern California
MS 113, DRB 155
Los Angeles CA 90089-1113

Dr. X. Q. Bao
Penn State University
Dept Engrng Science & Mechancs
Materials Research Lab
College Park PA 16802

Dr. Roshdy S. Barsoum
Department of the Navy
Office Chief of Naval Research
Code 1132SM
Arlington VA 22217-5000

Dr. Dean Batha
Fiber Materials, Inc.
5 Morin Street
Biddeford Industrial Park
Biddeford ME 04005

J. G. Batterson
NASA Langley Research Center
MS: 231
Hampton VA 23665

Prof. Amr Baz
Catholic University
620 Michigan Avenue NE
Dept. Mechinical Engineering
Washington DC 20064

Mr. Michael C. Beale
Newport News Shipbuilding
4101 Washington Avenue
E23-B600
Newport News VA 23607

Mr. Charles Beauchamp
Naval Underwater Systems Ctr
Code 8214
Newport RI 02841

Dr. Andrew J. Bell
Ecole Polytechnique Federale
Laboratoire de Ceramique
MX-D Edublens
1015 Lausanne Switzerland

Mr. Terry C. Bell
Lockheed Aeronautical Systems Co.
86 South Cobb Drive
D/73-73, Zone 0160
Marietta GA 30063-0160

Dr. William H. Bennett
Techo-Sciences Incorporated
7833 Walker Drive
Suite 620
Greenbelt MD 20770-3211

LTC Joseph H. Beno USA
DARPA
1400 Wilson Blvd.
Arlington VA 22209-2308

Mr. Timothy A. Berkoff
Sachs/Freeman Associates
Naval Research Lab
Code 6574
Washington DC 20375

Dr. Robert Betros
TRW Electronics & Space Division
1 Space Park
Redondo Beach CA 90278

Mr. Indar L. Bhatia
The MITRE Corporation
7525 Colshire Drive
MS Z380
McLean VA 22102-3481

Dr. Andrew S. Bicos
McDonnell Douglas Space Systems C
5301 Bolsa Avenue
M/S A3-Y857-13/3
Huntington Beach CA 92647

Dr. Joseph E. Bishop
Texas A&M University
Dept Aerospace Engineering
Room 744C
College Station TX 77843

Dr. Gilmer L. Blankenship
Techno-Sciences Inc
7833 Walker Drive
Suite 308
Greenbelt MD 20770-3211

Mr. James E. Bongardt Jr.
Newport News Shipbuilding
Airport Plaza I, Suite 1100
2711 S Jefferson Davis Highway
Arlington VA 22202

Ms. Leslie J. Bowen
53 Hillcrest Road
Concord MA 01742

Mr. Keith B. Bowman
USAF
ATTN: WL/FIBAC
Wright Patterson AFB OH 45433

Dr. J. Breakwell
Lockheed Missile & Space Company
3251 Hanover Street
MS 0/92-30B/250
Palo Alto CA 94304

Ms. Kimberly A. Britton
ADPA
2101 Wilson Boulevard
Suite 400
Arlington VA 22201-3061

Mr. Russell P. Brodeur
Pennsylvania State University
259 Materials Research Lab
University Park PA 16802

Dr. Alan Bronowicki
TRW Electronics & Space Division
One Space Park
Redondo Beach CA 90278

Mr. Richard Brown
ATOCHEM
3 Parkway
Philadelphia PA 19102

Ms. Jane Brumfield
ATOCHEM
Three Parkway
Philadelphia PA 19102-1321

Dr. A. P. Buckley
NASA
Marshall Space Flight Center
Huntsville AL 35812

Dr. Leonard J. Buckley
Naval Air Development Center
Polymar Physics Division
Code 6064
Warminster PA 18974

Dr. Steven J. Bullock
Purdue University
Sch Aeronautics & Astronautics
West Lagayette IN 47907

Mr. Robert Bupp
University of Maryland
Dept of Mechanical Engineering
College Park MD 20742

Dr. Shawn E. Burke
Charles Stark Draper Lab Inc.
555 Technology Square
Mail Stop 7C
Cambridge MA 02139

Prof. John Burns
VPI & SU
648 Whittemore Hall
Blacksburg VA 24061

Mr. Alper K. Caglayan
Charles River Association
55 Wheeler Street
Cambridge MA 02138

Dr. Patrick J. Cain
MTS Systems
14000 Technology Drive
Eden Prairie MN 55344

Dr. Hengchu Cao
University of California
College of Engineering
Santa Barbara CA 93106

Mr. Gregory T. Carman
Virginia Tech
225 North Hall
ESM Department
Blacksburg VA 24061-0219

Mr. Robert L. Carman
Rockwell International
Rocketdyne Division
6633 Canoga Avenue, MC IB47
Canoga Park CA 91303-2790

Dr. Thomas G Carne
Sandia National Labs
PO Box 5800
Org 2743
Albuquerque NM 87185

Mr. Bernie C. Carpenter
Martin Marietta
Defense Space & Communications
PO Box 179, B3085
Denver CO 80201

Prof. William N. Carr
New Jersey Institute of Technolog
323 King Boulevard
Newark NJ 07102

Mr. Kushi L. Chandiramani
Atlantic Applied Research Corp.
4 A Street
Burlington MA 01803

Dr. K. Chandrashekhara
University of Missouri-Rolla
Dept of Mechanical Engineering
Rolla MO 65401

Mr. Mark Chase
Potomac Research Incorporated
6121 Lincolnia Road
Alexandria VA 22312

Dr. C. S. Chen
Yale University
4305 Oldfield Drive
Arlington TX 76016

Mr. Peter Chen
University of Maryland
Dept of Mechanical Engineering
College Park MD 20742

Mr. C. C. Cheng
University of Maryland
Dept of Mechanical Engineering
College Park MD 20742

Mr. Liang-Chin Chin
Penn State University
CEEAM
227 Hammond Building
University Park PA 16802

Dr. Inderjit Chopra
University of Maryland
Dept of Aerospace Engineering
Room 0151 Building 088
College Park MD 20742-3015

Dr. Cheng-Chih Chu
Jet Propulsion Laboratory
4800 Oak Grove Drive
MS 198-326
Pasadena CA 91109

Dr. Y. T. Chung
McDonnell Douglas Space Systems C
5301 Bolsa Avenue
Huntington Beach CA 92647

Dr. Arthur E. Clark
Naval Surface Warfare Center
10901 New Hampshire Avenue
Code R43
Silver Spring MD 20903-5000

Mr. Ron L. Clark
Virginia Tech.
Dept. Mechanical Engineering
Blacksburg VA 24061-0238

Mr. Allan H. Clauer
Battelle Memorial Institute
505 King Avenue
Columbus OH 43201-2693

Prof. Richard O. Claus
Virginia Tech
Dept of Electric Engineering
648 Whittemore Hall
Blacksburg VA 24061

E. M. Cliff
VPI & SU
648 Whittemore Hall
Blacksburg VA 24061

Mr. Jeffrey D. Cohen
Du Pont Compnay
Fibers Department
PO Box 80702
Wilmington DE 19880-0702

Mr. Joe Colvin
Virginia Polytechnic Institute
Mechanical Engineering Dept
ATTN: H.H. Cudney
Blacksburg VA 24061-0238

Ms. Teresa R. Colvin
ADPA
2101 Wilson Boulevard
Suite 400
Arlington VA 22201-3061

Dr. Joseph A. Corrado
David Taylor Research Center
Code 1702
Bethesda DC 20084-5000

Lt. Derek F. Cossey
Phillips Laboratory
OLAC PL/STSS
Edwards AFB CA 93523

D. E. Cox
NASA
MS: 231
Hampton VA 23665

Dr. Edward F. Crawley
MIT
Space Eng Research Center
37-341
Cambridge MA 02193

Mr. Erik Cross
Penn State University
Materials Research Lab
Room 167
University Park PA 16802

Dr. C. Robert Crowe
DARPA
1515 Wilson Boulevard
Code 6320 Suite 705
Arlington VA 22209

Mr. Andrew Crowson
Army Research Office
Materials Science Division
PO Box 12211
Durham NC 27709-2211

Dr. Harley H. Cudney
Virginia Polytechnic Institute
Mechanical Engineering Dept
ATTN: H.H. Cudney
Blacksburg VA 24061-0238

Dr. Shirley D. Darrah
Fiber Materials, Inc.
5 Morin Street
Biddeford Industrial Park
Biddeford ME 04005-4497

Dr. Alok Das
Phillips Lab (AFSC)
Structures & Control Division
Edwards AFB CA 93523

Dr. Abhijit Dasgupta
University of Maryland
104 John Street
Perryville MD 21903-2630

Dr. R. Davidson
AEA Technology
Harwell
Oxfordshire OXII ORA United Kingd

Mr. Matthew A. Davies
Cornell University
105 Upson Hall
Mechanical & Aerospace Engring
Ithaca NY 14853

Mr. Christopher L. Davis
Penn State University
Dept. Aerospace Engineering
233 Hammond Building
University Park PA 16802

Mr. Hugh W. Davis
Ball Electro-Optics
Cryogenic Division
PO Box 1062
Boulder CO 80306-1062

Mr. L. Porter Davis
Honeywell Incorporated
PO Box 52199
M/S 2513
Phoenix AZ 85072-2199

Mr. Michael W. Dean
Ball Space Systems Division
MS-CO-10
PO Box 1062
Boulder CO 80303

Dr. Pete D. Dean
Lockheed Corporation
Dept. ORE93-60, Bldg 255
1111 Lockheed Way
Sunnyvale CA 94089-1212

Dr. Marc DeGraef
University of California-Santa Ba
College of Engineering
Materials Department
Santa Barbara CA 93106

Dr. Gary L. Denman
DARPA
1400 Wilson Boulevard
Arlington VA 22209

Dr. William Ditto
Naval Surface Warfare Center
10901 New Hampshire Avenue
Code R43
Silver Spring MD 20903-5000

Mr. Scott W. Doebling
Purdue University
Department of AAE
West Lafayette IN 47907

Mr. Ted A. Doederlein
U.S. Air Force
OLAC PL/STSS
Edwards CA 93523-5000

Dr. Richard L. Donovan
APTEK
1275 Lake Plaza Drive
Colorado Springs CO 80906

Dr. Jeffrey J. Dosch
SUNY at Buffalo
Dept. Mechanical Engineering
321 Jarvis Hall
Buffalo NY 14260

Dr. Joseph P. Dougherty
Penn State University
144 Materials Research Lab.
State College PA 16802

Mr. Bruce Douglas
University of Maryland
Electrical Engineering Dept
College Park MD 20742

Mr. James F. Dray
Mechanical Technology Incorporate
2361 Jefferson Davis Highway
Suite 1110
Arlington VA 22202-3876

Dr. Mark A. Druy
Foster Miller Incorporated
350 Second Avenue
Waltham MA 02154

Dr. Carolyn M. Dry
University of Illinois
1304 West Greet Street
106 Architecture Bldg. MC624
Urbana IL 61801

Dr. Pieter S. Dubbelday
Naval Research Laboratory
PO Box 568337
Code 5993
Orlando FL 32856-8337

Mr. Travis V. Dubois
Old Dominion University
Mechanical Engineering Dept
Norfolk VA 23529-0247

Dr. John V. Dugan Jr.
Cortana Corporation
520 N Washington Street
Suite 200
Falls Church VA 22046-3549

Mr. Huy T. Duong
Old Dominion University
Mechanical Engineering Dept
Norfolk VA 23529-0247

Dr. Mark Ealey
Litton/Itek Optical Systems
10 Maguire Road
Lexington MA 02173-3199

Dr. Donald L. Edberg
McDonnell Douglas Space Systems C
5301 Bolsa Avenue
M/S A3-Y857-13-3
Huntington Beach CA 92647-2048

Dr. S. L. Ehlers
Naval Avionics Center
6000 E. 21st St.
Indianapolis IN 46219-2189

MG William E. Eicher USA (Ret.)
ADPA
2101 Wilson Boulevard
Suite 400
Arlington VA 22201-3061

Dr. Azim Eskandarian
Penn State University
1031 Edgecomb Avenue
York PA 17403

Mr. Michael F. Evans
SFA Incorporated
1401 McCormick Drive
Landover MD 20785

Dr. Mark S. Ewing
Wright Lab
Wright Patterson AFB
Dayton OH 45433-6553

Dr. James A. Fabunmi
AEDAR Corporation
8401 Corporate Drive
Suite 460
Landover MD 20785

Mr. Russell Fairles
McDonnell Douglas
Dept. -357 MC 1021310
P.O. Box 516
St. Louis MO 63166

Ms. Andrienne Fenton
IOP Publishing LTD
1411 Walnut Street
Suite 200
Philadelphia PA 19102

Mr. Joseph A. Ference
General Dynamics Convair Division
PO Box 85357
MZ 41-6850
San Diego CA 92186-5357

Ms. Lorna Finch
TREK, Incorporated
3932 Saltworks Road
Medina NY 14103-9546

Mr. Shalom Fisher
Naval Research Lab
Code 8241
Washington DC 20375-5000

Mr. Robert H. Fitzgerald
Innovative Dynamics
243 Langmuir Labs
95 Brown Road
Ithaca NY 14850

Dr. Christopher C. Flanigan
SDRC
11995 El Camino Real
Suite 200
San Diego CA 92130

Dr. Anthony Fontana
NASA
Langley Research Center
Hampton VA 23665-5225

Mr. Philip A. Frank
SFA Incorporated
1401 McCormick Drive
Landover MD 20785

Mr. Bernard A. Frankpitt
University of Maryland
Systems Research Center
Building 115, Room 2164
College Park MD 20742-3311

Dr. Stephen W. Freiman
NIST
Materials Building 223, A257
Gaithersburg MD 20899-0001

Dr. P. L. Fuhr
University of Vermont
College of Engineering
Burlington VT 05405

Dr. Christopher R. Fuller
Virginia Tech
Dpt. of Mechanical Engineering
333 Norris Hall
Blacksburg VA 24061-0238

Dr. George W. Game
British Aerospace
Earth Observation/Science Div.
P.O Box 5, Filton
Bristol - England Bs12 7QW Unit

Prof. Mukesh Gandhi
Michigan State University
Dpt. of Mechanical Engineering
Room C130
East Lansing MI 4882[illegible]-1326

Dr. Ephrahim Garcia
Vanderbilt University
Smart Structures Lab.
Dept of Mechanical Engineering
Nashville TN 37202

Mr. Peter T. Gardiner
University of Strathclyde
Smart Structures Research In.
204 George Street
Glasgow Scotland G11XW United Kin

Dr. John J. Gassner
Foster-Miller Incorporated
350 Second Avenue
Waltham MA 02154

Mr. William A. Geideman
McDonnell Douglas Electronic Sys
1801 East St. Andrew Place
M/S A208
Santa Ana CA 92705

Dr. Fred G. Geil
Westinghouse Oceanic Division
P.O. Box 1488
M/S 9845
Annapolis MD 21404

Dr. Zheng Geng
George Washington University
704 Ivy Lane
Rockville MD 20850

Mr. Joseph J. Gerardi
Innovative Dynamics
244 Langmuir Labs
95 Brown Road
Ithaca NY 14850-1296

Dr. Warren C. Gibson
CSA Engineering
560 San Antonio Road
Suite 101
Palo Alto CA 94306

Dr. J. Gilheany
Catholic University
Dept of Mechanical Engineering
620 Michigan Avenue NE
Washington DC 20064

Mr. George C. Gillespie
University of Maryland
4312 Knox Road
College Park MD 10742

Mr. Tom Glennon
Anatrol Corporation
4120 Birch Street
Suite 108
Newport Beach CA 92660-2228

Mr. Davis Goldstein
10217 Deakins Hall Drive
Adelphi MD 20783

Mr. Theodore C. Goodenow
Epoch Engineering, Inc.
806 West Diamond Avenue
Suite 200
Gaithersburg MD 20878

Mr. Mel J. Goodfriend
Edge Technologies Incorporated
306 South 16th Street
Ames IA 50010-8102

W. L. Grantham
NASA Langley Research Center
MS: 231
Hampton VA 23665

Dr. Scott D. Gray
Sandia National Laboratories
PO Box 5800
Albuquerque NM 87185

Mr. David S. Greeley
Atlantic Applied Research Corpora
4 A Street
Burlington MA 01803

Mr. John L. Gubser
McDonnell Douglas Missile Systems
PO Box 516
Bldg. 304 MC: 3064025
St. Louis MO 63166

Ms. Rebeca M. Guerricagoitia
ADPA
2101 Wilson Boulevard
Suite 400
Arlington VA 22201-3061

Dr. Nesbitt Hagood
MIT
Space & Engineering Res. Ctr.
77 Massachusetts Avenue
Cambridge MA 02139

Dr. Sathya V. Hanagud
Georgia Tech
Institute of Technology
Atlanta GA 30332-0150

Ms. Tricia C. Hankinson
PRC Incorporated
1515 Wilson Boulevard
Suite 705
Arlington VA 22209

Mr. Brantley R. Hanks
NASA Langley Research Center
Mail Stop 230
Hampton VA 23665

Dr. John T. Harduvel
McDonnell Douglas Corporation
5301 Bolsa Avenue
Huntington Beach CA 92647-6237

Dr. Ronald T. Harrold
Westinghouse
1310 Beulah Road
Pittsburgh PA 15235

Dr. Timothy K. Hasselman
Engineering Mechanics Associates,
3820 Del Amo Boulevard
Suite 318
Torrance CA 90503-2155

Ms. Kristl B. Hathaway
Office of Naval Research
800 North Quincy Street
Code 1131
Arlington VA 22217

Mr. Michael J. Haun
Colorado School of Mines
Dept. of Metallurgical &
Materials Engineering
Golden CO 80401

Ms. Jennifer Heeg
NASA Langley Research Center
Code 231
Hampton VA 23365-5225

Dr. J. J. Helferty
Temple University
Dept of Electrical Engineering
Philadelphia PA 19122

Dr. John P. Henderson
Materials & Vibration Engineering
383 Ridgewood Drive
Fairborn OH 45324-4022

Mr. Chris Henry
Atochem Sensors, Inc.
950 Forge Avenue
Norristown PA 19403

Mr. Peter C. Herdic
SFA Incorporated
1401 McCormick Drive
Landover MD 20785

Ms. Gail A. Hickman
Innovative Dynamics
243 Langmuir Labs
95 Brown Road
Ithaca NY 14850-1296

Prof. Harry H. Hilton
University of Illinois
104 South Mathews
104 Transporation Building
Urbana IL 61801-2997

Dr. D. Hogg
University of Toronto
4925 Dufferin Street
Downsview Ontario M3H 5T6 Canada

W. Bruce Holt
ADPA
Two Colonial Place
2101 Wilson Blvd. Suite 400
Arlington VA 22201-3062

Mr. Suk-Yoon Hong
Penn State University
CEEAM
227 Hammond Building
University Park PA 16802

Mr. Ernst Hornung
MBB/ERNO Raumfahrttechnik GmbH
Postbox 10 59 09
Huenefeldstreet 1-5
2800 Bremen 1 Federal Rep Germany

Dr. Warren Hoskins
Lookheed Missiles & Space
PO Box 504
Dept. 62-18 Bldg. 551
Sunnyvale CA 94088-3504

Dr. Thomas R. Howarth
Naval Research Laboratory
P.O. Box 568337
Orlando FL 32856-8337

Dr. Chien Y. Huang
Grumman Corporation
MS A08-35
Bethpage NY 11021

Dr. James E. Hubbard Jr.
OPTRON Systems Incorporated
3 Preston Court
Bedford MA 01730

Mr. Gerard H. Hufstetler
US Army
AATD (AVSCOM)
Fort Eustis VA 23604-5577

Dr. David C. Hyland
Harris Corporation
MS 22/4847
Melbourne FL 32902

Dr. Roy Ikegami
Boeing Defense & Space Group
PO Box 3999
MS 82-97
Seattle WA 98124-2499

Prof. Daniel Inman
SUNY at Buffalo
Mechanical Systems Lab.
Buffalo NY 14260

Dr. Stephen Jacobsen
University of Utah
3176 Merrill Engineering Bldg
Salt Lake UT 84112

Dr. Dean Jacot
Boeing Aerospace Co.
PO Box 3999 JMS 82-24
Seattle WA 98124-2499

Mr. Robert Jacques
MIT
Aeronautics & Astronautics
77 Massachusetts Avenue
Cambridge MA 02139

Mr. Adesh Jain
University of Maryland
Dept of Mechanical Engineering
College Park MD 20742

Dr. Andrew P. Jardine
State University of New York
Dept of Materials Science
SUNY at Stony Brook
Stony Brook NY 11794

Mr. David W. Jensen
The Pennsylvania State University
233 Hammond Bldg.
University Park PA 16802

Dr. Conor D. Johnson
CSA Engineering
560 San Antonio Road
Suite 101
Palo Alto CA 94306

Dr. James D. Jones
Purdue University
School of Mechanical Enginring
1077 Ray W. Herrick Labs.
West Lafayette IN 47907-1077

Prof. Shiv P. Joshi
University of Texas.
500 West First Street
Box 19018
Arlington TX 76019

Dr. Laurie Judd
AEA Industrial Technology
Culham Laboratory
E5/150
Abingdon Oxon OX1430B United King

Dr. Jer-Nan Jung
NASA
M/S 297
Hampton VA 23665-5225

Mr. Kim Kahl
University of Maryland
Dept of Mechanical Engineering
College Park MD 20742

Dr. Daniel C. Kammer
University of Wisconsin
Dpt of Engineering Mechanics
Madison WI 53706

Dr. S. R. Kasiviswanathan
Michigan State University
Intelligent Materials Lab.
Dpt Of Mechanical Engineering
East Lansing MI 48824-1326

Mr. Mehdi Kaveh
Lord Corporation
1635 West 12th Street
Erie PA 16506

Mr. James Kelly
Office of Naval Technology
1515 Wilson Boulevard
Suite 705
Arlington VA 22209

Dr. Fred Kern Jr.
Atlantic Applied Research Co
4-A Street
Burlington MA 01803-3405

Dr. Farshad Khorrami
Polytechnic University
Dept of Electrical Engineering
333 Jay Street
Brooklyn NY 11201

Mr. Jae H. Kim
Purdue University
1282 Grissom Hall
Room 350
West Lafayette IN 49707-1282

Mr. Jinho Kim
Naval Surface Warfare Center
10901 New Hampshire Avenue
Code G43
Silver Spring MD 20903-5000

Dr. Sung J. Kim
Purdue University
1077 Ray W. Herrick Labs
School of Mechanical Enginring
West Lafayette IN 49707-1077

Dr. George C. Kirby
Naval Research Lab
Code 6382
Washington DC 20375

Dr. Gareth J. Knowles
Grumman Corporate Research Ctr
MS: A08-35
Bethpage NY 11714

Dr. Barry Koepke
Alliant Techsystems
Underseas Sys Div/Ceramic Ct
5121 Winnetka Avenue North
Minneapolis MN 55428-4256

Mr. Stewart M. Kohler
Snadia National Laboratory
PO Box 5800
Div 2334
Albuquerque NM 87185

Dr. Gary H. Koopmann
Penn State University
Center Acoustics & Vibration
157 Hammond Building
University Park PA 16802

Mr. Utpal Kopptkar
University of Maryland
Dept of Mechanical Engineering
College Park MD 20742

Dr. Leslie R. Koval
University of Missiouri-Rolla
Dept Mech & Aero Engineering
Rolla MO 65401

Mr. Larry A. Kraus
SRL Incorporated
1401 McCormick Drive
Landover MD 20785

Dr. N. Jayanth Kudva
Northrop Corporation
Dept 3853/82
One Northrop Avenue
Hawthorne CA 90250-3236

Dr. John P. Kurmer
Battelle Institute
505 King Avenue
Columbus OH 43201-2681

Dr. Drew M. L'Esperance
University of California
Department of Chemistry
Riverside CA 92521

Mr. Dimitris C. Lagoudas
Rousselaer Polytechnic Institute
Department of Civil Engineer.
Troy NY 12180

Mr. Brian M. Lampriere
Boeing Commercial Airplane Group
PO Box 3999
Seattle WA 98124

Mr. Eric J. Lang
University of Delaware
126 Spencer Laboratory
Newark DE 19716

Ms. Christina Lapeter
Virginia Polytechnic Institute
Mechanical Engineering Dept
ATTN: H.H. Cudney
Blacksburg VA 24061-0238

Mr. John Laudo
Battelle
505 King Avenue
Columbus OH 43201-2693

Dr. Chung E. Lee
Texas A&M University
Dept of Electrical Engineering
College Station TX 77843

Mr. Moti M. Leibowitz
Unitersity of Delaware
Dept of Mechnical Engineering
Newark DE 19716

Mr. Larry G. Leka
SFA Incorporated
1401 McCormick Drive
Landover MD 20785

Dr. Spiro Lekoudis
Department of the Navy
Office Chief of Naval Research
Code 1131
Arlington VA 22217-5000

Mr. George A. Lesieutre
Penn State University
Dept Aerospace Engineering
233 Hammond Building
University Park PA 16802

Mr. Chen Liang
Virginia Polytechnic Institute
840 University City Boulevard
Blacksburg VA 24061-0261

Prof. Doug Lindner
Virginia Tech
Dept of Electrical Engineering
Blacksburg VA 24061

Dr. K. Liu
University of Toronto
Institute of Aerospace Studies
4925 Dufferin Street
Downsview, Ontario M3H 5T6 Canada

Mr. Y. L. Lo
University of Maryland
Dept of Mechanical Engineering
College Park MD 20742

Dr. Richard Longman
Columbia Universtiy
Dept of Mechanical Engineering
New York NY 10027

Dr. B. J. Lurie
NASA JPL
4800 Oak Grove Drive
M/S 198-326
Pasadena CA 91109

Dr. Donald Lyons
Grumman Aerospace
Research Center
M/S A01-26
Bethpage NY 11714-3580

Lt. John R. Mackaman USAF
US Air Force
WL/FIBG
Wright Patterson AF OH 45433-6553

Mr. Brian J. MacLean
Martin Marietta
2288 Foothill Drive
#D209
Salt Lake City UT 84109

Dr. Walter M. Madigosky
Naval Surface Warfare Center
10901 New Hampshire Avenue
Silver Spring MD 20903-5000

Mr. A. Malik
University of Maryland
Dept of Mechanical Engineering
College Park MD 20742

Dr. Edward L. Marek Jr.
Sandia National Labs.
P.O. Box 5800
Division 1545
Albuquerque NM 87185

Mr. Kenneth R. Marken
Battelle Memorial Institute
505 King Avenue
Columbus OH 43201-2693

Dr. Charles W. Marschall
BATTELLE
505 King Avenue
Columbus OH 43201

Dr. Michael J. Martin
Michigan State University
Industrial Development Inst.
East Lansing MI 48823

Dr. David Martinez
Sandia National Laboratories
Division 1524
P.O. Box 5800
Albuquerque NM 87185

Mr. Bill Maryniak
TREK, Incorporated
3932 Salt Works Road
Post Office Box 231
Medina NY 14103-0231

Mr. Sami F. Masri
University of Southern California
Civil Engineering
M/C 2531
Los Angeles CA 90089-2531

Mr. Craig Mathews
University of Maryland
Dept of Mechanical Engineering
College Park MD 20742

Dr. Hiroshi Matsushita
National Aerospace Laboratory
6-13-1, Ohsawa
Mitaka
Tokyo 181 Japan

Prof. Yuji Matsuzaki
Nagoya University
Dept of Aerospace Engineering
Chikusa
Nagoya Japan

Mr. Russell G. May
Virginia Tech
Fiber & Elec. Op. Research Ctr
Electrical Engineering Dept.
Blacksburg VA 24061

Dr. James M. Mazzu
Charles River Analytics, Inc.
55 Wheeler Street
Cambridge MA 02138

Dr. Alaster McDonach
Strathclyde University
Smart Structures Rsch Inst.
204 George St, Glasgow
G1 1XW Scotland United Kingdom

Mr. John T. McHenry
Virginia Tech
Bradley Dept of Elec. Engring
340 Whittemore Hall
Blacksburg VA 24061

Prof. Ray Measures
University of Toronto
Aerospace Institute
4425 Dufferin Street
Downsview, Ontario M3H 5T6 Canada

Dr. Mohammed H. Megherhi
Piezo Kinetics, Incorporated
Post Office Box 756
Bellefonte PA 16823

Mr. Raymond J. Meilunas
Grumman Aerospace
Grumman Research Center
M/S A02-026
Bethpage NY 11714

Dr. L. D. Melvin
NASA Langley Research Center
M/S 321
Hampton VA 23665-5225

Dr. Walter R. Mendes
Nichols Research Corp
4040 S. Memorial Parkway
MS 3-2-09
Huntsville AL 35802

Dr. W. Craig Michie
University of Strathclyde
Smart Structures Rsch Inst.
204 George Street
Glasgow, Scotland G1 1XW United

Dr. John Michopoulos
Naval Research Laboratory
Code 6380
Washington DC 20375

Dr. Scott F. Midkiff
Virginia Tech
Dept. of Electrical Engineer.
340 Whittemore Hall
Blacksburg VA 24061-0111

Mr. John E. Miesner
George Washington University
12497 Alexander Cornell Drive
Fairfax VA 22033-2438

Dr. A. Keith Miller
Sandia National Laboratories
PO Box 5800
Organization 9145
Albuquerque NM 87185

Admiral William C. Miller USN
Chief of Naval Research
800 North Quincy Street
Arlington VA 22203-1906

Dr. Jovan Moacanin
California Institute of Technolog
Jet Propulsion Laboratory
4800 Oak Grove Dr, MS 125/112
Pasadena CA 91109-8001

Mr. Nicholas Mocerine
New Business Opportunities Inc.
P.O. Box 461
Hawthorne CA 90251

Mr. David B. Mohl
Naval Air Development Center
Code 5013
Warminster PA 18974

Mr. William D. Monaghan
US Department of the Interior
Bureau of Mines
PO Box 18070, Cochrans Mill Rd
Pittsburgh PA 15236-0070

Mr. Kenneth J. Moore
Cortana Corporation
520 N. Washington St. Ste 200
Falls Church VA 22046-3549

Dr. Robert E. Morgan
Naval Avionics Center
6000 E. 21 Street
Indianapolis IN 46219-2189

Mr. J. D. Muhs
Oak Ridge National Laboratory
PO Box 2003
Oak Ridge TN 37831-7280

Mr. Kent A. Murphy
Virginia Tech
Fiber & Elec. Op. Research Ctr
Electrical Engineering Dept.
Blacksburg VA 24061

Dr. M. Mutya
Catholic University of America
Mechanical Engineering Dept
620 Michigan Avenue
Washington DC 20064

Dr. Nagi G. Naganathan
The Univeristy of Toledo
Mechanical Engineering Dept
2801 West Bancroft Street
Toledo OH 43606

Mrs. Juock S. Namkung
College of Willima & Maryter
MS 2311
Nasa Langley Research Center
Hampton VA 23665

Mr. Marcello R. Napolitano
West Virginia University
Eng. Science Bldg. G-60
West Va. University
Morgantown WV 26505-6101

Dr. Ahid Nashif
Anatrol Corporation
10895 Indeco Drive
Cincinnati OH 45241

Dr. Michihiro Natori (M.C.)
ISAS
Inst of Space & Astronautical
3-1-1 Yoshinodai
Sagamihara, Kanagawa 229 Japan

Mr. Craig D. Near
Morgan Matroc Incorporated
232 Forbes Road
Bedford OH 44146

Mr. Mehrad Negahban
University of Nebraska
Ctr/Materials Resch & Analysis
Dpt of Engineering Mechanics
Lincoln NE 68588-0347

Mr. Peter E. Neumeier
General Dynamics Convair Division
PO Box 85357
Mail Zone 44-6400
San Diego CA 92186-5357

Mr. Thomas R. Newbauer
Los Angeles AFB
ATTN:SSN/CNPI
Los Angeles AFB CA 90009-2960

Mr. Stephen E. Newfield
Office of Naval Research
800 North Quincy
Suite 528
Arlington VA 22217-5000

Prof. Robert Newnham
Penn State University
Materials Research Lab.
University Park PA 16802

Dr. Jerry Newsom
NASA
Langley Research Center
MS 298
Hampton VA 23665-5225

Mr. Chris Niggemeier
University of Maryland
Dept of Mechanical Engineering
College Park MD 20742

Dr. Thomas E. Noll
NASA Langley Research Center
M/S 231
Hampton VA 23665-5335

Mr. Christian V. O'Keefe
Martin Marietta Labs
1450 South Rolling Road
Baltimore MD 21227

Dr. Thomas J. O'Keefe
University of Missouri-Rolla
Dept of Metallurgical Engring
111 Straumanis Hall
Rolla MO 65401

Dr. Michael C. O'Neal
Jet Propulsion Lab
4800 Oak Grove Drive
Pasadena CA 91109

Mr. Edward J. O'Neill
Global Associates Ltd
2300 Clarendon Boulevard
Suite 205
Arlington VA 22201

Mr. Gene P. O'Neill
United Technologies
Silver Lane MS: 129-14
East Hartford CT 06108

LTC Mike Obal
U.S. Air Force
SDIO/TNK
Pentagon, Room 1E178
Washington DC 20301-7100

Mr. Len Ogborn
University of Delaware
4 Harvest Trail
Newark DE 19702

Dr. Gerard Orcel
SpecTran Corporation
50 Hall Road
Sturbridge MA 01566

Dr. R. A. Osegueda
University of Texas
Civil Engineering Dept
El Paso TX 79968

Ms. Angelica Ottova
Michigan State University
Department of Physiology
East Lansing MI 48824

Ms. Zoubeida Ounaies
Penn State University
College of Engineering
1605 Highland Court
State College PA 16801

Prof. Umit Ozguner
Ohio State University
Dept of Electrical Engineering
2015 Neil Avenue
Columbus OH 43210

Mr. Eugene Pak
Grumman Corporation
Corporate Research Center
111 Stewart Avenue, M/S A08-35
Bethpage NY 11714

Dr. Markku J. Palantera
Helsinki University of Technology
Laboratory of Light Structures
Pumichenhuja 5A
Espoo SF-02150 Finland

Mr. Jiqin Pan
University of Maryland
Systems Research Center
College Park MD 20742

Dr. Hagop V. Panossian
Rockwell International/Rocketdyne
6633 Canoga Avenue
Mail Stop JB11
Canoga Park CA 91303

Mr. Kyong Park
ATOCHEM
3 Parkway
Philadelphia PA 19102

Dr. Ned Patton
Southwest Research Institute
6220 Culebra Road
PO Drawer 28510
San Antonio TX 78228-1510

Mr. Edward T. Peters
Arthur D Little Inc
Acorn Park
Cambridge MA 02140-2390

Dr. Lee D. Peterson
University of Colorado
Center for Space Structures
Campus Box 429
Boulder CO 80309-0429

Mr. Sean Pidgeon
IOP Publishing Ltd
1411 Walnut Street
Suite 200
Philadelphia PA 19102-3187

Ms. Shelby Pierce
Honeywell Inc.
13350 US 19 North
M/S 360-4
Clearwater FL 34624

Dr. Steven M. Pilgrim
Martin Marietta Laboratories
Advanced Ceramics Group
1450 South Rolling Road
Baltimore MD 21227-3898

Dr. Robert C. Pohanka
Department of the Navy
Office Chief of Naval Research
Code 1131
Arlington VA 22217-5000

Mrs. Anne B. Price
Newport News Shipbuilding
4101 Washington Avenue
E24/B600
Newport News VA 23607

Mr. David Price
U.S. Bureau of Mines
712 North Maramec
St. James MO 65559

Mr. Daniel A. Pruzan
Innovative Dynamics
243 Langmuir Labs
95 Brown Road
Ithaca NY 14850

Ms. Cindi L. Pulver
Honeywell SRC
10701 Lyndale Avenue South
MN09-F100
Bloomington MN 55420

Dr. Robert Quattrone
U.S. Army Construction Eng. Res.
P.O. Box 9005
Champaign IL 61826-9005

Mr. Larry A. Rabellino
Hi-Shear Technology
Ordnance Division
24225 Garnier Street
Torrance CA 90505-5323

Dr. Jayant Ramakrishnan
Dynacs Engineering Company
34650 US 19 North
Suite 301
Palm Harbor FL 34684-2157

Dr. Vittal S. Rao
University of Missouri-Rolla
Intelligent Systems Center
320 Engineering Research Lab
Rolla MO 65401

Dr. A. W. Raskob Jr.
APTEK, Inc.
1257 Lake Plaza Drive
Colorado Springs CO 80906

Mr. Bhakta B. Rath
Naval Research Lab
4555 Overlook Avenue SW
Code 6000
Washington DC 20032-5340

Dr. John R. Red-Horse
Sandia National Laboratories
PO Box 5800
Org 1545
Albuquerque NM 87111

Dr. Garth M. Reese
Sandia National Laboratories
PO Box 5800
Org 1545
Albuquerque NM 87185

Mr. Marc E. Regelbrugge
Lockheed Missiles & Space Div
3251 Hanover Street
M/S 0/93-30 B/251
Palo Alto CA 94304

Mr. Stanley Reich
Grumman Corporation
Aircraft Systems
IIII Stewart Ave, M/S K01-14
Bethpage NY 11714

Mr. John Reilly
University of Maryland
Systems Research Center
A.V. Williams Building
College Park MD 20742

Ms. Pamela A. Reis
Grumman Corporate Research Ctr
MS: A08-35
Bethpage NY 11714

Mr. Paul E. Rempes
McDonnell Douglas Missile Systems
PO Box 516
MC 1066086
St. Louis MO 63166

Dr. Darrell H. Reneker
University of Akron
Institute of Polymer Science
Akron OH 44325-3909

Mr. George W. Rinard
GR Associates
76 Hilltop Road
Cumberland MD 21502-5829

Dr. Andrew Ritter
Martin Marietta Laboratories
1450 South Rolling Road
Baltimore MD 21227-3898

Dr. Ruyen Ro
Penn State University
Dept. of Engineering Science
& Mechanics
University Park PA 16802

Mr. Craig A. Rogers
Virginia Tech
Dept of Mechanical Engineering
Smart Material & Structure Lab
Blacksburg VA 24061

Dr. Lynn C. Rogers
US Air Force
WRDC/FIBG
Structural Dynamics Branch
Wrigth-Patterson AF OH 45433-6553

Dr. William Rogers
General Dynamics
Ft. Worth Division
PO Box 748, M/S 2824
Ft. Worth TX 76101

Mr. Robert Roglin
Georgia Tech
1938 Glen White Drive
Deluth GA 30136

Dr. Robert Rogowski
NASA/LARC
M/S 231
Hampton VA 23365-5225

Mr. Kevin P. Roscoe
Newport News Shipbuilding
Airport Plaza I, Suite 1100
2711 S Jefferson Davis Highway
Arlington VA 22202

Mr. Michael J. Rossi
Grumman Corporation
Research Center
A08-35
Bethpage NY 11714

Mr. Paul S. Rutherford
Boeing Defence & Space Group
P.O. Box 3707, MS 4C-67
Seattle WA 98124-2207

Mr. E. Alexander Saenz
ADPA
2101 Wilson Boulevard
2 Colonial Place, Suite 400
Arlington VA 22201

Dr. A. Safari
Rutgers University
Dept of Ceramic Engineering
PO Box 909
Piscataway NJ 08855-0909

Dr. James R. Salois
General Motors Corp.
Saginaw Division
3900 Holland Road
Saginaw MI 48605

Mr. O. L. Santa Maria
NASA Langley Research Center
M/S 231
Hampton VA 23365-5225

Ms. Janet M. Sater
Institute for Defense Analyses
2001 N. Beauregard St.
Alexandria VA 22311-1772

Mr. Stanley S. Sattinger
Westinghouse Science & Technical
1310 Beulah Road
Building 401, Room 2X9C
Pittsburgh PA 15235

Mr. L. McDonald Schetky
Memry Technologies, Inc.
57 Commerce Drive
Brookfield CT 06804

Dr. Jeffrey S. Schoenwald
Rockwell International Science Ct
1049 Camino Dos Rios
MC: A9
Thousand Oaks CA 91360

Mr. Jeffrey N. Schoess
Honeywell Systems & Rsch Ctr
10701 Lyndale Avenue South
M/S MN09-D200
Minneapolis MN 55420

Mr. Paul J. Schubring
Cornell University
Department of T&AM
Kimball Hall
Ithaca NY 14853

Mr. David I. Schwartz
State University of New York
212 Ketter Hall
Dept of Civil Engineering
Buffalo NY 14214

Dr. Daniel J. Segalman
Sandia National Labs
PO Box 5800, Org. 1545
Albuquerque NM 87185

Dr. George P. Sendeckyj
Wright-Patterson AFB
WRDC/FIBEC
Dayton OH 45433-6553

Mr. Pradeep Sensharma
Virginia Tech
Dept Aerospace & Ocean Engring
Randolph Hall
Blacksburg VA 24061

Dr. Nava J. Setter
Ecole Polytechnique Federale
Laboratoire de Ceramique
MX-D Ecublens
1015 Lausanne Switzerland

Dr. Mo Shahinpoor
Sandia National Laboratories
P.O. Box 5800
Division 1534
Albuquerque NM 87185-5800

Dr. Nisar Shaikh
University of Nebraska-Lincoln
212 Bancroft Hall
Lincoln NE 68588-0347

Mr. David Shapiro
US Department of the Interior
Bureau of Mines
PO Box 18070, Cochrans Mill Rd
Pittsburgh PA 15236-0070

Dr. Montgomery T. Shaw
University of Connecticut
97 N. Eagleville Road
Storrs CT 06269-3136

Dr. Stuart J. Shelley
University of Cincinnati
Dept of Mch, Indus & Nuc Engr
Mail Location #72
Cincinnati OH 45221

Mr. Weicheng Shen
Martin Marietta Laboratories
1450 South Rolling Road
Baltimore MD 21227-3898

Mr. Harry D. Shirey
Piezo Kinetics, Inc.
Post Office Box 756
Bellefonte PA 16823

Mr. Thomas R. Shrout
Penn State University
150 Materials Research Lab.
University Park PA 16802

Mr. Peter Shyprykevich
Grumman Corporation
MS B44-35
Bethpage NY 11793

Dr. Pavel Sindelar
FAA Aeronautic Rsch Inst of Swede
PO Box 11021
Bromma S-16111 Sweden

Mr. Harmeet Singh
University of Maryland
Dept of Mechanical Engineering
College Park MD 20742

Dr. James Sirkis
University of Maryland
Dept. of Mechanical Eng.
College Park MD 20742

Mr. Samuel W. Sirlin
Jet Propulsion Lab
4800 Oak Grove Drive
M/S 198/326
Pasadena CA 91109

Prof. Robert E. Skelton
Purdue University
1293 Potter Center
West Lafayette IN 47907-1293

Prof. Gary L. Slater
University of Cincinnati
Department ASE/EME
Cincinnati OH 45221-0070

Dr. Barton Smith
Oak Ridge National Laboratory
Applied Technology Division
PO Box 2003
Oak Ridge TN 37831-7280

Dr. Herb G. Smith
McDonnell Aircraft Company
Structural Research
P.O. Box 516
St. Louis MO 63166-0516

Mr. James F. Smith
EDO Corporation/Electro-Acoustic
2645 South 300 West
M/S 3
Salt Lake City UT 84115

Dr. Wallace A. Smith
Department of the Navy
Office Chief of Naval Research
Code 1131
Arlington VA 22217-5000

Dr. Rudeen Smith-Taylor
NASA
Langley Research Center
Hampton VA 23665-5225

Dr. Jan Smits
Boston University
Dpt of Electrical Engineering
44 Cummington Street
Boston MA 02215

Mr. Venkata Sonti
Purdue University
1077 Ray W. Herrick Labs
School of Mechanical Enginring
West Lafayette IN 47907-1077

Dr. David N. Soo
European Space Agency
Postbus 299
2200 Noordwijk Netherlands

Mr. Jonathan E. Spanier
SFA Incorporated
1401 McCormick Lane
Landover MD 20785

Dr. John T. Spanos
Jet Propulsion Lab
4800 Oak Grove Drive
Pasadena CA 91109

Dr. Brian E. Spencer
Spyrotech, Inc.
4930 Superior Street
Lincoln NE 68523

Dr. Jay R. Spingarn
Sandia National Labs
PO Box 969
Div. 8311
Livermore CA 94551

Mr. Dieter R. Stoeckel
Raychem Corporation
300 Constitution Drive
M/S 522/7263
Menlo Park CA 94025

Dr. Staffan Strom
Royal Institute of Technology
Dept. of Electromag. Theory
Stockholm S-10044 Sweden

Mr. Joseph Su
Grumman Corporation
1111 Stewart Avenue
MS A08-35
Bethpage NY 11714

Dr. Hartono A. Sumali
Virginia Tech
Dpt of Mechanical Engineering
218 Randolph Hall
Blacksburg VA 24061-0238

Mr. C. C. Sung
Penn State University
Materials Research Lab
University Park PA 16802

Prof. K. Takahashi
Tokyo Institute of Technology
Dpt of Physical Electronics
O-Okayama, Megura-Ku
Tokyo 152 Japan

Prof. Junji Tani
Tohoku University
Institute of Fluid Science
2-1-1 Katahira Aoba-Ku
Sendai-shi, Miyagi Japan

Ms. Nimfa Teneva
Naval Research Lab
Code 8241
Washington DC 20375-5000

Mr. Lawrence B. Tentor
Virginia Polytechnic Institute
Mechanical Engineering Dept
ATTN: H.H. Cudney
Blacksburg VA 24061-0238

Dr. Joseph P. Teter
Naval Surface Warfare Center
Code R43
10901 New Hampshire Avenue
Silver Spring MD 20903-5000

Dr. James C. Thi
AT&T Bell Laboratories
1919 South Eads Street
Arlington VA 22202

Mr. Sridhar R. Thirupathi
The University of Toledo
Mechanical Engineering Dept
2801 West Bancroft Street
Toledo OH 43606

Dr. T. C. Thompson
MIT
Building 9, Room 428
77 Massachusetts Avenue
Cambridge MA 02139

Dr. Michael H. Thursby
Florida Institute of Technology
150 W. University Blvd.
Dept Electric & Computer Eng.
Melbourne FL 32901-6967

Mr. Michael T. Tichvon
TRW Engineering & Test Division
One Space Park
R4/2190
Redondo Beach CA 90278

Mr. H. Ti Tien
Michigan State University
Giltner Hall
East Lansing MI 48824

Mr. Thomas N. Tinley
David Taylor Research Center
Code 2081
Annapolis MD 21401

Dr. Harvey Tobin
Rome Air Development Center
ATTN: RADC/EEAS
Boston MA 01731

Dr. Toda
ATOCHEM
3 Parkway
Philadelphia PA 19102

Dr. Iradj G. Todjbakhsh
Rousselaer Polytechnic Institute
Prof. of Civil Eng. & Mech.
Troy NY 12180

Dr. William Tse
Lockheed Missiles & Space Co.
PO Box 504
Sunnyvale CA 94088-3504

Dr. Charlie D. Turner
Nichols Research Corporation
4040 South Memorial Parkway
MS3-2-09
Huntsville AL 35802

Dr. R. C. Twiney
GEC-Marconi Materials Technology
Carswell
Towcester
Northants NN12 8EQ United Kingdom

Prof. H. S. Tzou
University of Kentucky
242 Anderson Hall
Mechanical Engineering Dpt.
Lexington KY 40506-0046

Prof. K. Uchino
Sophia University
Kioi-Cho 7-1
Chiyoda-ku
Tokyo 102 Japan

Dr. Eric Udd
McDonnell Douglas
1801 East St. Andrew Place
Santa Ana CA 92647

Dr. Erdal A. Unver
AT&T Bell Laboratories
1919 South Eads Street
Arlington VA 22202

Dr. Thomas Valis
MIT
77 Massachusetts Avenue
Room 37-327
Cambridge MA 02139

Dr. Vijay Varadan
Penn State University
Acoustics Materials & Devices
149 Hammond Building
University Park PA 16802

Mr. V. V. Vardin
Penn State University
Materials Research Lab
University Park PA 16802

Mr. V. K. Vardin
Penn State University
Materials Research Lab
University Park PA 16802

Dr. Samuel L. Venneri
NASA Headquarters
Materials & Structure Division
Mail Code RM B647
Washington DC 20546

Dr. Danute R. Ventriglio
David Taylor Research Center
Code 2801
Annapolis MD 21401

Mr. Mats Vikstrom
Defence Material Administration
Aircraft Division
Stockholm S-11588 Sweden

Mr. Josef W. Vilsmeier
Deutsche Aerospace
MBB-FE202
P.O. Box 80 11 60
8000 Munich 80 Federal Rep German

Prof. Andreas Von Flotow
MIT
Dept Aeronautics & Astronautcs
77 Massachusetts Avenue
Cambridge MA 02139

Mr. Edward J. Vydra
Pre Finish Metals
2300 East Pratt Boulevard
Elk Grove Village IL 60007

Dr. Benjamin K. Wada
Jet Propulsion Laboratory
4800 Oak Grove Drive
Bldg 157, Room 507
Pasadena CA 91109

Mr. Dave Walsh
Honeywell, Incorporated
13350 U.S. 19 North
Mail Stop 360-4
Clearwater FL 34624

Mr. Raymond L. Walters
Office of the National Council
123 Old Executive Office Bldg
Washington DC 20500

Mr. Curtis Walz
University of Maryland
Dept of Mechanical Engineering
College Park MD 20742

Dr. Frederick E. Wang
Innovative Technology Intl, Inc.
10747-3 Tucker Street
Beltsville MD 20705

Dr. K. W. Wang
Penn State University
Mechanical Engineering Dpt.
150B Hammond Building
University Park PA 16802

D. J. Warkentin
MIT
Aeronautics & Astronautics Dpt
77 Massachusetts Avenue
Cambridge MA 02139

Dr. John Warren
Brookhaven National Lab.
Building 535B
Upton NY 11973

Mr. Shoichi Watabe
LTV Aircraft Division
PO Box 655907
Mail Stop 194-26
Dallas TX 75265-5907

Mr. Steve E. Watkins
University of Missouri/Rolla
220 Engineering Research Lab.
Rolla MO 65401

Prof. C.M. Wayman
University of Illinois
115B Metallurgy & Mining Bldg.
1304 W. Green Street
Urbana IL 61801

Cpt. Stephen G. Webb
USAF Academy
Dept of Engineering Mechanics
Colorado Springs CO 80840

Mr. Mark Webster
Jet Propulsion Laboratory
MS 157316
Oak Grove Drive
Pasadena CA 91106

Mr. Shmuel Weinstein
Westinghouse Electric Corp
1310 Beulah Road
Pittsburgh PA 15235

Dr. Terrence A. Weisshaar
Purdue University
School of Aeronautic & Astron.
Grissom Hall
West Lafayette IN 47907

Dr. John A. Wellman
Litton/Itek Optical Systems
10 Maguire Road
Lexington MA 02173-3199

Dr. Garry Wells
Kobe Steel Europe Ltd.
10 Nugent Road
Surrey Research Park
Guildford, Surrey GU2 5AF United

Mr. John H. Wendell
Atlantic Applied Research Corpora
4 A Street
Burlington MA 01803

Mr. Mark D. Wetzel
Du Pont Company
Engineering Department
PO Box 80357
Wilmington DE 19880-0357

Dr. Grady S. White
NIST
Ceramics Division 852
Materials Building 223 A256
Gaithersburg MD 20899-0001

Mr. Steve Wieczorek
AT&T
1919 S. Eads Street
Suite 300
Arlington VA 22202

Ms. Nancy Wieland
Honeywell, Incorporated
13350 US 19 North
Mail Stop 3160-4
Clearwater FL 34624

Dr. Jill Wilson
Cortana Corporation
520 N Washington Street
Suite 200
Falls Church VA 22046

Mr. Kennedy B. Wilson
Los Alamos National Laboratory
PO Box 503
IT-2/MS B229
Los Alamos NM 87544-0503

Mr. Erik M. Winkler
Global Associates Ltd
2300 Clarendon Boulevard
Suite 205
Arlington VA 22201

Dr. Stephen R. Winzer
Martin Marietta Laboratories
Materials Research Division
1450 South Rolling Road
Baltimore MD 21227-3898

Dr. W. R. Witkowski
Sandia National Labs
Division 1425
PO Box 5800
Albuquerque NM 87815-5800

Mr. Patrick Wolke
Honeywell Satellite Systems
PO Box 52199
Phoenix AZ 85072-2199

Dr. John Wood
University of Utah
3176 Merrill Engineering Bldg
Salt Lake City UT 84112

Dr. Robert W. Woolfolk
Southwest Research Institute
1235 Jefferson Davis Highway
Suite 1104
Arlington VA 22202-3293

Dr. Spencer T. Wu
AFOSR
Building 410
Bolling AFB
Washington DC 20332-6448

Mr. Zhen Wu
Penn State University
CEEAM
227 Hammond Building
University Park PA 16802

Prof. Hiroaki Yanagida
University of Tokyo
Faculty of Engineering
7-3-1 Hongo, Bunkyo-ku
Tokyo 113 Japan

Dr. Gershon Yaniv
Simula Incorporated
10016 South 51st Street
Phoenix AZ 85044

Dr. Y. C. Yiu
Lockheed Missiles & Space Co.
1111 Lockheed Way
M/S 0/77-50, B551
Sunnyvale CA 94089

Mr. Brad Yost
Wyle Laboratories
Phillips Lab
Building 2852A
Edwards AFB CA 93523

Dr. K. David Young
Lawrence Livermore National Labs.
P.O. Box 808
Mail Code L-156
Livermore CA 94550

Mrs. Rose M. Young
SFA Incorporated
1401 McCormick Drive
Landover MD 20785

Dr. Wei-Ching Yu
University of Connecticut
Institute of Materials Science
U-136
Storrs CT 06269-3136

Dr. Jianping Zhong
University of Kentucky
Dept of Mechanical Engineering
Lexington KY 40506-0046

Mr. Yan Zhuang
University of Maryland
System Research Center
Building 115, Room 2164
College Park MD 20742-3311

Mr. Bernd Zimmerman
FIMOD Corporation
Post Office Box 11192
Blacksburg VA 24062-1192

Dr. David Zimmerman
University of Florida
231 Aerospace Building
Gainesville FL 32611

Total No. of Labels: 453

ACTIVE MATERIALS/ADAPTIVE STRUCTURES

SYMPOSIUM AND EXHIBITION

4 - 8 November 1991

The exhibits will be open in conjunction with the receptions, all published breaks, and for one hour before the first session each morning. Exhibits will close at 7:15 PM on Wednesday, 6 November.

GUIDE TO EXHIBITORS BY BOOTH NUMBER:

1. Piezo Kinetics, Inc.
2. IOP Publshing LTD
3. Charles River Analytics, Inc.
4. BATTELLE
5. Honeywell, Inc.
6. Atochem Sensors, Inc.
7. TREK, Inc.
8. Anatrol Corporation

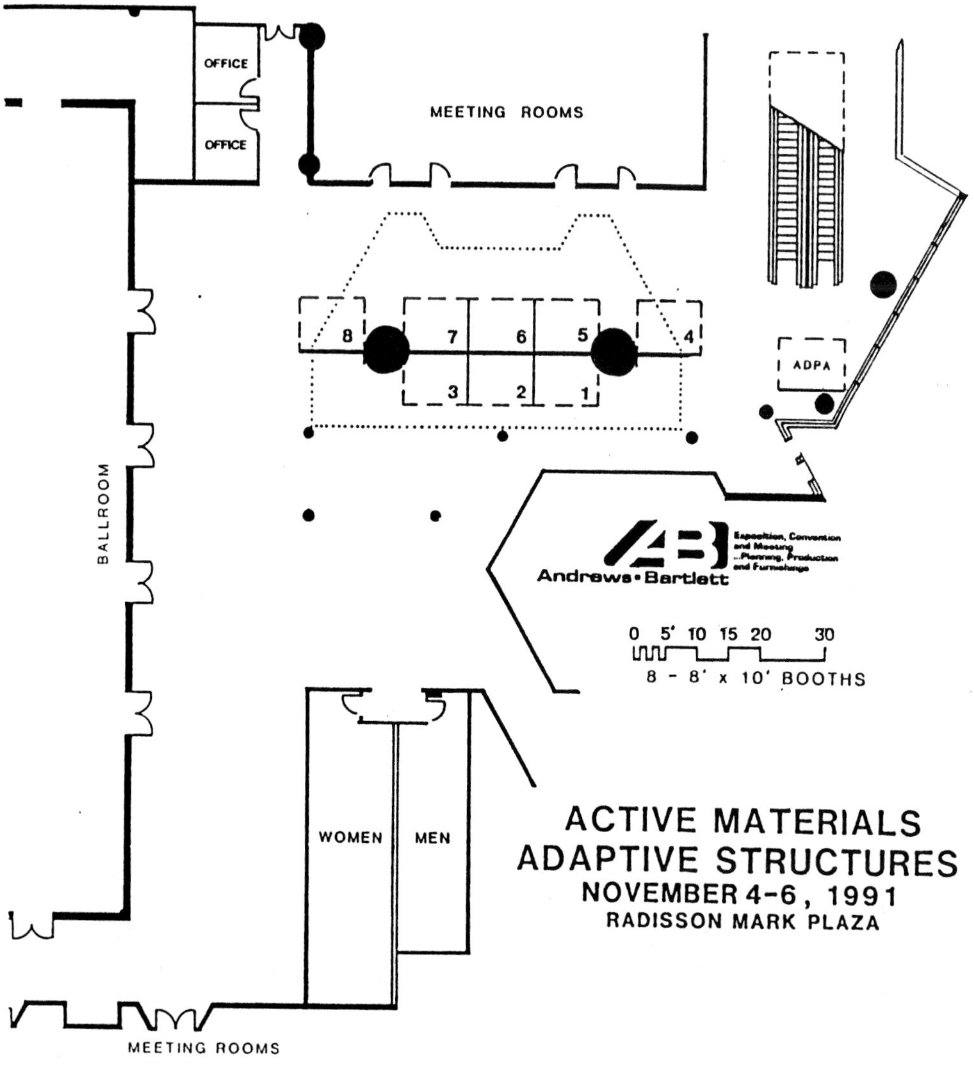

ABSTRACTS FROM
International Symposium & Exhibition on
Active Materials & Adaptive Structures
November 4-8, 1991
Alexandria, Virginia

TABLE OF CONTENTS

Paper Title | Page

Paper Title

Page

A MULTIAXIS ISOLATION SYSTEM FOR THE SPOT SATELLITE MAGNETIC BEARING RWA

BY
D. CUNNINGHAM, P. DAVIS AND F. SCHMIDT
Honeywell SSO

Design of a six degree of freedom isolation system using viscous damping is presented. Because of potential interaction with the control loops used to suspend the reaction wheel and control its output torque, the isolation system is required to provide very specific and tightly controlled values for the natural frequency and damping (Q) in all degrees of freedom. In addition, two of the translational degrees of freedom are required to have a very low natural frequency and the system must be testable under 1-G conditions. Consequently, the static deflections are unusually large for a space isolator application.

Systems trade studies showed that the only feasible means of providing these requirements is to employ an arrangement of six viscously damped tunable isolators in a "hexapod" configuration. Design equations were developed to predict the frequency and Q of each degree of freedom as a function of the mechanical impedance properties of the individual elements and the system geometry (mounting radius, angular skew, etc.) Iteration of the system design led to an optimum geometry and specification of requirements for the individual suspension elements.

Launch loads were predicted using a non-linear simulation that included the effect of compliant stops that limit overtravel of the reaction wheel into other spacecraft components. It was found that damping augmentation is necessary during the launch phase to prevent excessive impulse loads on the reaction wheel when the stops are contacted. This is provided by designing the viscously damped elements in an unusual fashion.

Normally spacecraft viscous dampers are extremely linear because they are designed to operate in laminar flow. Test results verify that their damping characterisics are invariant over at least 4 orders of magnitude. In order to significantly increase the damping during launch, the damper was redesigned to have a high Reynold's number when subjected to the flow rates which occur during launch. Predictions have been made showing that the required damping augmentation is provided with this scheme.

A prototype isolation system is currently being fabricated and preliminary test results will be available for inclusion in the final paper.

Passive Damping Design for Control System Stability on the SPICE Testbed

Y.C. Yiu
Lockheed Missiles and Space Company
1111 Lockheed Way, O/62-18, B551, Sunnyvale, CA 94089
408-742-4048

Eric M. Austin
CSA Engineering
560 San Antonio Road, Suite 101, Palo Alto, CA 94306
415-494-7351

Steven D. Ginter
Honeywell Inc., Satellite Systems Operation
P.O. Box 52199, Phoenix, AZ 85022
602-561-3244

Abstract

Prepared for the Active Materials and Adaptive Structures Conference, Nov 5-7, 1991, Alexandria, Virginia

Spacecraft structures that are required to hold precise alignments for accomplishing mission objectives often need additional augmentation. Induced structural vibration due to inevitable disturbances is always an important issue. Passive damping and active feedback control are two augmentation techniques that have been under serious investigation for many years with great potential to address induced vibration.

The *SPace Integrated Controls Experiment (SPICE)* testbed affords an opportunity to bring passive and active techniques together in an integrated application to a 6.5 meter, full scale, precision optical structure. The first experiment planned for the *SPICE* testbed is a *Precision Point Experiment (PPE)*. The *PPE* objective is to demonstrate a factor of 100 improvement in structural line-of-sight pointing stability in the presence of specified disturbances. The approach taken for the PPE was to achieve the primary improvement with active structure control and to use passive damping to ensure a robustly stable system.

Based on the baseline structural model, a preliminary control system was designed for the purpose of determining the performance requirements for both the actuators and damping in order to meet the goals of the experiment. For the disturbances specified, the active control system was designed with bandwith of a 100 Hz. In the control cross-over frequency region, however, the basic structure exhibits a number of lightly damped vibration modes. Therefore, the goal established for passive damping was to achieve 5% to 10% critical damping in these modes. The requirements for passive damping design consist of a list of target modes with accompanying damping values. The modal displacements and strain energy distributions of these modes were analyzed in detail to formulate passive design concepts. Three types of damping treatments were considered for passive damping design: damped struts, constrained layers, and tuned mass dampers. The most effective damping treatment was selected for each mode depending on its modal characteristics. The optimum locations, total number, and design parameters for each type of damping device were designed based on simplified design methods. For the control system stability, relatively large amounts of passive damping were required in a relatively small set of modes in the high frequency range. The system level finite element model was then updated to include these devices. The complex eigensolution was used to verify the system damping of integrated damping design and account for interaction between these damping devices. In order to track the modes, a special procedure that checks the cross-orthogonality between the undamped real modes and damped complex modes was used. Significant perturbations of mode shapes due to passive damping design were identified. Preliminary design specifications for the damping devices were prepared.

The objective of current risk reduction phase in the passive damping technology area is to design, fabricate, test, and verify the design and analysis methods of viscous and viscoelastic damping struts in the frequency range of interest. Future work will include a meticulous system level design, analysis and test program which includes an integrated active/passive system design and analysis, modal test of the damped structure, and precision point experiment of the integrated active/passive system.

Abstract for Active Materials and Adaptive Structures Conference, Arlington, VA, Nov. 5-7, 1991

The Role of Passive Damping in a Controlled Structures Testbed

Eric Anderson and Robert Jacques
MIT Space Engineering Research Center, Room 37-331
Cambridge, Massachusetts 02139

The importance of passive damping in the active control of structures will be reviewed. Approaches to the selection of distribution of damping in a large flexible structure will be described. Implementation of passive damping in one controlled structures technology testbed will be discussed. The emphasis will be on the benefits of damping for control.

Material and structural damping are extremely low in typical large flexible structures envisioned for space applications. The presence of additional passive damping is beneficial for a number of reasons[1,2,3]. For example, it provides a cushion for model accuracy for low frequency modes and is particularly important for reducing the sensitivity to poorly-modeled and unmodeled modes.

Approaches to predicting the distribution and characteristics of passive damping augmentation devices will be discussed. One approach to the CSI problem consists of a formal global optimization of structural configuration, passive damping, and active structural control. This has the advantage of linking damping and control, but can become extremely complicated. Such an optimization may itself become sensitive to parameters that are likely to change in an actual flight application. Other approaches mix some degree of engineering judgment with formal optimization. The most common approach is to locate dampers in areas where the concentration of strain energy for relevant modes is high. Local damping properties can be selected to give high loss factors in those modes.

In all approaches, the sensitivity of damping effectiveness to changes in the structural plant is likely to be greater when the number of damping devices is small. A highly distributed lower level of damping may be desirable in many cases. The advantages and difficulties associated with each approach will be discussed.

The MIT Interferometer Testbed is a controlled structures technology testbed designed to represent one concept for space-based optical interferometry[4,5]. The performance objective is the alignment of several mock siderostats mounted on a tetrahedral truss structure (See Figure 1). The goal is to maintain accurate pathlengths (such as AE and BE) in the presence of disturbances injected at point D.

The testbed contains active piezoelectric members and active piezoelectric and electrostrictive mirror mounts, as well as an on-board multi-axis laser metrology system capable of measuring nanometer-level displacements. The laser system is used in control and to judge performance. Accelerometers, strain gages, and load cells are also available for control.

In this paper, the passive damping components in the structure will be described. These include constrained layer viscoelastic struts[6], viscous D-Struts[7,8], and some use of shunted piezoelectric struts[9]. The role of each type of damping implementation will be described. Design approaches for achieving certain characteristics from different types of dampers will be reviewed. The procedure involves using damper component models to obtain desirable properties at global structural frequencies and in critical locations in the truss testbed. The rationale for choosing locations and individual damper characteristics will be detailed. The changes introduced in the frequencies and mode shapes will be discussed. The component-level and system-level performance of the D-Strut viscous dampers will be highlighted.

The control approach consists of two main areas: isolation of the disturbance and the payloads (mock siderostats); and distributed active control of the truss structure. The need for passive damping in both of these areas will be demonstrated. The isolation of the siderostats involves closing an optical loop in which the influence of structural modes is deliberately minimized. In this case, a small amount of passive damping is all that is needed to greatly improve the control[10]. The presence of damping can also greatly enhance structural control, particularly in helping to stabilize modes in the frequency range of controller rolloff.

References

1 von Flotow, A. and Voss, D.W., "The Need for Passive Damping in Feedback Controlled Flexible Strutures," Damping '91 Conference, San Diego, Feb. 1991.

2 Gueler, R., "The Virtues of Passive Damping for Feedback Controlled Structures," MIT S.M. Thesis, Dept. of Aeronautics and Astronautics, June 1991.

3 Morganthaler, D. R., "Passive and Active Control of Space Structures," 3 vols., WRDC-TR-90-3044, Sept. 1990.

4 Miller, D., Blackwood, G., Jacques, R., Hyde, T., and Kim, E., "The MIT Multipoint Alignment Testbed: Technology Development for Optical Interferometry," SPIE Conf. on Active and Adaptive Optical Components, San Diego, July 1991.

5 Hyde, T., Kim, E., Anderson, E., Blackwood, G., and Lublin, L., "MITs Interferometer CST Testbed," JPL Workshop on Technologies for Space Interferometry, Pasadena, CA, Apr. 1990.

6 Plunkett, R. and Lee, C. T., "Length Optimization for Constrained Viscoelastic Layer Damping," *J. Acoustical Society of America*, Vol. 48, No. 1, 1970.

7 Davis, L. P. and Wilson, J. F., "New Structure Design Criteria Offer Improved Pointing and Lower Weight", 59th Shock and Vibration Symposium, Albuquerque, NM, October 1988.

8 Anderson, E., Trubert, M., Fanson, J., and Davis, P., "Testing and Application of a Viscous Passive Damper for Use in Precision Truss Structures, AIAA Structures, Structural Dynamics, and Materials Conf., Orlando, April, 1991, Paper No. 91-0996.

9 Hagood, N. W., Chung, W. H., and von Flotow, A., "Modelling of Piezoelectric Actuator Dynamics for Active Structural Control," *J. of Intell. Mater. Syst. and Struct.*, Vol. 1, No. 3, pp. 327-354, July, 1990.

10 Garcia, J. G., Sievers, L. A., and von Flotow, A., "Broadband Positioning of 'Small' Payloads Mounted on a Flexible Structure," to be published in *J. Guidance, Control, and Dynamics.*

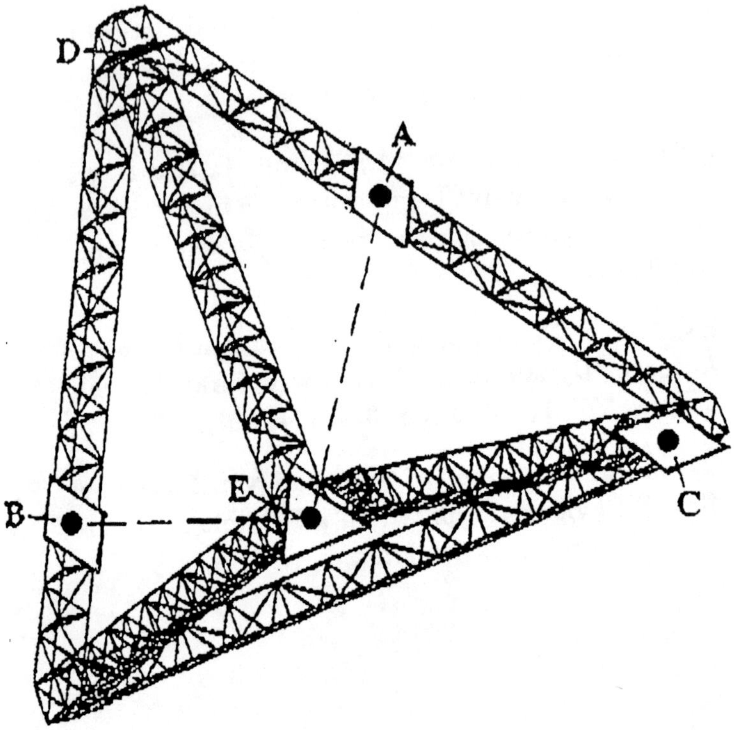

Figure 1

Smart Tuned-Mass Dampers

Conor D. Johnson, Joseph R. Maly, Kevin E. Smith
CSA Engineering, Inc.
Palo Alto, CA

Submitted to
ADPA/AIAA/ASME/SPIE Conference on
Active Materials and Adaptive Structures

ABSTRACT

The scenarios for many space systems require that, for their success, the effects of structural dynamics be reduced. Quite often, one vibrational mode plays a dominant role in the dynamic response, such as the first mode of a metering structure during a slew maneuver. Passive damping using tuned-mass dampers (TMD's) is a well-known, weight-efficient approach to suppress vibrations of a single mode (or a group of modes). Its main advantages over other types of passive damping treatments are that only a small amount of added weight is needed to achieve relatively high levels of damping, and it has minimal side effects on primary structure design. However, to be effective, TMD's must be kept tuned to the frequency of the offending mode. A prototype TMD that will tune itself to an offending mode has been designed, built, and tested. It can keep itself tuned to the offending mode, even if that mode changes frequency. The prototype confirmed that a "smart" TMD could be built.

The properties of viscoelastic materials used in TMD's are temperature dependent, which is usually a disadvantage. In the present effort, the temperature dependence has been exploited as a mechanism for tuning the TMD. This was accomplished by developing and implementing an active control system where the dynamics of the TMD and base structure are sensed and the temperature (and thus the stiffness) of the viscoelastic material in the TMD is controlled for optimum tuning. The thermal control will also allow the TMD to work in the harsh space environment.

For realistic applications, the control system was required to meet two conditions:

1. The natural frequency and damping of the mode of the base structure to be damped is initially unknown and can change with time in a continuous or noncontinuous manner. For instance, a slowly decreasing fuel load or articulated members would produce a slowly changing natural frequency, while the docking of an orbiter or release of a missile would produce a near instantaneous change in natural frequency.

2. The excitation is unknown in type and level, and both may change in time. The excitation (and thus the response) may also drop below the sensitivity of any sensor used by the TMD. This loss of excitation and response must not result in a wrong action by the controller.

These requirements provide the broadest, most realistic view of how a TMD might be required to interact with a large space structure. However, these requirements severely limit the types of information available to the controller. In effect, the controller has virtually no *a priori* knowledge of the base structure or how it will be excited. This means that the controller needs to work solely on the basis of response measurements and assumptions about the underlying structure, TMD, and the control objectives.

A prototype TMD was designed and tested. The study was successful in demonstrating all of the essential aspects of the self-tuning TMD. The controller can find and track a single structural mode and produce the maximum system damping for that particular TMD. Thin-foil heaters were embedded in the viscoelastic material for thermal control. The entire unit could be insulated to minimize the total amount of power required. Once the TMD is tuned, very little power is required to maintain the tuning or shift the stiffness to follow a mode.

SMART POLYMERIC MATERIALS FOR ACTIVE CAMOUFLAGE

Dr. L. J. Buckley and Mr. D. Mohl
Naval Air Development Center
Warminster, PA 18974

Abstract

Segmented polyurethanes that contain the diacetylene group in the molecular backbone were synthesized and studied as potential active visual camouflage materials. The polyurethanes have a segmented block copolymer structure consisting of crystalline hard domains containing the diacetylenes within amorphous regions. The materials varied by type of hard segment and soft segment molecular weight. The diacetylene groups can be reacted in the solid state to produce a chromic material that will change its absorption behavior with strain or temperature. The fully conjugated polydiacetylene structure with its extensive pi-electron delocalization enables this behavior. The morphology of these materials dictates the ultimate length of the conjugated polydiacetylene structure and was investigated with small-angle x-ray scattering. This enabled the selection of systems that were well phase separated with a strong chromic effect. The structure is typically lamellar and highly dependent upon processing.

UV-Visible studies were performed to evaluate the mechanochromic and thermochromic nature of these materials. The change in absorption behavior was quantified for each of the various polyurethane systems investigated. The soluble nature of these materials before forming the polydiacetylene structure enables efficient deposition as a coating.

In addition to the adaptive material, the active camouflage system consists of sensors and controls. The sensors identify the ambient wavelengths and intensity of the surroundings. The controller processes the information from the sensors and sends signals to modify color and intensity. Various sensor configurations were investigated such as: a diffraction grating with photodetectors; a prism with a linear charge-coupled device (CCD); primary color filters with photocells; and a color CCD camera. The color CCD camera provided the best overall features. The output of the camera contains all the information needed to determine color and intensity of the targeted area. The sensor outputs are fed through an anolog to digital converter into the microprocessor. In order to identify the perceived color of the scene, the microprocessor maps the sensor data into a three dimensional color space. Once the color of the scene is identified, the microprocessor will determine how to change the material to blend with the background.

The overall objective of this effort was to develop materials for an active camouflage system that is capable of changing both, color and intensity.

Dr. Leonard J. Buckley
Code 6064
Naval Air Development Center
Warminster, PA 18974
(215) 441-2823

SMART ELECTROMAGNETIC ABSORPTIVE AND SHIELDING MATERIALS

Vijay K. Varadan and Vasundara V. Varadan
Center for the Engineering of Electronic and Acoustic Materials
Pennsylvania State University, University Park, PA 16802

The need to achieve electromagnetic compatibility of electronic components, computers, etc., and also absorption of certain unwanted electromagnetic signals in the commercial and military environment has never been so great. Even Cable Television systems leak signals in excess of the 200 microvolt limit which is far in excess of the cumulative leakage index (CLI) of 64. The CLI standard was designed to prevent harmful interference to aeronautical communications in the 108 to 137 MHz and 225 to 400 MHz frequency bands. In order to achieve this goal, scientists and engineers are designing circuits which are less susceptible to signal errors, circuit designers are laying out their hardware with more understanding and substrates, cables and enclosures are pacaged to give better shielding. All these design and implementation are commonly pursued with materials such as carbon, ferrrite, graphite, conducting polymers, etc. .These materials, however, provide electromagnetic interference (EMI) shielding and microwave and radar absorption only over a limited frequency band. What is really needed for the safe environment are materials which will be active and adapt according to the radiating and interfering sources. In this talk, one family of such smart materials developed at Penn State using chirality and tunablity with d.c. bias field in dielectric composites will be presented.

It will be shown in this paper that chiral composites can be attractive as highly efficicent absorbers. Electromagnetic waves can discriminate between objects of different handedness due to their transverse nature, which implies that the origin of chirality need not necessarily be molecular as in the case of optically active media. Effective chiral composites may, therefore, be constructed by embedding chiral microstructures in non-chiral host media. Chiral polymer coatings for microwave application consist of chiral inclusions dispersed in a dielectric matrix material. In our Center, chiral coatings are realized by embedding a large number of miniature ceramic, polymer, carbon, graphite and other dielectric or dielectric and magnetic helical fibers in a lossy dielectric material or by randomly dispersing microballoons filled with chiral polymers in a lossy polymer host material. The chiral materials when irradiated with electromagnetic waves induce surface current and surfce charge desnsities which in turn produces artificial magnetization. In addition, when a linearly polarized wave hits a chiral scatterer, the scattered fiel has both left circularly polarized (LCP) and right circularly polarized (RCP) components travelling with different phase velocities. In chiral composites, multiple scattering between chiral inclusions enhances this phenomenon even further.

The induced surface current and charge densities in chiral materials depend on the pitch and diameter of the helical inclusions and also the dielectric properties of the helices and the host material. Active and adaptive composites may be made by controlling these parameters. In case one cannot change the parameters of the chiral inclusions, it is possible to achieve smart absorptive and shielding composites by having a tunable dielectric or frequency selective host material. The tunability may be acccomplished by a d.c. bias field which in turn will not affect electromagnetic wave absorption and shielding.

Another family of smart materials useful for high temperature EMI and RAM applications is a ceramic based magnetoelectric composites. These composites have both ferroelectric and magnetic properties and they can be grouped under anisotropic chiral materials.

The composite materials presented in this paper provides about -100 to 110 dB shielding and about -30 to -45 dB absorption from 30 MHz to 110 GHz. Applications ranging from shielding computer and electronic components, EMI shielding conformal coating on P.C. boards, composite shelters to microwave and radar absorbing coatings will be discussed.

THE NEW BLM SYSTEM: SELF-ASSEMBLING BILAYER
.IPID MEMBRANES (s-BLMs) H. Ti Tien, T.
Martynski* and A. Ottova*, Membrane Biophysics
aboratory (Giltner Hall), Department of Physiology,
Michigan State University, East Lansing, MI 48824
USA); #Institute of Physics, Poznan Technical
niversity, Poland; *Slovak Technical University,
ratislava, CSFR.

Artificial bilayer lipid membranes (planar BLMs) irst formed more than two decades ago have been idely studied as a model for biomembranes [1]. For ong-term studies as well as for biotechnological pplications, a common concern has been the chanical stability of the BLM, since rarely do ey last more than a few hours. For this reason, it s not only desirable but also impertive that a ethod be found to generate long-lasting BLMs. In is communication, we report a simple and novel thod for self-assembling BLMs on solid support. e new BLM system possesses not only the requisite chanical stability, but also dynamic properties]. Experimentally, the method for forming lf-organized solid supported bilayer lipid mbranes (s-BLMs) is based on interactions between phiphilic lipid molecules and nascent metallic rface. Advantages of the new method are several: particular the mechanical stability is improved orders of magnitude. Insofar as we can acertain, s-BLM, apart from its simplicity in formation and superior stability, behaves similarly in almost ry respect to that of a conventional BLM except transmembrane ion movement. The presence of id support on one side of the BLM precludes ion nsport. This is offset, however, by its potential electronic conduction (2).

The conventional BLM system and its development a model for biomembranes has been recounted ewhere (1,3). Together with 'black' soap films, Langmuir-Blodgett (L-B) technique provided the cial insight for its realization. It should be tioned, however, there is one major difference een the L-B layers on rigid substrate and the . Apart from its bimolecular thickness, a BLM is iquid-like, dynamic structure in a metastable e. On the other hand, a layered structure made he L-B technique is rigid and in a solid state. so-called 'solvent-free' BLMs made by the L-B nique required the pretreatment of the aperture petrolatum gel or squalene. From a -assembling point of view, we conclude that it ifficult to envision how a BLM separating two ous solutions can be made from two rigid, -like monolayers of lipids without a eau-Gibbs border. For biosensor development, it ur opinion that a fluid bilayer is of crucial tance.

LMs with diameters up to 10 mm can be formed by iety of methods, one of which is based on the uir-Blodgett technique introduced by Takagi et 4). Alternatively, a new type of BLM may be d by a method recently developed in our atory in which one side of a solvent-free BLM chored to a hydrophilic (either conducting or nducting) support while the other side is in t with aqueous solution in the usual manner.

Such a system could be useful in the development of biosensors and molecular electronic devices (1,5-9).

In the last decade or so, planar BLM has shown promise in the field of chemical/bio sensors for potential applications in medicine, industry, and clinical laboratories (1,5-8). The principal idea behind the development of BLM-based biosensors is quite simple. We envision that, for detection in biological environments, the sensing element should be biocompatible and be biomembrane-like. Thus, the new bilayer lipid membrane (s-BLM) has great potential and is an ideal choice upon which to develop a new class of electrochemical/bio sensors. In immunochemical reactions, the antigen (Ag) that corresponds to antibody (Ab) to be detected is incorporated into a BLM. This modified BLM then becomes a sensing element specific for the antibody. With techniques now available, the electrical properties of the BLM such as the membrane potential, capacitance, resistance, dielectric breakdown voltage and other electrical parameters can be readily measured. Biosensors based on s-BLMs, once successfully developed, would be very cheap with the added advantage of ease of measurements.

It is worthy to note that s-BLMs with minimeter to micrometer demensions have other attributes; they are essentially microelectrodes. As such, s-BLMs can be readily developed for a variety of practical applications. It seems highly likely that the new method for self-assembling BLM on solid support, like its predecessor, will have an impact on lipid bilayer-based research in the years to come.

REFERENCES

1. H. T. Tien, 1988, in "Thin Liquid Films," I. Ivanov, ed., Marcel Dekker, Inc., New York Chapter 14.
2. H. T. Tien and Z. Salamon, Bioelectrochem. Bioenerg., 22, 211 (1989).
3. G. Dryhurst and K. Niki, eds., 1988, "Redox Chemistry and Interfacial Behavior of Biological Molecules," Plenum, New York.
4. H. T. Tien, 1974, "Bilayer Lipid Membranes (BLM): Theory and Practice," Marcel Dekker, Inc., New York.
5. J. Janata, 1987, in: "Proc. Sym. Chemical Sensors," D. R. Turner, ed., The Electrochemical Society, Inc., Pennington, N. J., pp. 258.
6. D. B. Kell, 1987, in: "Biosensor Fundamentals and Applications," A. P. F. Turner, I. Karube, G. S. Wilson, eds., Oxford Press, Oxford, p. 427.
7. U. J. Krull, M. Thompson, and H. E. Wong, 1986, in: "Fundamentals and Applications of Chemical Sensors," D. Schuetzle, R. Hammerle, J. W. Butler, eds., Am. Chem. Soc., Washington D.C., p. 351.
8. W. M. Reichert, C. J. Bruckner, and J. Joseph, 1987, Thin Solid Films, 152:345.
9. H. T. Tien, Z. Salamon and A. Ottova, in: "Critical Reviews in Biomedical Engineering," CRC Press, in press (1991).
10. H. T. Tien and T. Martynski, Bioelectrochem. Bioenerg., in press (1991).

Extended abstract for CONFERENCE ON ACTIVE MATERIALS AND ADAPTIVE STRUCTURES

November 5-7, 1991, Alexandria, VA.

Controlled Formation and Properties of Responsive Polymers

Stephen G. Weber, Elizabeth A. Wise, Andrew D. Hamilton,
Stephen J. Geib, Fernando Garcia-Tellado

Department of Chemistry, University of Pittsburgh,
Pittsburgh, PA 15260

The design of 'smart materials' that show both a controlled structure and a specific response to exogenous chemicals is an important goal. We are presently taking two approaches to this problem, both centered on the use of non-covalent interactions to form the structural foundation of the material and to provide the origin of its specific response.

The first approach employs directed hydrogen bonding interactions to self-assemble two or more components into a well defined and predictable crystal lattice. Aliphatic dicarboxylic acids and the 2-aminopyridine diamides of rigid aromatic diacids will self assemble as shown into an alternating polymeric cocrystal with two hydrogen bonds between each acid and aminopyridine unit (Figure 1). We can control the precise structure of this

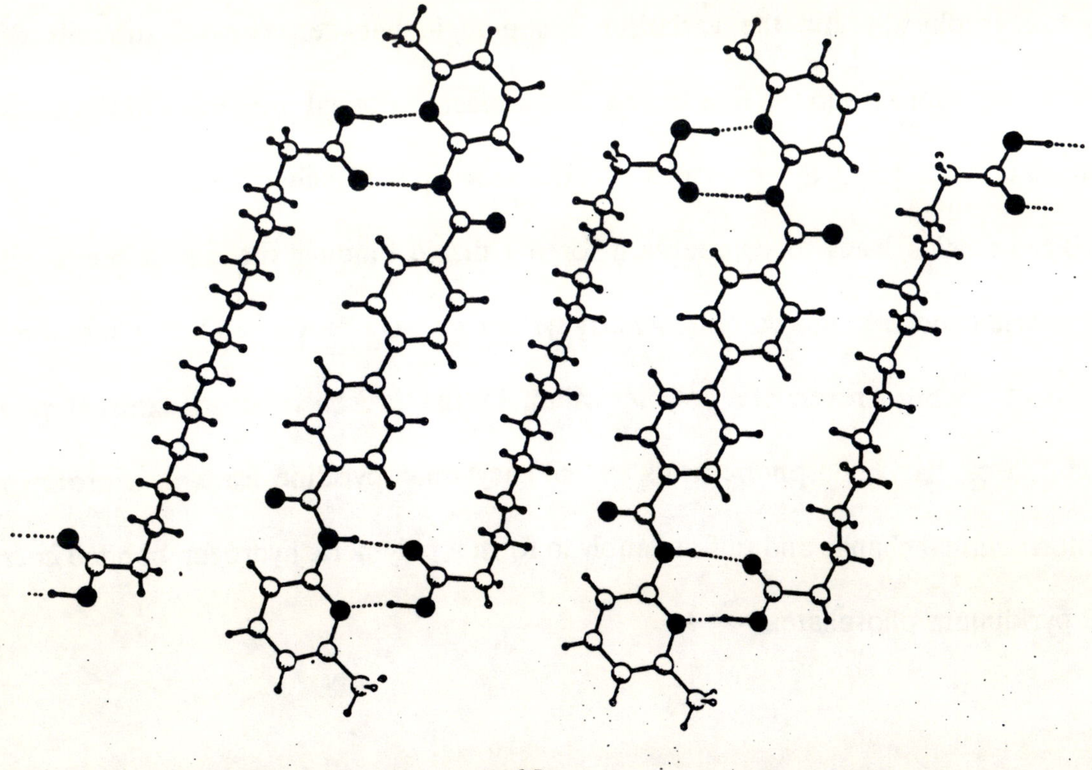

Figure 1

material by changing the relative length of the diacid and diamide. Shortening the diamide leads to a decrease in the angle between the components and the horizontal. Subsequent shortening of the receptor will increase this angle to the point where the two components are well-matched and the angle is 90° (Figure 2). These changes have all been characterized

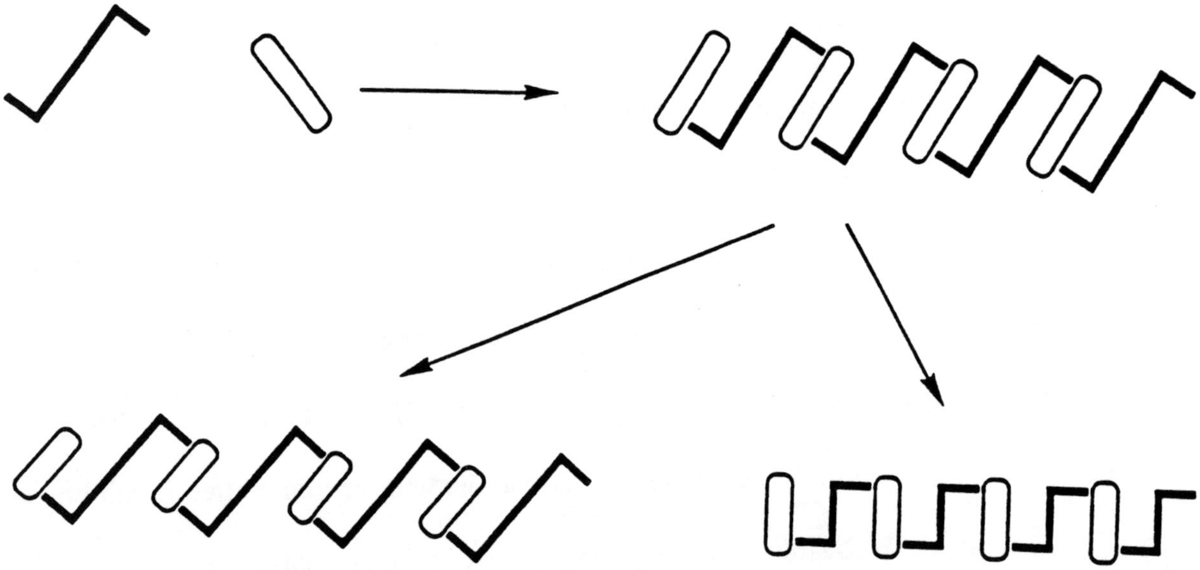

Figure 2

crystallographically and the variation in slip-angle between parallel subunits offers an interesting approach to the fine tuning of electrical or optical properties in the crystal. The materials respond to the presence of externally applied substrates. Addition of simple carboxylic acids leads to competition for the diacid binding site and a break up of the polymeric structure. In a complementary system we have developed a responsive monomer that will undergo reversible polymerization in the presence of an external phosphate. Addition of diaryl phosphoric acids to 2,6-diacylaminopyridine leads to a proton induced conformational change and self-assembly to form a polymeric hydrogen bonded cocrystal of the pyridinium phosphate.

The second system studied is commonly known as "slime"; it consists of aqueous sodium borate and polyvinylalcohol in a weakly basic solution. We have carried out an investigation of the phase diagram of this system at low borate concentration (moles of borate less than or equal to the moles of alcohol in the polyvinyl alcohol). Viscosity is dependent on the extent of borate crosslinking of the polyvinyl alcohol. The material properties of such a system depend on the activity of the borate which can be controlled with a second alcohol such as glycerol or mannitol. The added alcohol indirectly influences the material properties of the material by influencing the extent of crosslinking.

Just as crystal structures can be controlled, properties of monocrystalline structures can also be controlled through noncovalent interactions. Because such interactions are reversible, they can be interrupted by chemical alterations in the environment. Smart materials that respond to the presence of specific chemical compounds can be created based on these principles.

The extension of such concepts to materials based on molecular specific interactions will be discussed.

THREE-DIMENSIONAL PHASE-STRAIN MODEL FOR EMBEDDED OPTICAL FIBER SENSORS: EXPERIMENTAL VERIFICATION AND APPLICATIONS TO DIFFERENT SENSOR TYPES

by

J. S. Sirkis, C. Mathews, Y.L Lo, A. Dasgupta, and K. Kahl
University of Maryland
Department of Mechanical Engineering
College Park, Maryland 20742
301-405-5265

This paper reviews the mathematical development and provides experimental verification of a three-dimensional phase-strain model for structurally embedded Mach-Zehnder optical fiber sensors. Extensions of this phase-strain model to four additional smart structure fiber optic sensors is then provided. The first series of verification tests consist of embedding single mode optical fibers within isotropic uniaxial compression specimens of four different sets of material properties. The Young's modulii of the host materials range from 20 GPa to 2 GPa, and the host Poisson's ratios range from .36 to .24. Generalized plane strain theory of elasticity solutions based on Savin's plane strain displacement fields are developed for this loading and geometry, and are then used with the three-dimensional phase-strain model to calculate the strain induced phase shift in the fiber given the applied load history. The actual phase history is measured with a Mach-Zehnder interferometer and found in good agreement with the calculated values. The phase-strain model developed by Butter and Hocker, which has found common usage for structurally embedded optical fiber sensors, is also used to calculated the strain induced phase shifts. The agreement between these phase history predictions and the experiments is good only for very low stiffness host materials, and the agreement becomes progressively worse as the stiffness of host material approaches that of the optical fiber. Since Butter and Hocker's model does not include any information about the host material, this series of tests show that the optical fiber must be treated as an elastic inclusion, and a system level approach must be used when interpreting the scalar output signal from structurally embedded optical fiber sensors since that output will be a function of the entire strain tensor.

It is interesting to note that as the host stiffness becomes a factor of ten or so smaller that the fiber stiffness, the optical fiber will reinforce the host material and behave as though it were surface mounted. In such cases the less complicated Butter and Hocker theory applies, and thus provides a one-to-one map between a single strain component and the sensor output. This presents one possible benefit of the resin rich zones commonly encountered when optical fibers are embedded in laminated composite materials. This resin rich zone results in the optical fiber being surrounded by low modulus material except where the optical and reinforcing fibers are in contact at the optical fiber north and south poles. The resin rich zone/optical fiber material system and geometry has lead to a second set of the verification tests which are similar to the first, but this time the host material is a graphite-epoxy laminated composite, where the optical fiber is layed-up perpendicular to the natural reinforcing fibers, and the specimen is loaded in tension. Generalized plane strain finite element methods are used in this case to calculate the strain induced phase shifts since elasticity solutions do not yet exist for this geometry and loading.

Finally, extensions of the three-dimensional phase-strain model are given for the Fabry-Perot, Michelson, polarimetric, and dual-mode classes of optical fiber sensors. The extensions of the theory to the Fabry-Perot and Michelson sensors are straight forward. The extension of the theory to polarimetric sensors is accomplished by assuming that the fiber is optically anisotropic, but materially isotropic. The ramifications of these assumptions are fully discussed. Further, the influence of the changes in direction of the principal planes of the strain field in the polarimetric sensor is mathematically treated. The main conclusion of this treatment is that an all-fiber circular polariscope version of the polarimetric sensor must be used for structurally integrated applications where the host structure is experiencing multi-axial, time-varying load fields. These types of load fields are common in most projected smart structure applications. Finally, the extension of the three-dimensional phase-strain theory to dual-mode sensors is somewhat complex, and has been carried out by Lindler and Reichard at Virginia Tech. Their work is summarized here for completeness.

Spatially Weighted Fiber Optic Sensors for Smart Structure Applications

K. A. Murphy, B. R. Fogg, A. M. Vengsarkar, and R. O. Claus
Fiber & Electro-Optics Research Center, Bradley Department of Electrical Engineering
Virginia Tech, Blacksburg, VA 24061 - 0111

Currer. .esearch in vibration sensing and control has shown that variable sensitivity, spatially distributed transducers may be more suitable than point sensors for optimal control architectures. We have recently shown that fiber sensors can be used as vibration-mode filters by appropriate placement of the sensors on a structure. We also proposed that by adjusting the weighting sensitivity function of the fiber sensor one could achieve the same vibration-mode filtering effect without considering the issue of sensor placement.

In this paper, we present results obtained from spatially weighted optical fiber sensors. We will review the development of distributed, weighted fiber sensors through the following examples:

- The sensor acts as a vibrational-mode filter by appropriate placement along the vibrational antinodes, when the weighting function, $p(x) = 1$. We demonstrate this for a clamped-clamped beam and show an accurate correlation between theoretical analysis and experimental results.

- The weighting function is varied by using tapered fibers such that $\Delta\beta$ varies along the length of the sensing segment. We describe experimental results from the same and compare the sensor outputs with conventional-core fiber sensors.

- $p(x)$ is also varied by the exposure of germanium doped, e-core fibers to high optical powers that induce nonlinearities and result in periodic gratings within the fibers. Methods of writing the gratings to obtain specific $p(x)$'s are described and results are presented.

We will also describe recent efforts in developing spatially weighted single-mode fiber sensors and predict possible uses in vibration sensing as well as hydrophone applications.

Single-Fiber, Dual Modal-Domain Sensors

Christian V. O'Keefe
Martin Marietta Laboratories

Ashish Vengsarkar
~~Virginia Polytechnic Institute and~~ State University
AT&T Bell Laboratory

Single-fiber sensors, such as modal-domain sensors and polarimetric sensors, while useful as strain sensors, are also very susceptible to temperature effects. We have exploited this simultaneous response to temperature and strain to create a single-fiber sensor using a single light source that can measure both temperature and strain independently.

This sensor combines the principles of a polarimetric sensor, which detects strain based on the differential phase velocities along orthogonal axes in a polarization-preserving fiber, and a modal-domain sensor, which uses the differential phase velocities between two modes in a fiber. These two effects can be combined such that two modal-domain sensors are created along the two orthogonal axes of a two-mode polarization-preserving fiber. Since each sensor responds differently to temperature and strain, each generates an independent output, offering the possibility of extracting the strain and temperature.

A variety of fibers were tested for their response to temperatures ranging from 20 to 100° C and axial strains up to 0.36%. The responses were used to create a matrix whose condition number was evaluated to determine the robustness of the sensor to small measurement inaccuracies. This index provides an objective measure for assessing the comparative merits of various types of optical fibers for use in a multiparameter sensor.

Weighted Distributed-Effect Sensors for Smart Structure Applications

by

Douglas K. Lindner and Karl Reichard

Fiber & Electro-Optics Research Center
Bradley Department of Electrical Engineering
Virginia Tech
Blacksburg, VA 24061
(703) 231-4580

Abstract

Recently, there has emerged a new class of sensors for vibration sensing in large flexible structures. These sensors are novel in that they are configured to respond to strain along a significant gauge length but they have a scalar output. Hence, they are called distributed-effect sensors. The most well known example is the piezoelectric laminate PVDF film (Lee and Moon, 1990). Another sensor in this class is the modal domain optical fiber sensor (Murphy, et. al., 1990; Cox and Lindner, 1991). By physically altering the sensors along their gauge length or by placement of the sensor on the structure, these sensors can be configured to measure a wide variety of structural parameters. These measurands include modal amplitudes (Lee and Moon, 1990) and travelling waves (Collins, et al, 1991). In a sense, distributed-effect sensors are naturally suited for measuring signals on structures described by distributed parameter models. Clearly, these sensors offer new capabilities over point sensors for sensing vibrations in structures.

These sensors are attractive for Smart Structure applications because they are flexible enough to conform to complex surfaces and they add very little mass to the structure. Fiber optic sensors are particularly attractive because they can be embedded directly into composite materials, they are immune to EMI, are low power, and have excellent sensitivity and dynamic range. Furthermore, they can operate in harsh environments and they can measure a wide variety

physical parameters including chemical concentration, temperature, pressure and strain.

In this paper we consider the use of weighted distributed-effect sensors for vibration suppression control systems. By "weighted" sensor we mean the local sensitivity of the sensor to strain varies as a spatial function of its length. Considering the control system as a whole, we see that the weighting function acts as a spatial filter and as a distributed control gain. This weighting function can be chosen to reduce the order of the dynamic compensator (Lindner, et. al., 1990) or to improve the performance of a control system to suppress acoustic radiation (Lindner, et. al., 1991). Here we analyze the performance of control systems which contain a weighted distributed-effect sensor.

We develop a model of the weighting functions which can be implemented by distributed-effect sensors. This model incorporates the inaccuracies and limitations due to sensor fabrication and attachment. An explicit relationship with modal domain optical fiber sensors and PVDF film is drawn. This model is used to investigate the performance of these sensors in control systems. Two cases are studied. First, the ability of these sensors to act as modal filters is characterized in terms of the inaccuracies of the sensors. This characterization in turn determines the limitations imposed on the ability of distributed-effect sensors to measure shapes (as weighted sums of modes) and the to implement low order compensators (as state feedback of the mode shapes). Secondly, the weighting functions can be used to change the "degree" of observability of the system. We determine the extent to which the observability of the system can changed in terms of several well known measures of observability. The impact of this design variable on the performance of the closed loop system is investigated. The results are illustrated using a cantilevered beam.

References

Collins, S.A., D. W. Miller, and A. H. von Flotow, "Piezopolymer Spatial Filters for Active Structural Control," *Proceedings ot the Workshop on Recent Advances in Active Control of sound and Vibration*, Blacksburg, VA, 1991, pp. 219-236.

Cox, D. E. and D. K. Lindner,"Active Control for Vibration Suppression in a Flexible Beam Using a Modal Domain Optical Fiber Sensor," accepted for the ASME Journal of Vibration and Acoustics, 1991.

Lee, C.-K. and F. C. Moon, "Modal Sensors/Actuators," *Journal of Applied Mechanics*, Vol. 57, 1990, pp. 434-441.

Lindner, D. K., W. T. Baumann, F. Ho, and E. Bielecki, "Modal Domain Optical Fiber Sensors for Control of Acoustic Radiation," *Proceedings of the Workshop on Recent Advances in Active Control of Sound and Vibration*, Blacksburg, VA, April, 1991, pp. 839-850.

Lindner, D. K., K. M. Reichard, W. T. Baumann, and M. F. Barsky, "Measurement and Control of Flexible Structures Using Distributed Sensors," *Proceedings of the 29th IEEE Conf. on Decision and Control,* Honolulu, HI, December, 1990, pp. 2588-92.

Murphy, K. A., M. S. Miller, A. M. Vengasarkar, and R. O. Claus, "Elliptical-Core Two Mode, Optical Fiber Sensor Implementation Methods," *Journal of Lightwave Technology,* Vol. 8, 1990, pp. 1688-1696.

Vengsarkar, A., "Two-Mode, Elliptical-Core, Weighted Fiber Sensors for Vibration Analysis," to appear in the *Proceedings of the Fiber Optic Sensor-Based Smart Materials & Structures Workshop* , Blacksburg, VA, April, 1991.

ACTIVE STRUCTURAL CONTROL DEMONSTRATOR FOR SPACECRAFT APPLICATIONS

G.W. GAME : HEAD OF NEW ATTITUDE AND ORBIT CONTROL SYSTEMS.
EARTH OBSERVATION AND SCIENCE DIVISION
BRITISH AEROSPACE
FILTON,
BRISTOL
U.K.

THIS PAPER DESCRIBES A DEMONSTRATOR FOR A CONTROL SYSTEM WHICH MAINTAINS THE ALIGNMENT OF A STRUCTURE SUBJECT TO LOW FREQUENCY TEMPERATURE AND/OR LOAD VARIATIONS. THE STRUCTURE IS 3-DIMENSIONAL WITH SIGNIFICANT INTERACTIONS BETWEEN AXES AND IS TYPICAL OF A STRUCTURE USED IN SPACECRAFT APPLICATIONS. THE PAPER DESCRIBES THE CONTROL METHOD EMPLOYED WITH DETAILS OF THE CONTROLLER/ACTUATOR SELECTION AND POSITIONING. THE TRADE-OFFS CONSIDERED DURING THE DEVELOPMENT OF THE DEMONSTRATOR ARE SUMMARISED. THESE INCLUDE THE MODEL OF THE STRUCTURE EMPLOYED BY THE CONTROLLER, THE POSITIONING AND SELECTION OF THE SENSORS AND ACTUATORS WITH THE OBJECTIVE OF MAXIMISING OBSERVABILITY AND CONTROLLABILITY AND DETAILS OF THE OPTIMISATION OF THE CONTROLLER PARAMETERS.

EXPERIMENTAL RESULTS ARE PRESENTED AND COMPARED WITH SIMULATION RESULTS FROM A MATHEMATICAL MODEL OF THE SYSTEM.

Flows between Structural and Control Designs at the Example of the Extendable and Retractable Mast

J. Bals, Institut fur Dynamik der Flugsysteme, Oberpfafenhofen, FRG
W. Charon, Dornier GmbH, FRG

The frame of Active Structure design is the general integrated structure/control formulation which foresees a decomposition into system design requirements and separate structure and control requirements and finally designs. The authors intend to advise of their experience about the Extendable and Retractable Mast (ERM) which passive version has been developed by DORNIER as prime contractor under ESTEC. Its imagined active version is an elastic 20 m long mast fitted with piezoelectric local actuators and an offset rigid antenna which pointing stability must be improved.

Since the structural design as part of the Active Structure design is reported in several papers, the control design is emphasized in the paper (choice of controller type, stability, verification).

The information flows betwenn the two independent (structural and control) designs are hierarchically managed. Very few experiences have been gathered on these flows which are therefore not well defined. This very important aspect is also practically emphasized at the example of the ERM.

BRITE/EURAM Project ASANCA (Advanced Study for Active Noise Control in Aircraft)

F. Mornal, MATRA-MS2I, Centre "Les Quadrants"
3 Avenue du Centre, Guyancourt BP 235
78052 St Quentin en Yvelines, France

ABSTRACT :

Initial assessments indicate that active noise control, the introduction of secondary-noise or anti-noise canceling the original noise, has a promising potential for solving the critical low frequency interior noise problem of current and future aircraft.

The main goal of the ASANCA project is to investigate this technique in greater detail and to identify possible optimum active noise control systems for future practical applications. The study will largely enhance the current technology in this field and with this strongly improve the competitivenessof future European aircraft. In addition, a significant technology spin-off to other industrial branches can be expected which can be fruitfully used by all but not only aircraft manufacturer partners.

The objectives will be met by performing extensive well balanced theoretical and experimental research including advanced interior noise calculations and related noise and vibration measurements in aircraft in flight and in a fullscale fuselage test section on the ground. Extensive work wil be performed on the development of optimum prototype active noise control systems, which in the later phase of the study will be initially critically evaluated by flight testing.

Control of Multiflex Systems

A.Silva*   R.Franco†   J.Ramakrishnan‡

Abstract

Control-Structure Interaction effects are dominant in highly flexible space systems with high gain controllers. The integration of structures and controls for a class of space structures is addressed in this paper. The integrated structure-control design methodology seeks to design space systems with a high degree of robustness. Multibody flexible (Multiflex) models are used to describe the dynamics of the spacecraft. The use of model reduction techniques coupled with different controller design strategies are presented. The proposed approaches are demonstrated using a multibody spacecraft model.

*Agusta S.p.A., Milan, Italy
†ESA-ESTEC, Noordwijk, The Netherlands
‡Dynacs Engineering Co.Inc., Palm Harbor, FL.

SPATIALLY DISTRIBUTED SHELL CONVOLVING SENSORS : Theory and Applications

H. S. Tzou [1] and J. P. Zhong [2]

[1] Department of Mechanical Engineering

[1] Center for Robotics and Manufacturing Systems
University of Kentucky
Lexington, KY 40506–0046

[2] Conmec Inc.
Allentown, PA 18103

ABSTRACT

Observation spillover can introduce system instability to undamped distributed structural systems. This problem can be prevented via modal filtering using distributed piezoelectric modal sensors which are spatially shaped and convoluted such that they are only sensitive to specific modal mode(s). In this paper, detailed electromechanics – ***sensor mechanics*** – of spatially distributed piezoelectric shell convolving sensors are analyzed and results presented. It is observed that, sensor sensitivity can be classified into two components: 1) the transverse modal sensitivity and 2) the membrane modal sensitivity in which the former is primarily contributed by bending strains and the later is by membrane strains. Design of spatially distributed cosine–shaped convolving sensors for ring structures is proposed and evaluated. Parametric studies suggest that the transverse sensitivity increases and the circumferential sensitivity remained constant when the ring becomes thicker. Both transverse and circumferential sensitivities increase when the piezoelectric layer becomes thicker or with higher piezoelectric constants.

---

† Supported by NSF, Army Research Office, and Kentucky EPSCoR.

(C/CnvlScr–ADPA.PzShl3.AIAA–91)

ABSTRACT

FOR

"ACTIVE MATERIALS AND ADAPTIVE STRUCTURES"

FRACTURE BEHAVIOR OF PIEZOELECTRIC/ELECTROSTRICTIVE MATERIALS

Stephen W. Freiman
Ceramics Division
NIST
Gaithersburg, MD

Piezoelectric (e.g., barium titanate, lead zirconate titanate) and electrostrictive (e.g., lead magnesium niobate) ceramics are key components of active and smart materials and structures. Being ceramics, these materials are subject to brittle failure. There are no mechanisms of plastic flow which can substantially reduce stresses at crack tips. Failure typically takes place from flaws arising during processing (inclusions, voids) or machining (surface cracks). Critical flaw sizes are usually quite small, of the order of 20 to 50 micrometers. Linear elastic fracture mechanics provides a tool by which we can model and quantify the relations between flaw sizes and the mechanical reliability of these materials. Because flaw sizes are of the order of grain sizes, local perturbations in the isotropic nature of the material must be accounted for in a fracture mechanics approach.

This paper reviews the effects of composition and processing variables on the microstructures of these materials. Microstructure is a major factor determining their fracture behavior. In addition, phase transformations from the non-ferroelectric to the ferroelectric state are also important in determining the fracture toughness of these materials. The twin structures generated by the phase transformation interact with the crack to produce an approximately 30% increase in critical fracture toughness. These materials are also quite sensitive to moisture enhanced crack growth in which flaws extend at stresses well below those at which rapid failure occurs. The effect of moisture is known to be due to a stress-accelerated chemical reaction between a water molecule and the chemical bonds in the solid at the crack tip. These bonds undergo large strains due to the application of stresses far from the crack. Such subcritical crack has been modeled; these models allow one to construct so-called "design diagrams" for components made from these materials.

Electric fields applied to these materials can have a direct effect on crack growth behavior. These fields are locally enhanced at crack tips, giving rise to field concentrations near the tip. This increase in local electric field in turn leads to locally enhanced strains which arise due to either the piezoelectric or electrostrictive nature of the material. It has been shown experimentally that such local strains lead to reductions in flaw growth for

cracks oriented perpendicular to the direction of the applied field. These results are in agreement with theoretical models. However, it has also been observed that delaminations between metal electrodes and the ceramic can also result from the application of an electric field.

Recent work has focussed on the effects of cyclic loading on failure of piezoelectric material (lead zirconate titanate) of interest for high power transducers. This work has shown that entirely new mechanisms of failure can come into play under the application of resonant cyclic loading. This loading, while not necessarily resulting in degradation of the static strength and fracture toughness of the material, does lead to extensive microcracking and other local damage sites. Continuous application of cyclic fields leads to specimen heating and to ultimate failure in relatively short times. Transmission electron microscopy has been used to illustrate the types of damage resulting from cyclic loading under various conditions.

In summary, the materials of prime interest for smart structures and components are brittle ceramics whose fracture behavior is dependent on their microstructure and the external environment. More work is needed to fully characterize these materials in terms of their response to cyclic loading under applied electric fields.

TEM Study for Domain Wall Structures in Ferroelectric Materials

by
M DeGraef and D. R. Clarke
Materials Dept.
University of California, Santa Barbara 93106

The structure and dynamical properties of domain walls in ferroelectric and ferroelastic materials are of fundamental importance for "Smart Applications". As a support for theoretical models of domain wall movement and interactions, a detailed transmission electron microscopic (TEM) study of 90° and 180° boundaries in ferroelectric perovskites was undertaken. High resolution TEM observations and in-situ low temperature TEM observations of domain superstructures in PSZT and related materials will be presented. A novel specimen stage design, allowing low temperature observations combined with in-situ compression and electric field measurements on domain boundaries will also be discussed.

Deformation and Breakdown of Ferroelectric Ceramics Under Applied Mechanical and Electrical Fields

Hengchu Cao and Anthony G. Evans, Materials Department, University of California, Santa Barbara, CA 93107

Ferroelectric ceramics have been employed both as sensors and as actuators. The mechanical behavior of these ceramics dictates the overall mechanical performance when used as single actuator and when incorporated in a smart system. A systematic experimental investigation has been undertaken to study the mechanical and electrical response. Specifically, polarization charges and strains are measured throughout the loading history. Severe nonlinear and hysteretic behaviors are observed. Preliminary attempts have been made to establish the constitutive laws. A novel fracture specimen is devised to study the breakdown of the material. Both cracking and electric breakdown are observed. This information is used to provide guidelines for designing multilayer ferroelectric actuators.

CONCEPTUAL DESIGN, KINEMATICS AND DYNAMICS OF SWIMMING ROBOTIC STRUCTURES USING ACTIVE POLYMER GELS

M. Shahinpoor

Department of Mechanical Engineering

University of New Mexico

Albuquerque, new Mexico 87131

ABSTRACT

Discussed are the structural design, kinematics and dynamics of swimming of autonomous swimming robotic structures which utilize an arrangement of electrically controlled polymeric ionic gel muscles. The general structural design of such swimming robotic structures is considered to be in the form of a submarine structure which is partially encapsulated in an elastic or flexible membrane filled with a counterionic electrolyte such as water+acetone. In such an encapsulated portion of the robotic swimming structure are specifically arranged polyacrylamide or PVA–PAA polymeric cylindrical fibres or bundles. The arrangement of, say, polyacrylamide fibres is such that it is capable of generating microprocessor–based electrically controlled propagating transverse waves to propel the partially encapsulated membrane structure in any direction and in any desired manner.

INTRODUCTION

Kuhn and Katchalsky [1, 2] originally reported on the possibility that certain co–polymers may be chemically contracted and expanded like a synthetic muscle.

As originally reported by Kuhn, Horgitay, Katchalsky and Eisenberg [3] a three

A study on control of a light weight robotic system using piezoelectric motor, sensor and actuator

Zhen Wu, Xiaoqi Bao, Vijay K. Varadan and Vasundara V. Varadan (Department of Engineering Science and Mechanics, Research Center for the Engineering of Electronic and Acoustic Materials, The Pennsylvania State University, 227 Hammond Building, University Park, PA 16802)

One of the trends in the evolution of the robot is the goal of light weight, compliance (or flexible) and fast response in the robotic manipulators and end effecters in which the drivers/motors are self-contained. But, there are two major technical difficulties in reducing the weight: (1) reducing the mass and bulk of conventional motor/driver; (2) heavy and stiff structure must be used for prevention of vibration interference during operation. New Approaches proposed in this paper are using piezoelectric motors as light weight driving sources and embedding piezoelectric sensor/actuator in the light weight flexible frame structure to actively control the undesired residual vibrations.

Piezoelectric motor is an unconventional motor operation on an utterly new principle of obtaining rotation by ultrasonic vibrations. Compare with conventional electromagnetic motor, it has light weight, simple structure, quick response in starting and braking, direct adaptability to low speed rotation. Piezoelectric motors can be applied for automation, robotics, audio and visual appliances, and automotive electrical components. Several prototype piezoelectric motors are made under the guidance of comprehensive theoretical modeling using the finite element method. To seek the applications of the piezoelectric motor in the robotic system which requires quick and precise responses, experimental studies on the response characteristics as well as position and speed control methods of the piezoelectric motor are conducted and a computer controlled servo driver using piezoelectric motor is made.

The natural vibrations become to be low frequency and lightly damped in the light weight, flexible structure, and hence become more difficult to control by using passive damper. An active damper using piezoelectric sensors/actuators and electronic feedback loop can successfully damp out the residual vibration of the structure in a very short period of time. The control of the residual vibration is achieved by picking up the bending vibration signal from the piezoelectric sensors and feeding back to the piezoelectric actuators. A linear rate velocity feedback control algorithm is used in this experiment.

Finally, A light weight single link robotic arm including a flexible arm with the piezoelectric sensors and actuators attached at the sides and a servo controlled piezoelectric motor as the driver is constructed. With the control of computer, the arm tip can move to programed destination quickly and smoothly, the settling time is significantly shortened. This "pizeo-robot" successfully achieves the goals of light weight, flexibility and dexterity in response.

Submitted to the Conference on
Active Materials and Adaptive Structures,
Alexandria, Virginia, Nov. 1991.

Experimental Verification of a Nonlinear Based Controller for Slewing of Flexible Multi-Body Systems

Farshad Khorrami

Control/Robotics Research Laboratory
School of Electrical Engineering & Computer Science
Polytechnic University, 333 Jay Street
Brooklyn, NY 11201
E-Mail: Khorrami@puee1.poly.edu

Abstract

In this paper, experimental results for slewing of a class of flexible multi-body systems (i.e., flexible multi-link manipulators) are given. A two-link flexible arm has been developed at the Control/Robotics Research Laboratory at Polytechnic university. The arm has been designed to study different configurations (i.e., rigidity) of the arms. A two-stage controller is utilized for vibration damping and end-effector trajectory tracking of a two-link flexible manipulator. The controller design strategy is based upon an inner-loop controller attained through asymptotic expansions and an outer-loop controller for further vibration damping and robustness enhancement of the closed-loop dynamics to parameter variations and unmodelled dynamics in the system. The outer-loop controller is a *linear output feedback* designed according to a *quadratic cost criterion.* The measurement used in the outer-loop controller is obtained through an accelerometer mounted on the flexible forearm. The real-time computing power is provided by a digital signal processing board (TMS320C30 based) capable of 33 Mflops.

I. Extended Summary

Robustness to parameter variations and unmodelled dynamics are central issues in design of feedback control systems. Many approaches have been proposed to satisfy and enhance the robustness of the closed-loop systems. One approach is the introduction of frequency dependent weightings in the linear quadratic regulator problem [1, 2]. The choice of frequency dependent weightings will permit a greater flexibility in shaping the loop transfer function. At the same time, control actions with high frequencies may be penalized more than the low frequency ones in order not to excite the unmodelled high frequency dynamics. This is especially important for the class of flexible structures and flexible-link manipulators since low-order models describing the low-frequency dynamics are usually utilized. In this paper, this approach is advocated in addition to a nonlinear controller for flexible-link manipulators.

The complication in controller synthesis for multi-link flexible manipulators is due to the fact that the input/state map of flexible-link manipulators is not externally feedback linearizable [3, 4]. In addition, the dynamics of flexible-link manipulators are much more complicated than the corresponding rigid-link manipulators. Not only the distributed parameter nature of the dynamics is a complication, but also the moving boundary conditions at the tip of the flexible links connected to the next link are major difficulties. Several modeling techniques and different control algorithms have been proposed for flexible-link manipulators ([5, 6, 7, 8, 9], to name a few).

In this paper, an inner-loop nonlinear feedback controller is designed first. At the second stage, an outer-loop controller is applied for suppression of vibration due to the rigid body motion and also to enhance the robustness of the controller to parameter variations and unmodelled dynamics in the system. The inner-loop controller is based on the asymptotic expansion techniques reported earlier [10, 5]. In the inner-loop design the $\mathcal{O}(1)$ terms in the dynamics are feedback linearized and also the vibrations induced on the links by this controller are taken into consideration. The outer-loop design is an *output feedback linear quadratic design.* This advocated controller is implemented on a two-link flexible arm recently developed at Polytechnic University.

II. Experimental Setup

A two-link robot arm with replaceable links has been developed. Different configurations (i.e., rigidity) can be studied by varying the lengths and the thickness of the arms. The actuator for the first link is a direct drive DC motor and the second link is actuated by a geared DC motor through anti-backlash gears. Each arm is instrumented with piezoelectric type accelerometer at the tip and several strain gages along the link. Furthermore, the angular positions and velocities of the arms are measured by optical encoders and DC tachometers respectively. The second joint has been designed to be lightweight; however, the joint is being supported on a granite table with air cushion. All the sensors are signal conditioned and filtered for anti-aliasing. A CCD camera is also being mounted above the setup to provide information on the end-effector tracking performance of the robot manipulator. The real-time computing power for the experiments is a digital signal processing board (i.e., TMS320C30 based) capable of 33 Mflops under supervision of a '386 based machine with a clock speed of 33 MHz.

At this point, we have tested the individual subsystems in the setup and have identified the models of the actuators. A careful model validation of the actuators have been performed taking into account the effects of stiction. The actuator models have been developed through time-domain (step responses) and frequency-domain (frequency response) techniques. For frequency response purposes, low frequency current commands were given to the motors and the tachometer outputs (the velocity) were recorded. All the gains in the system are evaluated. At this point, we are developing an actual simulation of the experimental setup by including the effects of the anti-aliasing filters in addition to all the actuator and sensor dynamics. We have also implemented several simple feedback control strategies.

References

[1] N. Gupta, "Frequency-shaped cost functionals: Extension of linear-quadratic-gaussian design methods," *Journal of Guidance and Control*, vol. 3, no. 6, pp. 529–535, 1980.

[2] F. Khorrami and Ü. Özgüner, "Frequency-shaped cost functional for decentralized systems," in *Proceedings of the 27th Conference on Decision and Control*, (Austin, Texas), pp. 417–422, Dec. 1988.

[3] X. Ding, T. J. Tarn, and A. K. Bejczy, "A novel approach to the dynamics and control of flexible robot arms," in *Proceedings of the 27th Conference on Decision and Control*, (Austin, Texas), pp. 52–57, Dec. 1988.

[4] F. Khorrami, "Dynamical properties of manipulators exhibiting flexibilities," in *Proceedings of the IEEE International Conference on Systems Engineering*, (Pittsburg, PA), pp. 1–4, Aug. 1990.

[5] F. Khorrami, "Analysis of multi-link flexible manipulators via asymptotic expansions," in *Proceedings of the 28th Conference on Decision and Control*, (Tampa, FL), pp. 2089–2094, Dec. 1989.

[6] C. Oakley and R. Cannon, "Initial experiments on the control of a two-link manipulator with a very flexible forearm," in *Proceedings of the 7th American Control Conference*, (Atlanta, GA), pp. 996–1002, 1988.

[7] E. Schmitz, "Modeling and control of a planar manipulator with an elastic forearm," in *Proceedings of the 6th IEEE International Conference on Robotics and Automation*, (Scottsdale, Arizona), pp. 894–899, Apr. 1989.

[8] A. Tzes, M. J. Englehart, and S. Yurkovich, "Input preshaping with frequency domain information for flexible-link manipulator control," in *Proceedings of the AIAA Guidance, Navigation, and Control Conference*, (Boston, MA), pp. 1167–1175, Aug. 1989.

[9] R. H. Canon and E. Schmitz, "Initial experiments on the end-point control of a flexible one-link robot," *The International Journal of Robotics Research*, vol. 3, no. 3, pp. 62–75, 1984.

[10] F. Khorrami and Ü. Özgüner, "A singular perturbation analysis of a distributed parameter model of flexible manipulators," in *Proceedings of the 7th American Control Conference*, (Atlanta, Georgia), pp. 1704–1709, June 1988.

A SEMI-SMART CAPACITIVE SKIN FOR ROBOT COLLISION AVOIDANCE IN SPACE APPLICATIONS

John M. Vranish, NASA/Goddard Space Flight Center
R.L. McConnell, University of West Virginia
Edward Cheung, Jackson and Tull
Wadi Rahim, DSTI

Abstract

Introduction

Telerobotic serving of Orbital Replacement Units (ORU) presents formidable problems. These modules must be handled in a damage-free manner and a high degree of safety. At the same time, because of the premium of platform space on-orbit, the modules will necessarily be located in close proximity to each other. However, there is mounting evidence that astronauts will not have the time to perform the maintenance [1]. It must be done robotically (or telerobotically). From inside a pressurized module, the astronaut can use cameras on the telerobot to guide it to the vicinity of the ORU in need of replacement. Unfortunately, due to occlusion and human error there is always the danger of a collision(s) enroute, either on the robot arms or the payload. In addition, once in the area of the ORU attachment point, precise control and collision avoidance is needed due to the constrained environment.

The problem of occlusion can be overcome to a limited extent by using many camera views of the task site. Unfortunately, the operator can become confused and slowed down if several video monitors need to be continuously examined. Moreover, despite all the efforts by the system builders and human operators, collision-free motion of the arm is still not guaranteed.

Clearly, a proximity sensing skin is needed both on the arm and the payload. The data of the sensitive skin can serve as input to the collision avoidance programs and provide the precise data needed for the docking manuever. Only a sensing skin on the ORU can actually ensure that collisions do not occur during the docking/removal process. It will be shown that a semi-smart sensing skin can perform the entire task, including those elements traditionally requiring machine vision.

In this paper, the implementation strategies will be presented. Following this, the basic characteristics and performance of the "Capaciflector" skin will be described in the context of the generic task of ORU replacement. Afterwards, the motion planning algorithms will be discussed.

Implementation Strategies

There are three distinct environments in which the sensing must operate. Collision avoidance with the robot arm presents the least burden on sensing requirements where accurate position information is not required, but sufficient range must be available. Television cameras will be used to guide the operator in robot motion, but will not cover all parts of robot arms. In this case, resolution of a few centimeters is sufficient to prevent collisions.

Actual servicing of the ORUs where the containers are removed and reinserted in a cluttered environment will place severe strain on the operator. Robot mounted cameras will not be able to see around the container which must be placed accurately between existing containers. In all probability, physical contact between containers during replacement will be highly discouraged in order to eliminate damage. In this case, proximity sensors will be required on the ORU surface to aid the operator during this operation. At these close quarters, resolution on the order of 1 cm, will be required. Sensors will also be desirable on the surfaces of the ORU to prevent collisions during gross placement. Therefore, power and communication lines will be required to the ORU container while attached to the robot.

In the third case, docking of the end-effector of the robot arm to attach to the ORU and final placement of the ORU to the locking position on the station fixture, will require placement to within a fraction of a centimeter where mechanical docking devices can take over. Again, this operation will be complicated by the fact that the view by the camera will be obscured during the final docking maneuver.

Capacitive Sensor System

The capacitive sensors, termed "Capaciflectors", have been described in several papers [2], and consist of a simple planar sensor with a shield to reflect the fields from local grounds. Objects the size of a human hand can be detected at sufficient range. The capacitive sensor is a "proximity" sensor with essentially a hemispherical pattern, allowing fairly wide separation while accommodating substantial overlap. The electronic circuitry to drive the sensor is presently located on a circuit board about 6x6 cm. Ultimate packaging will likely be a single silicon chip fully integrated within the sensor package. Therefore, sensors and electronics will not significantly impact the dimensions of the robot arms or the payload. As there are no protrusions, physical damage is unlikely in normal operation. With thin films for the conducting surfaces, shorts due to perforations from micrometeorites are unlikely to occur. Small perforations have virtually no effect of the sensor.

The proposed configuration will consist of several sensors on each robot arm for collision avoidance, a few sensors on the end-effector of the robot arm to allow docking with attachment devices and ORU attachment points, and sensors on the sides and end of the ORU for gross movement, insertion into storage/operation compartment, and for final docking with the station attachment. Each sensor will contain driver and detection electronics which will preprocess the information and present that information to the robot controlling computer as proximity and range information.

Robot Motion Planning

The robot controller will interpret the sensor data in terms of direction of movement allowed. The obstacle avoidance approach will be to identify for each period in time the point on the manipulator that is closest to an obstacle with the help of the sensor skin. This point, termed the *obstacle point*, is then assigned a velocity away from the obstacle surface. In this manner, the robot arm is able to stop motion towards the obstacle (regardless of the configuration of the arm), and is able to move away from or along the surface of the obstacle. The robot controller will move to the desired position of the arm if this is possible, in the case of an obstruction, the arm automatically and smoothly departs from the desired end-effector position maintaining safety at all times.

In addition, semiautomatic alignment and positioning is proposed for final insertion and docking. The data of the sensors on the ORU will be fed back to the human operator in such a manner to facilitate and semiautomate the docking procedure.

References

1. W. Fisher, C. Price, Space Station Fredom External Maintenance Task Team, Final Report, National Aeronautics and Space Administration, Johnson Space Center, TX, July 1990.

2. John M. Vranish, Robert L. McConnell, and Swami Mahalingam, "Capaciflector" Collision-Avoidance Sensors for Robots, Journal of Electrical Engineering and Computers, to be published.

ABSTRACT - Active Materials and Adaptive Structures
November 5th-7th 1991,
Alexandria VA

DO EMBEDDED SENSOR SYSTEMS DEGRADE MECHANICAL PERFORMANCE OF HOST COMPOSITES?

Dr. R. Davidson
AEA Technology, Harwell

The development of smart composite materials relies upon building into the composite structure during fabrication a suitable nervous system which can sense the localised environment. In order to produce an adaptive structure, actuators are also necessary which respond to the sensor signals to effect shape change or vibration control. These actuators would be either embedded, as in the case of shape memory alloy wires or surface bonded as in piezo electric ceramics.

For embedded sensor or actuator systems concern has been expressed as to the possible structural strength degradation which may result. This paper considers the modelling of stress and strain fields around and within optical fibres embedded in carbon reinforced composites. The saturation is complex since fibre optics generally consist of a silica based core/cladding combination designed to give the waveguiding and sensing characteristics of the fibre plus a polymeric protective coating to reduce handling damage and increase environmental resistance. The optical fibre package of this type will always be significantly larger in diameter than the reinforcing fibres which are typically 10µm in diameter. When such optical fibres, typically 100-300µm in diameter, are embedded in composite laminates there will be an inevitable disruption of the reinforcing fibres in the vicinity of the fibre optic. the nature of this disruption will not only be dependent upon the diameter of the embedded fibre but also on the relative orientation of the optical fibre and the neighbouring reinforcing fibres.

In order to be acceptable the fibre sensor must:

- produce a minimum perturbation in the distribution of reinforcing fibres;
- not significantly reduce the strength or stiffness characteristics of the composite;
- not suffer from excessive attenuation such that sensing techniques can not be applied;
- include suitable fibre pigtails for the input and output laser light.

The work presented here aims to determine the effect of an embedded optical fibre on laminate strength as a function of fibre properties and orientation, such that general recommendations for the manufacture of composite components containing optical fibres can be made. Several laminate/optical fibre geometries have been studied using finite element analysis to determine the magnitude and extent of stress concentrations caused by the embedded fibre when the laminate is under uniform thermal, tensile or shear loads applied in the plane of the fibre cross section. For these simple load cases a two dimensional analysis was used. Initial studies were concentrated on geometries where the optical fibre lay parallel to ply reinforcement fibres. The more complex case where the optical fibre lay perpendicular to the reinforcement was then studied.

The analytical results will be compared with experimental data derived from the fracture of CFRP plates containing embedded fibre optics having different sizes and coatings.

The work shows that it is possible to design a sensor fibre which can be successfully embedded which will not significantly degrade the static mechanical properties of the host composite.

TENSILE STRENGTH AND STIFFNESS REDUCTION GRAPHITE/BISMALEIMIDE LAMINATES WITH EMBEDDED FIBER OPTIC SENSORS

David W. Jensen, Jesús Pascual, James A. August

The Pennsylvania State University
Department of Aerospace Engineering
233 Hammond Building
University Park, Pennsylvania 16802

ABSTRACT

Embedded optical fibers introduce geometric discontinuities in their host composite structure, due to their relative large size and reduced mechanical properties in comparison to typical reinforcing fibers. The objective of this investigation was to quantify the effect of optical fibers embedded on varying orientations and locations on the tensile mechanical properties of graphite/bismaleimide (Gr/BMI) laminates. Experimental strength and stiffness data from seven configurations are compared to a control configuration without embedded optical fibers. All specimens were fabricated from G40-600/5245C Gr/BMI pre-preg tyape to form a $[0_3/90_2/0]_s$ stacking sequence. The orientation with respect to loading direction and adjacent graphite fibers, and the location within the laminate was varied to form different configurations. Uniaxial tension testing was carried out in accordance with ASTM standards. Catastrophic failure was prompted by graphite fiber fracture. The results show that optical fibers do have a detrimental effect on the tensile behavior of Gr/BMI laminates; however, tensile properties are only modestly reduced. For the particular configurations investigated, the effects on tensile strength ranged from 1% to 9%, while the stiffness reduction ranged from 1% to 11%.

INTRODUCTION

The primary purpose of fiber optic smart structures is to enhance the material performance and survivability of aerospace structures. However, embedded optical fibers have been shown to introduce geometric discontinuities in their host composite structure, due to their relative large size and reduced mechanical properties in comparison to typical reinforcing fibers. Specifically, the diameters of typical fiber optic sensors are relatively large in comparison with those of typical structural reinforcing fibers (e.g., on the order of 125-500 microns, while the diameter of graphite fibers is approximately 5-10 microns). By comparison, a single layer of an advanced composite material is typically only about 120-140 microns thick. The objective of this investigation was to establish the significance of the orientation of embedded optical fibers on the tensile behavior of

MICROMECHANICS OF FIBER OPTIC SENSORS

Eugene Pak[1], Vicente DyReyes[2], and Eugene S. Schumter[3]

1. Grumman Corporate Research Center
2. Grumman Aircraft Systems
3. Grumman Data Systems

Coated optical fiber embedded in a host matrix is analyzed in the framework of linear elasticity. Closed form solutions are obtained for the case of out-of-plane longitudinal shear load and inplane uniaxial load. Optical fiber and its coating is modeled as isotropic concentric circular inclusions. It is assume that the host matrix is infinite, isotropic, and homogeneous elastic material. This study is useful in determining the effects of the coating and its elastic constants on the strain transfer to the optical fiber. Also studied is the effect of the elastic constants on the strain concentration in and around the optical fiber.

Boundary element analysis is performed to study the interaction between the optical fiber and the near by ply interface. Feasiblility of optical fiber sensor detecting delamination of composite laminate is also studied.

In order to address the micromechanics issues for elliptical-core optical fiber sensor, a closed form solution is obtained for an antiplane elliptical inclusion problem using comformal mapping technique. The effects of ellipticity on strain transfer and the stress concentration are studied and comparisons are made with the circular case. This study can also be useful in understanding stress distribution around elliptical resin pockets forming around an optical fiber embedded perpendicular to the composite fibers.

MICRO-MECHANICS OF SENSOR-HOST INTERACTIONS

IN FIBER-OPTIC SENSORS

EMBEDDED IN LAMINATED "SMART" COMPOSITES

A. Dasgupta, J. Sirkis and Y. Wan
Mechanical Engineering Department
University of Maryland
College Park, MD 20742

ABSTRACT

It is well known that embedding fiber-optic sensors in laminated composites causes a lenticular resin-rich interlaminar pocket to be created around the fiber-optic inclusion, except for the special case when the native reinforcing fibers in the adjacent layers of the host composite are parallel to the optical fiber. The specific geometry of this lenticular pocket has significant implications in terms of the micromechanical issues of the problem. This study investigates both the geometry of the resin pocket as well as the resulting stress and strain fields.

The micromechanical issues are related to the local interactions between the host and the sensor and include strain and heat transfer mechanisms between the host and the fiber-optic transducer, as well as damage induced due to the presence of the fiber and the resin pocket, under external applied loads. Experimental as well as analytical evidence indicates the presence of stress/strain concentrations in the host structure in the immediate vicinity of the resin pocket and of resulting damage under cyclic thermo-mechanical loads. The regions of high stresses can be at the interface between the optical fiber and the resin, in the host adjacent to the optical fiber or in the host at the tip of the resin pocket. Similar gradients are expected in temperature fields under thermal loads. The stress/strain/temperature concentrations can not only perturb the values of the very parameters being sensed by the fiber-optic transducer, but may also cause damage to the sensor-host assembly. The damage may take the form of debonding between the sensor and the surrounding resin and/or microcracking in the resin and/or interlaminar crack propagation at the sharp tip of the lenticular pocket. The damage mechanisms are important because they can affect the reliability/durability of the sensor by degrading the strain/heat transfer mechanisms over a period of time, and in extreme cases, by threatening the very integrity of the host structure. Clearly, the importance of the micromechanical issues are integral to the performance and reliability of the sensor.

In this paper, we present first, a quantitative investigation of the geometry of the lenticular resin-rich pocket using both analytical prediction tools and experimental verification. This information is of critical importance to subsequent studies of the micromechanical interactions between the sensor and the host. Energy methods are used

to obtain the shape and size of the pocket as a function of critical parameters such as properties and size of the optical fiber; properties and size of the host composite laminae and reinforcing fibers; and finally lamination parameters such as stacking sequence and lamination pressure. As a first order approximation, a linear elastic analysis is performed by ignoring the changes occuring during the resin vitrification process. The interactions between the resin and the fibers of the host composite are only approximately modeled. Higher order perturbations of the resin pocket geometry due to stresses generated by the resin vitrification process are ignored. Predictions show excellent agreement with photomicrographs of the actual geometry observed by fabricating and sectioning composite samples containing embedded optical fibers. Parametric studies are performed by varying important parameters such as the diameter of the optical fiber, material system and stacking sequence of the host composite and lamination pressure.

The second part of this paper contains results of finite element investigations of the stress and strain fields in and around the predicted resin-pocket. This information forms the basis of our understanding of the micro-mechanical interactions between the sensor and the host. The predicted resin pocket geometry is automatically discretized and analyzed by the finite element method under different loads states imposed on the sensor-host assembly. Commonly encountered load states in practical situations are in-plane extensional/compressive loads and transverse shear loads due to vibrational loads; and thermo-mechanical loads due to temperature excursions. Stress results are presented for these load states for a $[0/90]_s$ graphite-epoxy laminate with an embedded 80 micron diameter optical fiber.

This study provides a simplified approach to modeling the geometric intrusiveness of a fiber-optic sensor on its host composite and provides important insights into the influence of different physical parameters on the geometry of the resin-pocket, and on the resulting stresses, under several different loads. The implications on the micromechanical issues in embedded fiber-optic sensors are significant and are clearly discussed in this paper.

COMPRESSIVE STRENGTH AND STIFFNESS REDUCTION IN GRAPHITE/BISMALEIMIDE LAMINATES WITH EMBEDDED FIBER-OPTIC SENSORS

David W. Jensen, James A. August, and Jesús Pascual

The Pennsylvania State University
Department of Aerospace Engineering
233 Hammond Building
University Park, PA 16802

March 1991

The research presented in this paper examined the change in mechanical properties of Graphite/Bismaleimide (Gr/BMI) laminates due to the presence of embedded fiber-optic sensors. The specific objective of this investigation is to quantify the effects of embedded optical fibers on the compressive behavior of composite laminates. The approach taken was to compare experimental strength and strain gage data from seven test groups to a control group. The specimens in the test groups had optical fibers embedded parallel and perpendicular to the loading direction and/or the adjacent fiber direction, while the control group had no embedded fibers (optical fiber configurations shown in fig. 2). All specimens were fabricated from G40-600/5245C Gr/BMI pre-preg with a $(0_3\ /90_2\ /0\ )_s$ stacking sequence. The optical fibers were precisely positioned during hand layup using a special fixture. Specimen manufacture, uniaxial compression testing using an IITRI fixture, and data reduction were performed in accordance with ASTM Standard D3410-87. Preliminary results indicate embedding of optical fibers reduces the ultimate compressive strength anywhere from 1.4% to 70.1% while reducing the elastic compressive stiffness anywhere from 6.9% to 20.5%, depending on the configuration of the embedded sensor. It was concluded that optical fibers should be embedded parallel to adjacent graphite fibers and parallel to the loading direction, if possible.

PIEZOELECTRIC CERAMIC-POLYMER COMPOSITES WITH 0-3 CONNECTIVITY

K.H. Han, R.E. Riman and A. Safari
Department of Ceramic Science and Engineering
Rutgers - The Sate University of New Jersey
Piscataway, NJ 08855-0909

Composites with 0-3 connectivity consist of a three-dimensionally connected polymer matrix loaded with discrete ceramic particles. This type of composite is easy to fabricate and amenable to mass production. However, the properties of piezoelectric 0-3 composites are strongly dependent on the piezoelectric ceramic filler material and polymer phases, as well as the fabrication method employed. In most of the earlier works, PZT, pure or modified $PbTiO_3$ or $(Pb,Bi)(Ti,Fe)O_3$ ceramic powder were used as the ceramic filler materials. Ceramic powders were prepared by solid state reaction or chemical solution techniques such as sol-gel and coprecipitation. Various kinds of polymers such as epoxy, rubber, and thermoplastic or thermosetting polymers were used as the polymer matrix. To fabricate composites, most of the methods were based on the mechanical mixing of ceramic powder with polymer. Using this technique, the mixture is compounded by shear mixing and formed into a sheet by hot-rolling or pressing. Recently, we developed a new method, colloidal processing, to prepare the flexible piezoelectric ceramic-polymer composites with 0-3 connectivity. In this method, the polymer was dissolved in a solvent and ceramic powder was dispersed to form a suspension. Powder-polymer coacervates were precipitated from the suspension by the addition of a nonsolvent. Colloidal filtration and subsequent pressing consolidated the coacervates to form composites. The excellent microstructural uniformity imparted by this method combined with the improved powder characteristics of chemically derived ceramic powder significantly increased the polability of 0-3 composite. The largest d_{33} (65 pC/N) and $d_h g_h$ figure of merit (5950×10^{-15} m^2/N) were obtained from the composite composed of coprecipitated Mn-doped $(Pb,Bi)(Ti,Fe)O_3$ ceramic powder and epoxy polymer. The effect of important processing parameters on the dielectric and piezoelectric properties will be discussed in this presentation.

FLEXIBLE PIEZOELECTRIC MATERIALS -
APPLICATION TO PRESSURE AND VIBRATION SENSING

Fred G. Geil*

Piezoelectric materials can be attached to structures for the purpose of sensing sound pressures or structure vibration. If the materials are flexible they easily may conform to a curved surface, or to a flexing surface. The signals that may be sensed by piezoelectric materials must be AC by nature, as in sound pressure and vibration, not DC by nature, as in barometric/hydrostatic pressure or static strain. The AC signals may be used as is, as in a hydrophone application, or they may be inputs to an active cancellation system. In either case the area of the sensor is an important design parameter. In the case of hydrophone arrays the area determines the beam pattern. In the case of vibration sensing, the area determines the spatial frequencies or modes that may be sensed, and those that will be rejected because two or more spatial half cycles are being sensed by one sensor.

The two flexible piezoelectric materials of greatest availability are a piezoelectric composite tile composed of lead titanate powder in a neoprene matrix, and polyvinylidene flouride film. Both materials have no natural piezoelectric behavior but are polarized during manufacture by subjecting them to a high electric field imposed across the thickness dimension, while cooling from a high temperature. The resulting piezoelectric constants of greatest interest are the hydrostatic "g" constants in the case of sound sensing, and the lateral "g" constants in the case of vibration.

This paper also covers the correct application of each of these sensing materials, and presents experimental data showing the performance attained. One attractive feature of these new materials is the forming of arrays by simply dividing the electrodes into smaller areas. In this way a single tile or film may form an array of any number of elements.

Part of the successful application involves choosing an appropriate preamplifier which must be sufficiently close to the sensor that there will be no signal loss due to capacitive loading by the connecting cable. When the signal to be sensed is weak, electrical noise may predominate, and the choice of preamplfier circuit becomes critical. A computer model is described which enables optimization of the sensor-preamplifier interface.

*Westinghouse Oceanic Division
Box 1488. M/S 9845

LARGE AREA PIEZOELECTRIC-POLYMER COMPOSITES

Darrah, S.D.[1], Batha, H.D.[1],
Damjanovic, D.[2], Cross, L.E.[3]

Large area 1-3 fiber reinforced piezoelectric polymer composites have potential use a sensors, acoustic generators and sound and vibration damping materials on large structures. Small area composites are used successfully as medical transducers, but fabrication limitations prevented consideration for large systems operating at lower frequencies. Advanced composite manufacturing techniques now allow fabrication of these large area composites with precise location of PZT rods and continuous glass fibers in the transverse plane.

Plates up to 6.3 mm thick have been made with square PZT 5H rods, 0.5 to 1 mm cross-sectional width, normal to the surface. The ceramic concentrations have varied from 1.3 to 20% by volume with transverse fiber reinforcement concentrations of 5 to 11%. both flexible and rigid epoxy resins have been used in the matrices. Different methods of applying metallic electrodes to the large area surfaces have been investigated and appropriate poling methods have been studied. The ability to repair defective or damaged areas has also been demonstrated.

Precise placement of piezoelectric rods and transverse fibers is necessary for reproducibility of composite manufacture and for meaningful characterization. Resin impregnation processing is critical to avoid residual thermal stress. These requirements were satisfied by use of the FMI® Ultraloom® and resin impregnation procedures developed in the manufacture of very large structures.

FMI® is a Registered Trademark of Fiber Materials, Inc.
Ultraloom® is a Registered Trademark of Fiber Materials, Inc.

The presented data will show the effect of rod size, concentration and spacing, reinforcement concentration, resin type; and overall specimen dimensions on piezoelectric properties. Measured properties include K, d_h, k_t, and the impedance spectrum.

[1] Fiber Materials, Inc.
5 Morin Street
Biddeford Industrial Park
Biddeford, ME 04005-4497

[2] EPFL
Departement des materiaux
Laboratoire de ceramique
MX-D Ecublens
CH-1015 Lausanne
Switzerland

[3] Penn State University
Materials Research Lab
University Park, PA 16802

PIEZOELECTRIC AND ELECTROSTRICTIVE COMPOSITE ACTUATORS

R.E. Newnham, Q.C. Xu and S. Yoshikawa
Materials Research Laboratory
The Pennsylvania State University
University Park, PA 16802

Several types of actuators composed of metal end plates and PZT (or PMN) ceramics have been developed recently. Shallow cavities between the ceramic disk and the metal electrodes have been designed to convert and amplify a radial displacement of the piezoelectric ceramic into a large axial motion of the metal end caps. Large d33 and dh coefficients in excess of 2500 pC/N were obtained with the new metal-ceramic composite transducers. The behavior of the electrically-induced strain with pressing force, frequency, and electrical field will be described, along with the important geometrical and material considerations, and some interesting applications.

FINE SCALE PZT FIBER / POLYMER COMPOSITES

A. Safari and D. J. Waller,
Department of Ceramic Engineering,
Rutgers - The State University of New Jersey,
Piscataway, New Jersey 08855-0909

PZT ceramic material was fabricated in woven fiber form using a 'relic' process. Piezoelectric composites incorporating the woven PZT within a polymer matrix were evaluated for potential usage in low and high frequency transducer applications, such as hydrophone and biomedical imaging (ultrasound). The advantages of relic processing are that it allows the production of large area composites with a fine-scale distribution of PZT and polymeric phases.

Carbon fabrics consisting of activated carbon fiber woven in two dimensions were impregnated with PZT precursor material after soaking in an alkoxide stock solution. After drying, fabrics were stacked 15-20 layers high on a bed of coarse PZT powder, which allows shrinkage to occur freely. Heat treatment in air at 550°C / 10hrs and 700°C / 4hrs served to burn out the carbon and calcine the Niobium doped PZT. The result was PZT ceramic relics which attained the form of the original carbon fiber weave. After sintering in a Pb-rich atmosphere, ceramic relics were vacuum backfilled with polymer to yield PZT / polymer composites.

Composites were fabricated with a number of different types of epoxy polymer* and were poled and tested in two distinct orientations. In the first composite orientation, PZT fibers were parallel to the electroded surfaces, whereas in the second, the fibers were perpendicular. Poling of composites was accomplished either conventionally or by the corona discharge technique.

Electromechanical characterization included the measurement of K, tan δ, d_{33}, g_{33}, d_h and g_h coefficients. Critical processing parameters such as stock solution concentration, number of soakings, sintering schedule and polymer method were varied and processing-property relationships were investigated and will be presented.

* Eccogel (1365-0, 1365-45, 1365-80) epoxy resins;
Spurrs standard composition epoxy

Acknowledgement: The authors are grateful for financial support received from the Office of Naval Research.

ABSTRACT

Benefits of Controls-Structures Interaction Technology for Future NASA Missions

William L. Grantham
NASA Langley Research Center

To be presented at the
Conference on Active Materials and Adaptive Structures
November 5-7, 1991
Alexandria, Virginia

Example "case studies" are presented in this paper to show how Controls-Structures Interaction (CSI) technology, when used in the design and control of large space structures, can increase mission performance and enable certain future NASA missions. Many future NASA missions have common CSI technology needs due to their inherent flexibility which requires the lower structural frequencies and spacecraft control system to occupy the same spectral bandwidth. These studies have been used to help formulate and direct the CSI technology development program being jointly pursued at Langley Research Center (LaRC), Jet Propulsion Laboratory (JPL), and Marshall Space Flight Center (MSFC).

Several CSI benefit studies have been completed to date as part of an ongoing assessment process: 1) missions requiring large antennas, 2) missions requiring large optical systems, and 3) missions requiring the use of closed-loop controlled flexible remote manipulator arms for in-space assembly. Also reported are the recent results concerning Space Station Freedom user accommodations and the influence of routine in-orbit disturbances and other effects such as astronaut movement which critically affect precision pointing and microgravity experiments.

ABSTRACT

Research on the Structural Dynamics and Control of Flexible Spacecraft

Brantley R. Hanks
NASA Langley Research Center

To be presented at the

Conference on Active Materials and Adaptive Structures
November 5-7, 1991
Alexandria, Virginia

This paper presents a brief explanatory overview of the technical nature and importance of structural dynamics and control for proposed future spacecraft structures. Requirements and needs for spacecraft which are assembled or constructed in space will be emphasized. The paper will also describe on-going test and analysis activities in a continuing interdisciplinary technology program at the NASA Langley Research Center to develop structural dynamics and control methods for application to future missions. The following topics will be included:

1. Ground Verification of Large Flexible Spacecraft - Difficulties and risks in ground verification of large spacecraft will be overviewed. The possibility of using scale models for ground tests of spacecraft which are too large or too flexible to test accurately on Earth or which change during their lifetime will be discussed. A recently constructed dynamic test model of Space Station Freedom and an accompanying in-progress test-analysis program will be described. Suspension approaches for supporting gravity-sensitive spacecraft during dynamic tests in Earth-gravity will be discussed. Experimental results on candidate devices for some categories of tests will be described.

2. On-Orbit Dynamic Response Measurements - Analytical investigations of proposed on-orbit dynamic tests of Space Station Freedom will be described. The primary purpose of these tests would be to complete verification but they would also provide data for calibration of ground tests and analysis in order to develop improved methods for future applications.

3. Controls-Structures Interaction (CSI) - Needs and difficulties in the control of flexible spacecraft will be briefly reviewed. Experimental results of CSI tests conducted on a 20-meter foldable truss-beam known as the Mini-Mast, developed for ground test studies of CSI, will be described. A second CSI experiment, also about 20 meters in dimension and based on science platform topology, will be described. Initial control results will be presented.

4. System Identification - The ability to develop mathematical models based on experimental data is an important subtask in spacecraft dynamics for future missions. Some recent studies in this technology area will be discussed.

ABSTRACT

Integrated Analysis and Design for Flexible Space Structures

James G. Batterson
NASA Langley Research Center

To be presented at the

Conference on Active Materials and Adaptive Structures
November 5-7, 1991
Alexandria, Virginia

A major objective of the analytical design methods development activity at the Langley Research Center (LaRC) Controls-Structures Interaction (CSI) Program is to develop and verify unified modeling and analysis techniques which provide the accuracy, speed, and computational efficiency needed to design integrated controls-structures systems that meet dynamic requirements on configuration, pointing, and operations for 21st century NASA missions. Progress has been made in both integrated design methodology and computational efficiency. Integrated design methods have been developed for two nonmodel-based controllers. Control optimized and integrated designs have been completed using both static and dynamic dissipative controllers with significant damping gains resulting from the integrated designs. Model-based controllers, including LQG and H-infinity, have been investigated and found to be very sensitive to assumed model structure. This presentation will summarize the background, objectives, and accomplishments to date. This includes analysis, simulation and laboratory test data for optimized controllers, as well as analysis and simulation results for several integrated designs, one of which is currently being fabricated for spring 1992 testing.

ABSTRACT

Summary of LaRC CSI Flight Experiment Activities

Anthony Fontana
NASA Langley Research Center

To be presented at the
Conference on Active Materials and Adaptive Structures
November 5-7, 1991
Alexandria, Virginia

The technical objectives, configurations, and development plans will be presented for four Controls-Structures Interaction (CSI) on-orbit experiments/demonstrations.

The Middeck Active Control Experiment (MACE) is being developed by the Massachusetts Institute of Technology (MIT) and is scheduled for launch in 1994. MACE investigates the CSI phenomena of a multi-instrument science platform (multiple interacting control systems) via closed-loop experiments in the middeck area of the Shuttle crew quarters.

The Jitter Suppression Experiment (Jitter) will be developed by McDonnell Douglas and is also scheduled for launch in 1994. Jitter is a Shuttle bay experiment which investigates the CSI issues associated with vibration suppression in optical benches through the application of passive and active damping treatments.

Langley Research Center (LaRC) and Johnson Space Center (JSC) are conducting a joint activity under the Technology Bridging Program for improving the operational performance of the Shuttle Remote Manipulator System (RMS). LaRC is designing control laws for rapid damping of RMS oscillations. These controllers will be tested on the crew training simulator at JSC in 1992. The simulation results will be used to decide whether or not to propose a flight demonstration on a subsequent Shuttle mission.

LaRC, MSFC, and Jet Propulsion Laboratory (JPL) are jointly defining the requirements for a CSI laboratory in space. This facility will allow validation of a broad range of CSI technologies to meet the needs of the CSI community including NASA and DOD, industry, and universities. A Phase A definition is planned for 1992.

ABSTRACT

The Controls-Structures Interaction Guest Investigator Program

Rudeen Smith-Taylor
NASA Langley Research Center

To be presented at the
Conference on Active Materials and Adaptive Structures
November 5-7, 1991
Alexandria, Virginia

The Guest Investigator (GI) Program is one of the five elements of the NASA Controls-Structures Interaction (CSI) Program. Through the GI program, researchers from industry and academia use government testbeds to validate advanced control techniques and integrated controls-structures designs. The objective of the GI program is to support CSI technology advancement by 1) involving CSI researchers from academia and industry, 2) providing the researchers with the most advanced CSI test facilities for conducting experimental validation, and 3) disseminating the experimental results to the research community.

Phase I of the CSI GI program was initiated in 1988 with the selection of eight researchers to conduct CSI experiments in two NASA test facilities. Phase I was a 2-year program that was completed in April 1991. Phase II, which will be awarded in mid-summer of 1991, is a joint NASA/DoD program. The five GI's selected for Phase II will conduct experiments in three newly developed government testbeds. The presentation will give an overview of the CSI GI program, describe the test facilities used in Phase I and to be used in Phase II, present some experimental results from Phase I, and briefly discuss the research thrusts of Phase II.

SHAPE MEMORY ALLOYS:
PROMISE AND PROBLEMS

by

Charles W. Marschall

and

Terrance Hill

Battelle
505 King Avenue
Columbus, Ohio 43201

For presentation at Active Materials
and Adaptive Structures Conference
November 5-7, 1991
Radisson Mark Plaza Hotel
Alexandria, Virginia

ABSTRACT

Because of their unique properties, shape memory alloys are strong candidates for use in "smart" structures and devices. In this paper, the authors review the main types of shape memory alloys, the characteristics of each type (including both one-way and two-memory), and the mechanisms responsible for their unique behavior, to better enable designers of smart structures and devices to utilize the alloys to their best advantage.

Despite their promising features, shape memory alloys have encountered problems in finding suitable applications, due in part to their metallurgical and mechanical complexity. The authors will review some of the problems that have been encountered in applications that involve shape memory alloys, to illustrate that maximum utilization of the their capabilities will require the contributions of designers, materials scientists and engineers, and alloy producers.

Actuation And Control With Ni-Ti Shape Memory Alloys

Dieter Stoeckel
Tom Waram

Raychem Corp., Menlo Park

"Shape Memory" describes the effect of restoring the original shape of a plastically deformed sample by heating it. This phenomenon results from a crystalline phase change known as "thermoelastic martensitic transformation". The shape memory effect in Ni-Ti alloys can be used to generate motion and/or force in actuators, fasteners and couplings. At temperatures below the transformation temperature, Ni-Ti alloys are martensitic. In this condition they are very soft and can be deformed easily (like soft copper). Heating above the transformation temperature recovers the original shape and converts the material to its high strength, austenitic, condition (like steel).

The transformations from austenite to martensite and vice versa do not take place at the same temperature. The hysteresis is an important characteristic of the heating and cooling behavior of shape memory alloys and actuators made from these alloys. Depending on the alloy used and/or its processing, the transformation temperature as well as the shape of the hysteresis loop can be altered in a wide range. Transformation temperatures (As) between -100°C and +100°C can be achieved; some Ni-Ti alloys show a pronounced premartensitic (R-phase) transformation which is characterized by a very narrow hysteresis of 0 to 5°C. On the other hand, a very wide hysteresis of over 150°C can be realized in Ni-Ti-Nb alloys after a particular thermomechanical treatment.

The standard thermomechanical processing of Ni-Ti alloys generates a very steep hysteresis loop (a greater shape change with a lesser change in temperature) which generally is desirable in applications where a certain function has to be performed upon reaching or exceeding a certain temperature. Special processing can yield a hysteresis loop with a more gradual slope, i.e. a small shape change with temperature. This behavior is preferred in applications where proportional control is required.

Ni-Ti shape memory actuators respond to temperature changes with a shape change. The change in temperature can be caused by a change in the environment or by electrically heating the Ni-Ti element. In the first case, the shape memory alloy acts as a sensor and an actuator (thermal actuator). In the second case, it is an electrical actuator that performs a specific task on demand. Thermal as well as electrical Ni-Ti actuators combine large motion, rather high forces and small size, thus they provide high work output. They usually consist of only a single piece of metal, e.g. a straight wire or a helical spring, and do not require sophisticated mechanical systems.

Shape memory thermal actuators have been used successfully in the areas of thermal compensation, thermal actuation and thermal protection. They often fit into tight spaces in given designs, where other thermal actuators, like thermostatic bimetals or wax actuators, would require a major redesign of the product. In flow-control or oil pressure control valves, for example, helical springs can be placed in the fluid path, without restricting the flow. Thus, they provide fast response to changes in temperature.

Electrical actuators have been used to replace solenoids, electric motors etc. in applications where quiet operation, small dimensions, small or large forces and simplicity of the design is required. By controlling the power during electrical actuation, specific levels of force and/or specific positions can be maintained.

This paper describes the shape memory effect in Ni-Ti alloys with emphasis on processes to achieve particular shapes of the hysteresis loop. Design principles for shape memory actuators will be discussed and some applications of thermal and electrical actuators will be presented.

THE SHAPE MEMORY EFFECT AND RELATED PHENOMENA: C. M. Wayman, Dept. of Materials Science and Engineering, University of Illinois, 1304 W. Green St., Urbana, IL 61801, USA.

The shape memory effect (SME) is associated with a martensitic transformation in which a material undergoes a displacive, shear-like phase change when cooled below a certain temperature, M_s. The transformation is complete when a lower temperature, M_f, is reached. When the martensite is deformed (below M_f) it undergoes a strain which is completely recoverable upon heating, which begins at temperature, A_s, and is completed at a higher temperature, A_f. The M_s, M_f, A_s and A_f temperatures depend on the alloy system involved and recovery strains range from 2-10%. The martensite in shape memory alloys (SMAs) may also be isothermally induced above M_s by an applied stress, known as stress-induced martensite (SIM) which disappears (reverses) when the stress is removed, resulting in a mechanical type (as opposed to thermal type) shape memory. The formation and reversion of SIM gives rise to superelastic behavior. The "two way" SME combines aspects of the "one way" SME and SIM formation.

ACTIVE CONTROL OF BUCKLING OF NITINOL-REINFORCED COMPOSITE BEAMS

A.BAZ, M.MUTUA and J.GILHEANY

Mechanical Engineering Department
The Catholic University of America
Washington, DC 20064

ABSTRACT

The buckling characteristics of flexible fiberglass composite beams are actively controlled by activating optimal sets of a shape memory alloy (NITINOL) wires which are embedded along the neutral axes of these beams. With such control capabilities, the beams can be manufactured from light weight cross sections without compromising their elastic stability.This feature will be invaluable in building light weight structures that have high resistance to failure due to buckling.

The present study is an extension of a study by A.Baz and L.Tampe (Proc. of ASME Conference on Failure Prevention and Reliability, Vol. DE-16, Montreal, Canada, Sept. 1989, pp.211-218, edited by S.Sheppard) where the shape memory actuators are placed external to the beams. Embedding the shape memory wires inside the beams results in smart composite beams with distributed actuation capabilities. Such capabilities produce more favorable and uniform control action than that produced by the discrete external actuators described by Baz and Tampe.

The underlying phenomena influencing the behavior of the NITINOL-reinforced composite beams when subjected to external compressive loading will be presented. The individual contributions of the fiberglass-resin laminate, the NITINOL wires, and the shape memory effect to the overall performance of the composite beam are modeled using the finite-element method. In the model, the effect of the operating temperature produced by activation of the NITINOL wires and the initial strain of the NITINOL wires are accounted for. The model yields an equivalent stiffness matrix for the NITINOL-reinforced beam. Model resultsindicate that the flexural stiffness of the beam is reduced considerably due to the application of the external compressive loads and by the thermal loading resulting from the temperature rise accompanying the activation of the NITINOL wires. However, the beam stiffness is enhanced by the strain energy imparted to the NITINOL wires as they undergo their phase transformation upon activation. The stiffness enhancement due to the strain energy imparted by the shape memory effect can be tailored not only to compensate for the reduction of stiffness due to the axial and thermal loading, but to exceed them by more than twofold. In this manner, the external and thermal loads can keep increasing without reducing the stiffness to its buckling limit. In other words, activating the NITINOL wires can increase the critical buckling load of a NITINOL-reinforced beam several folds over that of an unreinforced beam of the same geometry.

A closed-loop computer-controlled system is built to validate the finite element model. The system is used to control the buckling of a fiberglass-polyester resin beam which is 75 cm long and 0.45 cm thick. The beam is reinforced with 8 NITINOL-55 wires that are 0.55 mm in diameter and have a matrensitic transformation temperature of about 40°C. The beam is subjected to compressive loading using a pneumatically powered cylinder which is coupled directly to one end of the beam. The other end of the beam is fixed. The NITINOL wires are fixed at one end and their other end is connected to a load cell to monitor the initial pre-load as well as the the phase transformation load developed when the wires are activated by the control system. The deflection of the mid-span of the beam, during the occurrence of the buckling, is sensed by two non-contacting sensors which also serve as physical stops to prevent the beam from undergoing excessive deflections as it buckles under load. The position signals are fed through an analog-to-digital (A/D) converter to a micro-processor to control the action of an ON-OFF controller with an adjustable dead-band. The computed control action is sent via a digital-to-analog (D/A) converter to a power amplifier to provide the power necessary to heat the NITINOL wires above their transition temperature. Upon phase transformation, the wires strain energy will increase to counterbalance the effect of the applied axial and thermal loads in order to bring the beam back to its undeflected position.

The effect of the initial pre-loads of the NITINOL wires on the critical buckling loads of the beam is determined with and without the activation of the wires. Also the effect of varying the control system parameters, i.e. the controller gain and the width of the dead band, is investigated. The effect of activating several strands of the wires on the system performance is also considered. The applied load, beam deflection, wire load, wire temperature and beam temperature are sampled and analyzed by the computer during all the experiments.

The results obtained confirm the developed theoretical model and indicate that the critical buckling load can be increased between two to three times when compared to the uncontrolled beam.

The present investigation demonstrates successfully the high buckling characteristics of NITINOL-reinforced composites which makes them suitable for a wide range of critical applications where strength to weight ratio is important e.g. in aircraft panels and large structures. [Work supported under a grant from ARO]

Development of a Low Atomic Number, Sensitive Strain Gage

R. L. Donovan
A. W. Raskob, Jr.

APTEK, Inc.
1257 Lake Plaza Drive
Colorado Springs, CO 80906

Introduction. Radiation environments such as exist in underground nuclear tests (UGTs) present harsh physical and electrical conditions to transducers and instrumentation alike [1]. There are two chief problems which combine to limit the usefulness of standard resistive strain gages: 1) radiation deposition in the gage and its leads induces heating, and 2) severe electrical noise in the instrumentation creates a high level background signal.

Conventional strain gages cannot be used in regions where significant x-ray shine through exists. Typical strain gages are made of high atomic number (high-Z) metals such as copper, constantan, nichrome, etc. When these gages are used to measure strain on irradiated low-Z materials (such as carbon-carbons, carbon-phenolics, aluminum, etc.) they are susceptible to radiation-induced heating. This heating will compromise the gage's ability to yield an accurate, calibrated strain signal. Precautions may be taken, such as shielding the gage, but these measures may compromise the effectiveness of the gage or the conditions of the experiment [4,2].

The long cable lengths (150 to 1000 feet) typically needed, in conjunction with the UGT environment, can give rise to significant radiation induced electrical noise currents. Very small strains (~ 10's of μstrain) produce only μvolts from a standard resistive gage bridge. This small signal requires a large degree of amplification and often the use of special buffers, circuitry and cabling in an attempt to attain an acceptable signal to noise ratio.

A study was conducted to evaluate the feasibility of using polyvinylidene flouride (PVDF or PVF_2),a piezoelectric polymer, as a highly sensitive and low atomic number (Z) dynamic strain sensing element. Such a gage is attractive for two reasons: 1) it possesses a high level electrical response to strain, and 2) it is relatively transparent to x-rays. These gages were tested for correlation to resistive strain gages. Modelling of the gage and electrical circuit was performed to design parametric experimental studies to determine effects of various gage parameters on performance.

Description of Work and Results. A lumped parameter electrical response model was derived for the gage and associated circuitry. Solutions for dynamic amplitude and phase behaviour were obtained for a number of combinations of gage area, aspect ratio, and PVDF thickness. This was used to design a parametric study test matrix. A number of runs were made using PSPICE [3] (a circuit analysis code) to validate the simpler lumped parameter model. Heating effects due to x-ray radiation were computed using an energy deposition code (GRAD [5]).

Standard resistive gages were compared to PVDF gages for a variety of dynamic strain inputs. The gages were attached to prismatic, cantilevered beams of various dimensions. Their positions on each beam were chosen so they would experience similar strains (for at least the lowest beam modes). A fixture was used to provide repeatable and analytically verifiable excitations to a wide variety of beams. In this way, many different strain frequencies and magnitudes, representative of various UGT data as collected in the past with resistive gages, could be modelled.

Calculations using GRAD showed that the heating expected in a typical gage installation as a result of energy deposition will not raise the gage above its Curie temperature (the point at which the material looses its piezoelectric quality). Analytical modelling of the PVDF gage and recording instrumentation circuit gave good correlation with actual gage performance. PSPICE circuit analysis results were consistent with the lumped parameter model (thus validating it).

A range of different strain frequencies and magnitudes, representative of various UGT data (collected in the past with resistive gages), were input to the beams using the test fixture. Recorded strain data showed excellent frequency and amplitude correlation between the standard resistive and PVDF gages as shown in Figure 1.

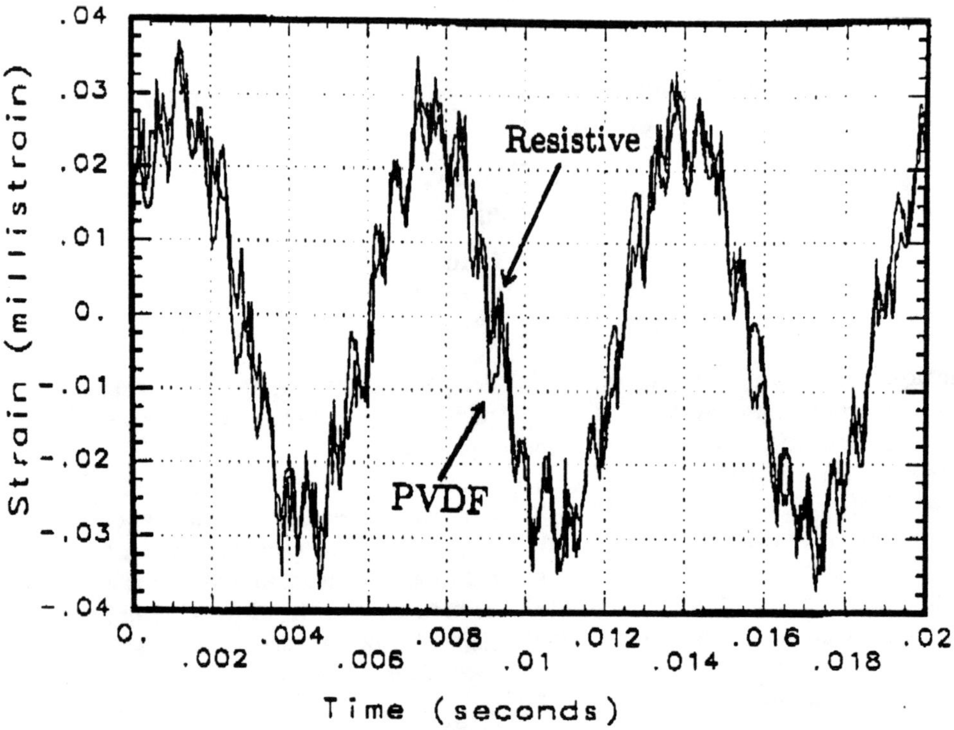

Figure 1: Comparision of PVDF and resistive strain gage outputs.

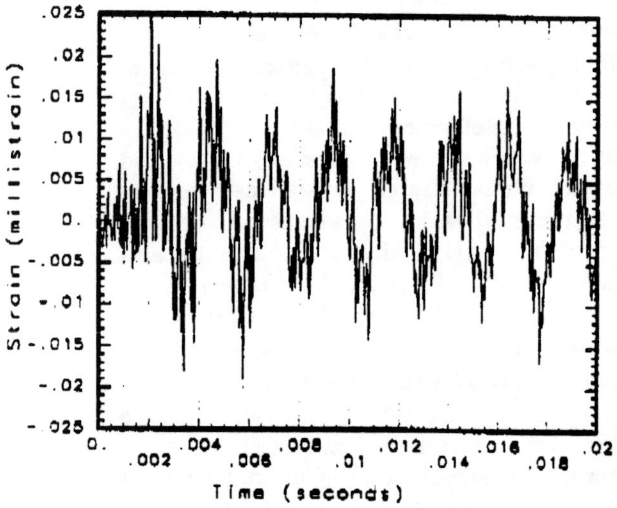

Figure 2: Resistive strain gage output.

The greater sensitivity of the PVDF gage enabled it to reliably measure lower amplitude strain signals which were obscured in the resistive gage records by signal noise. This is shown in Figures 2 and 3. The resistive gage in Figure 2 exhibits an noise level which is roughly the equivalent of a 5 μstrain peak-to-peak signal, whereas the PVDF gage noise level in Figure 3 is substantially lower. The PVDF and resistive gages

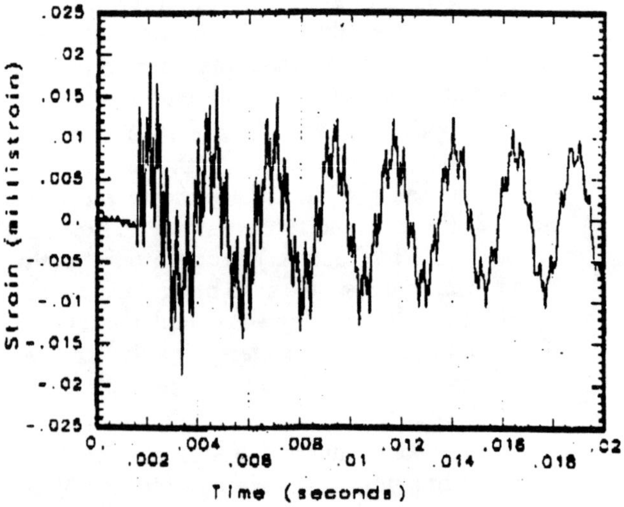

Figure 3: PVDF strain gage output.

Table 1: Comparison of Predicted Resistive and PVDF Gage Outputs

Input Strain (μstrain)	Resistive Gage Bridge Output[1] (mV)	Low-Z Gage (mV) No-Load Gage[2]	Low-Z Gage (mV) 100' Cable[3]
2000	3.84	44.9×10^3	2.59×10^3
1000	1.92	22.5×10^3	1.29×10^3
500	0.96	11.2×10^3	$.646\times10^3$
100	.192	2.25×10^3	$.129\times10^3$
50	.096	1.12×10^3	$.065\times10^3$
10	.019	$.224\times10^3$	$.013\times10^3$

All outputs are prior to amplification

[1] Resistive gage bridge excitation voltage 4.0V, gage factor 2.05

[2] PVDF film without any external capacitance

[3] RF-21 cable (@ 10pF/foot), 0.175×0.25 inch gage area

agreed over all strain magnitude ranges tested which included peak-to-peak strains between 10 μstrain to 1.5 millistrain.

The analytical model provided a basis for extrapolating laboratory performance to actual test/instrumentation environments. Table 1 presents a comparison of predicted output levels for a resistive strain gage and a PVDF gage with and without cable loading.

Conclusions. The feasibility of using PVDF as a low-Z, sensitive strain gage was demonstrated. The lower-Z composition lead to reduced thermal response to x-rays, and allows for forward placement in radiation environments. The gage works well with standard UGT instrumentation.

PVDF is excellent for measuring dynamic response and extremely low strains. Its output is at least two orders of magnitude greater than that of resistive gages, resulting in a much higher signal-to-noise ratio. PVDF response is linear over a wide range of strain amplitudes and over a wide dynamic range of frequencies. Its sensitivity makes it ideal for use in the study of precision structures, such as space structures.

Acknowledgements. This work was sponsored by the Defense Nuclear Agency (DNA) under a Phase I Small Business Innovative Research (SBIR) project.

References

[1] DeMuth, N., Meiers, D., "Impulsive Response Investigation of S-200-E Beryllium Rings," Kaman Sciences Corp., AFWL-TR-72-223, May 1973

[2] Oscarson, J., Seitz, D., "Impulsive Load Response of Soft-bonded Rings and Arcs," Kaman Sciences Corp., DNA3862F, 12 Dec. 1975

[3] PSPICE, MicroSim Corp., Jul. 1989

[4] "Experimental-Theoretical Correlations of Impulsively Loaded Rings," Kaman Nuclear Report KN-71-89(R), 25 Feb. 1971

[5] Williams, G.C., **Program GRAD User's Guide**, APTEK, Inc., 1989

Title: Composite Cure Monitoring Using Optical Fiber Sensors

Authors: Bernd Zimmermann
Marten DeVries
Richard Claus

ABSTRACT

We report on the performance of a novel fiber optic composite cure monitor which has been developed for advanced composite fabrication process control. The monitor is based on a sensing fiber made out of the composite resin itself. The objective is to avoid incorporation of foreign materials into the composite for cure monitoring purposes. Embedded silica fibers have been known to weaken the composite structure, or cause inhomogeneities which have been attributed to result in composite delamination. By choosing the material of the monitoring optical waveguide to be the same as that of the resin of the composite, the inhomogeneities and associated negative effects within the composite are anticipated to be minimized. Feasibility was demonstrated with neat resin specimens and pre-impregnated graphite/epoxy coupons.

INTRODUCTION

The motivation for this work is based on the need to monitor the cure state of composites during the fabrication process. In addition to being able to perform an in-situ measurement on the state of cure, it is desirable to do so by using sensors which can be embedded deep within the composite without affecting the integrity of the finished component. This is especially critical for thick composites, where conditions in the center of the specimen may be very different from those on the outside surfaces where temperature and pressure are usually monitored. The sensor elements to be embedded must therefore be as non-intrusive as possible either by minimizing their size or by manufacturing them out of materials which are either identical or very similar to those of the composite itself. Embedded fiber optic sensors have been proposed [1], and used for cure monitoring applications. Many of these sensors, however, are somewhat intrusive due to their large size and/or "incompatible"material type, often deteriorating the composite's structural integrity. It has been suggested to utilize optical fiber sensors made out of the composite resin itself such that the cure can be monitored in the composite without affecting its structural integrity [2]. Preliminary experiments were conducted with industrial grade fast cure epoxies, and feasibility was demonstrated. We have shown that this concept can be used not only with fast cure epoxies, but also with typical composite resins and pre-impregnated laminates.

Bernd Zimmermann, FIMOD Corporation, P.O. Box 11192, Blacksburg, VA 24062.
Marten DeVries, Richard Claus, Fiber & Electro-Optics Research Center, The Bradley Department of Electrical Engineering, Virginia Polytechnic Institute & State University, Blacksburg, VA 24061.

BACKGROUND

The organic matrix composite cure monitor is based on correlating the resin refractive index to its state of cure. The refractive index change of the resin is monitored using optical fiber waveguide techniques. A waveguide fiber is manufactured using the resin material of the organic matrix composite itself. After insuring that the resin fibers have been allowed to cure completely, they are embedded in the resin to be monitored. An optical signal launched into the cured resin fiber will excite a number, M, of optical "modes" given approximately by:

$$M = \frac{v^2}{2}, \quad (1)$$

where v, the waveguide normalized frequency, is given by :

$$v = \frac{2\pi a}{\lambda}(n_1^2 - n_2^2)^{\frac{1}{2}}. \quad (2)$$

In Equation (2), a is the resin fiber diameter, λ the operating wavelength, n_1 the refractive index of the cured resin fiber, and n_2 the refractive index of the resin to be monitored. As the resin cures, the optical power, P, transmitted through the resin fiber changes according to:

$$\frac{dP}{dt} = \frac{dP}{dn_2}\frac{dn_2}{dt}, \quad (3)$$

where dP/dn_2 is assumed to be proportional to dM/dn_2, that is:

$$\frac{dP}{dn_2} = K\frac{dM}{dn_2}. \quad (4)$$

K depends on several factors including optical launch conditions into the lead fiber, fiber interaction length, cure temperature, and cure pressure. As will be explained in the experimental section, a reference signal, P_{ref}, will be tapped from the optical source to compensate for source output drift. The Normalized Transmitted Power (NTP), which will be related to the state of cure, is given by:

$$NTP = \frac{P}{P_{ref}}. \quad (5)$$

EXPERIMENTS

The experimental part consisted of first determining the resin's optimum operating wavelength, i.e., the wavelength at which optical loss was minimum. For the particular resin of interest a low attenuation "plateau" was observed between 800 and 1100 nm. We therefore opted to conduct our experiments using a CW 816 nm laser source.

After the resin fiber sensors were manufactured and fully cured, they were placed in a heating assembly as shown in Fig. 1. This assembly allowed cure monitoring using neat resin specimens. The assembly consisted of a laser, a fiber optic splitter, two photodetectors, an optical power meter, two lead fibers, a resin fiber sensor, a heating block, and a GPIB interface to a personal computer. The output of the laser was split into two arms; the output of one of the splitter arms was sent directly to photodetector # 1 as reference signal, while the output of the other arm was launched into the resin fiber through a lead-in fiber. The output of the resin fiber was picked up by a lead-out fiber and directed to photodetector # 2. The power levels of the two detectors were processed by the optical power meter and passed on through a GPIB bus to a computer for data acquisition. An A/D board in the computer also allowed acquisition of the temperature in the heating block.

The results of monitoring the cure of neat resin specimens using the assembly shown in Fig. 1 is given in Fig. 2. This figure shows a graph of Normalized Transmitted Power (NTP) and its numerical derivative (dNTP/dt) versus time of cure. As can be observed, NTP decreases steadily as the resin cures while dNTP/dt, which is indicative of the rate of cure, approaches zero.

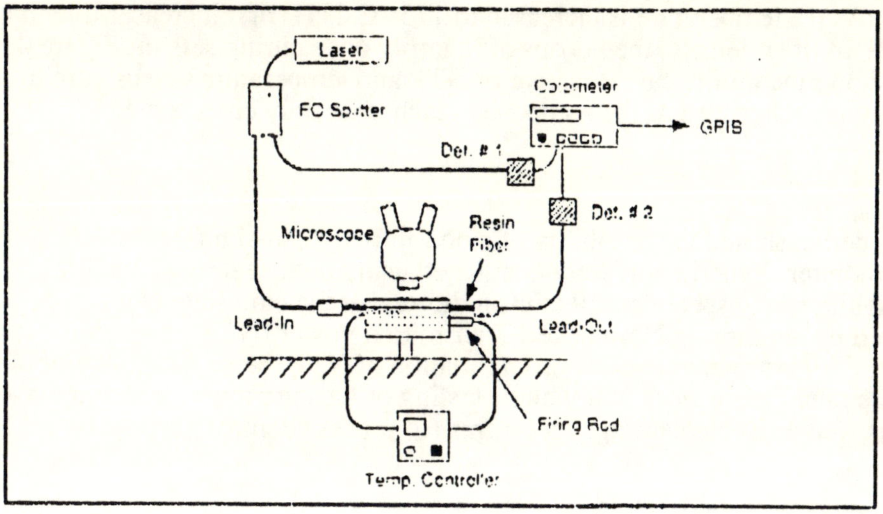

Figure 1. Fiber Optic Cure Monitoring Assembly.

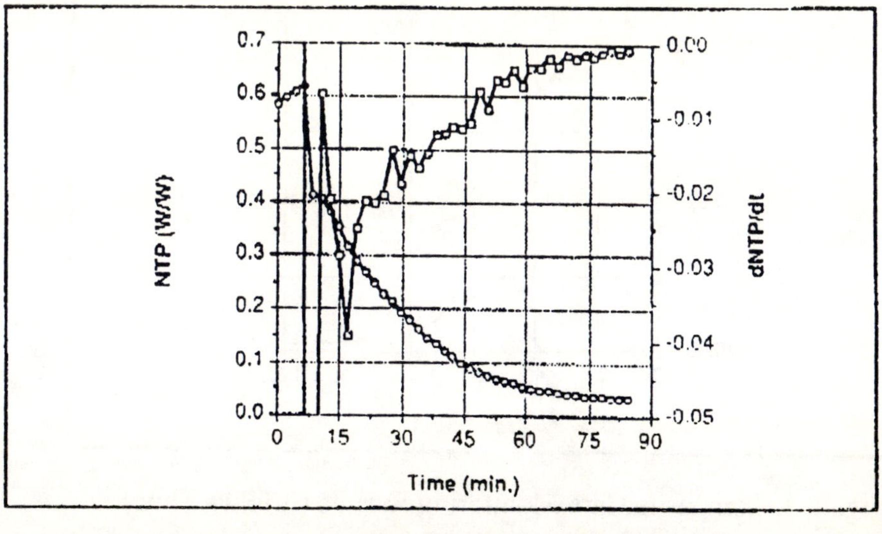

Figure 2. Neat Resin Cure Monitoring Results.

Similarly, we used the assembly shown in Fig. 1 to monitor the cure of a pre-impregnated laminate composite coupon. The prepreg laminates were placed in an aluminum mold which contained access ports through which the resin fiber sensor was routed. A total of 32 prepreg plies (0°-90° orientation) was used; the resin fiber fiber was placed between the 12th and 13th ply, parallel to the matrix fibers. Pressure and temperature to the mold were applied using a 1 ton hot press. A pre-cure temperature of approximately 125 °C was used for 45 minutes to condition the specimen before ramping up the temperature to 175 °C.

The results of monitoring the cure of the composite coupon are shown in Figures 3 and 4. Figure 3 shows NTP and temperature versus cure time, and clearly demonstrates the sensor's behavior both during the pre-cure conditioning cycle as well as the cure cycle. During the 45 minute conditioning period NTP falls slowly, already indicating the presence of resin gelation. As the temperature is increased to 175 °C, NTP first increases due to a decrease in the resin fiber density, then drops off sharply once curing sets in. Figure 4, on the other hand, shows the numerical derivative of NTP and temperature versus cure time. Again, dNTP/dt approaches zero as the composite reaches its fully cured state.

CONCLUSION

We have demonstrated the feasibility of a non-intrusive, in-situ fiber optic composite cure monitor. Experiments were conducted using both neat resin and pre-impregnated graphite/epoxy specimens. Results indicate that the cure state of a composite can be determined by monitoring Normalized Transmitted Power (NTP) through the resin fiber waveguide. Furthermore, numerical differentiation of NTP seems to allow monitoring of composite cure rate. Future work will include testing of the cure monitor in autoclave environments, as well as implementing algorythms for process control purposes.

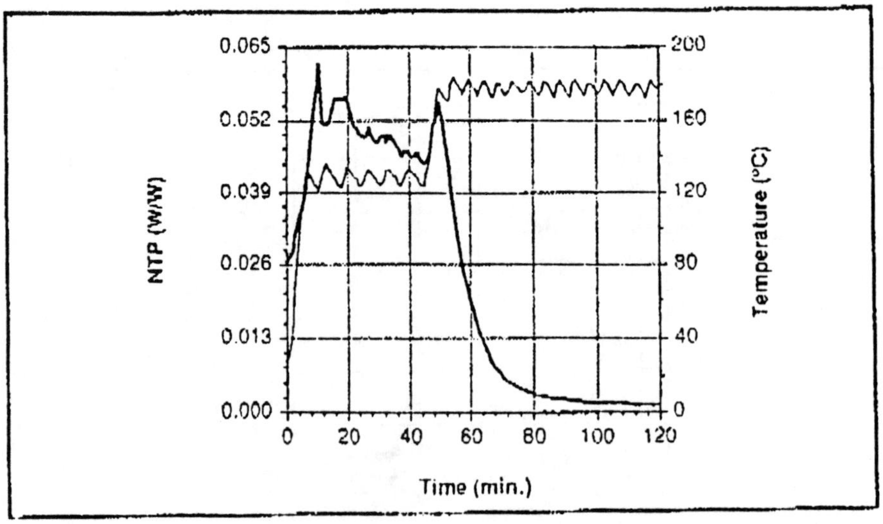

Figure 3. Composite Cure Monitoring Results (NTP vs Time).

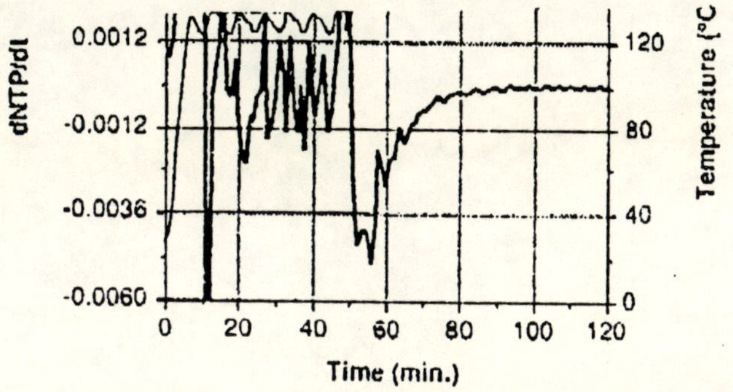

Figure 4. Composite Cure Monitoring Results (dNTP vs Time).

ACKNOWLEDGEMENTS

This work was sponsored by the U.S. Army Materials Technology Laboratory (USAMTL), Watertown, MA under contract # DAAL04-90-0013 (SBIR Phase I).

REFERENCES

1. M.A. Druy, L. Elandjian, and W.A. Stevenson, SPIE Proceedings Vol. 986, 130 (1988).

2. M.A. Afromowitz, J. of Lightwave Technology, Vol. 6, No. 10, (October 1988).

Smart Materials for Sensing and/or Remedial Action to Reduce Damage to Materials

Carolyn Dry, Assoc. Prof.
University of Illinois

SENSING AND REMEDIAL ACTION

The sensing of cracking or corrosion of a settable material by a chemical or physical sensor which, in the process of sensing, starts the activation of a remedial process is the topic of this research. Specifically, materials containing various types of hollow porous fibers filled with a chemical which release into the matrix at appropriate times, or over time, are being developed to address the issues of material brittleness and durability; i.e. cracking and corrosion.

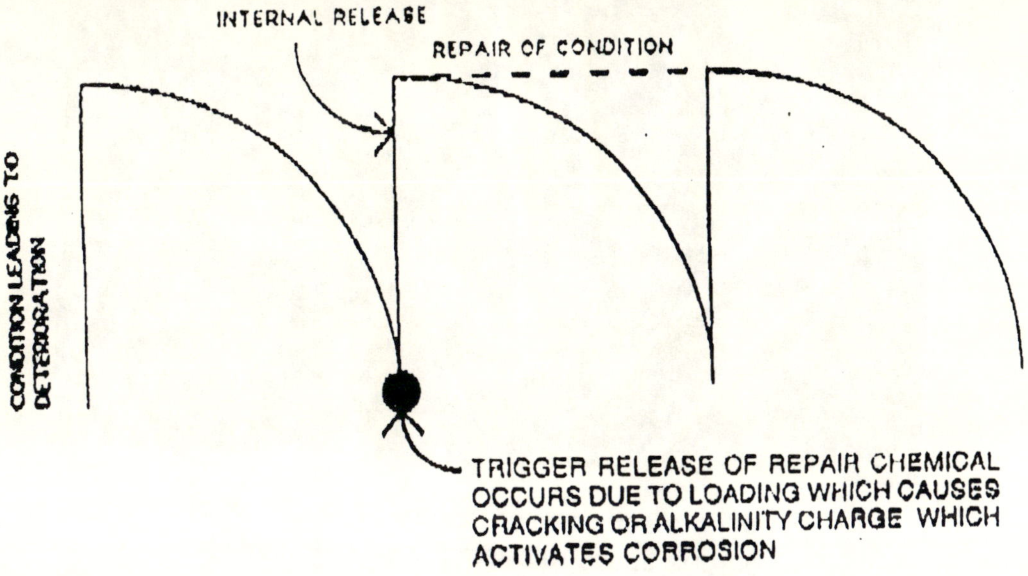

Figure 1

Cracking

Cracking, due to loading, is a major problem in materials durability because the cracks increase permeability. Loading over time has a cumulative effect which can lead finally to complete deterioration of the component or structure. The design to alleviate this problem consists of filling hollow porous glass fibers containing cross-linking crack closing chemicals which close the cracks. The chemicals are released from the fibers when the fibers flex due to loading. This is the ideal situation in which the agent of environmental degradation, namely loading, is the stimulus to release the repair chemical.

Corrosion

Corrosion is an electrochemical process that requires an electrical current in a moist and oxygenated condition. The entry of chloride ions or carbonation can cause the pH to be reduced to 11.5 pH, at which point corrosion begins. The addition of calcium nitrite, an anticorrosion chemical, protects the steel. In this design calcium nitrite is put into hollow porous fibers coated with polyol. Polyol dissolves in an alkalinity of appropriate pH (11.5), thus, releasing the anticorrosion chemical. The cause of deterioration, reduction of alkalinity due to chloride or carbonation which causes the corrosion is the sensor (coating dissolution) also which activates the remedial action (release of calcium nitrite).

Scanning electron microscope photograph of porous fibers releasing their chemical

REMEDIAL ACTION

Internal time release of chemicals from hollow porous fibers can offer remedial and repair action to reduce damage without the sensing function.

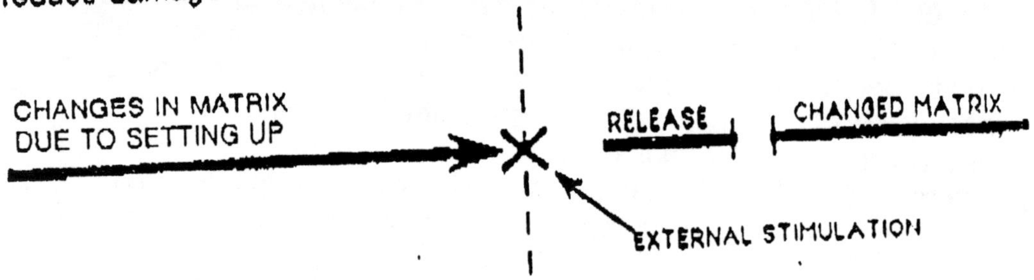

Figure 2

Chemical Attack Due to Permeability

The ability of water and chemicals to intrude because the matrix is permeable is the major problem in durability. In this design permeability is addressed by the release of sealants into the body of the matrix from hollow, porous fibers. Methyl methacrylate is contained in the fibers for later release by coating the fibers with wax. To be effective, the methyl methacrylate has to be released from the fibers after the matrix is set up and then it has to be polymerized. Heat is the stimulus which not only releases the methyl methacrylate from the porous fibers by melting the wax coating, but it also dries out the matrix to receive the methyl methacrylate and polymerizes the methyl methacrylate in place in the matrix. The heat stimulus has that threefold effect of accomplishing release, matrix drying, and polymerization.

OTHER DESIGNS

Other designs on which research is commencing are:

1.) internal timed release of charged ions from hollow metal fibers by application of an electrical current to move sealant to the desired location or to move anticorrosion chemicals toward the metal.

2.) coat fibers containing anticorrosion chemicals with oxygen or moisture sensitive materials which degrade as the matrix contents of O_2 and H_2O charges.

ISSUES

The question of refilling the fibers with chemicals for very long remedial action is being researched. A network of hollow fibers vessels capable of being refilled, larger volume porous ceramic plugs, and surface accessible timed release patches are candidate vehicles.

A Self-Sensing Piezoelectric Actuator For Collocated Control

Jeffrey J. Dosch, Daniel J. Inman, and Ephrahim Garcia
Mechanical Systems Laboratory
Department of Mechanical and Aerospace Engineering
State University of New York at Buffalo
Buffalo, New York 14260

ABSTRACT: A technique has been developed which allows a single piece of piezoelectric material to concurrently sense and actuate in a closed loop system. The motivation behind the technique is that such a self-sensing actuator will be truly collocated and has applications in active and intelligent structures, such as vibration suppression. A theoretical basis for the self-sensing actuator is given in terms of the electromechanical constitutive equations for a piezoelectric material. In a practical implementation of the self-sensing actuator an electrical bridge circuit is used to measure strain. The bridge circuit is capable of measuring either strain or time rate of strain in the actuator.

The usefulness of the proposed device was experimentally verified by actively damping the vibration in a cantilever beam. A single piezoceramic element bonded to the base of the beam functioned both as a distributed moment actuator and strain sensor. Using a rate feedback control law, the first mode of vibration was suppressed, reducing the settling from 35 seconds to 2.5 seconds. Using a positive position feedback law the first two modes of vibration were suppressed; the first mode settling time was reduced from 35 to 0.3 seconds and the second mode settling time was reduced from 7 seconds to 0.9 seconds.

A Novel Sensor for Adaptive and Smart Structures

Nisar Shaikh
Center for Material Research and Analysis
University of Nebraska-Lincoln

State-of-the-art sensing for adaptive and smart structures is accomplished with embedded sensors. Today, optical fibers and Nitinol sensors are being used where strain gages and accelerometers once were. Each of these methods has advantages and disadvantages which make it more suitable for specific applications.

A fiber coated with piezoelectric material offers a viable alternative to some existing sensors. These sensors are active and do not require power input or an external signal. They also make it possible to directly measure the induced electrical voltage, without the conversion required for optical signals. The sensors offer great advantages in fibrous composites, where they can be easily incorporated into the strands. The sensors can be conveniently dispersed or distributed along the entire structure.

The feasibility of inducing strain sensing properties in a structure has been demonstrated [1] with beam samples made from stainless steel strips deposited with a thin film of piezoelectric ZnO. The technique of sputtering fibers with ZnO was developed by coating Ni-Cr wires of 143 micron diameter. The vibration sensing experiments were carried out on beams made of brass strips embedded with ZnO coated Ni-Cr wires. The tests consisted of forced vibration through a shaker as well as natural vibration by impulse. The work is progressing now with making filament wound composites which carry the sensing fibers.

(This work is supported by the Material Science Division of the U.S. Army Research Office. Their encouragement and support are greatly appreciated.)

REFERENCE

1. Shaikh, Nisar and Dillon, Rod," Smart Structural Composites", U.S./Japan Work Shop on Smart/Intelligent Materials and Systems, pp 287-293, Hawaii, 1990.

Time–Scale Effects in the Dynamics of Shape–Memory Alloys

Eugene M. Cliff

Aerospace and Ocean Engineering Department
and
Interdisciplinary Center for Applied Mathematics
Virginia Polytechnic Institute and State University

Abstract

A dynamical model for a shape-memory alloy has been described by a system of coupled partial differential equations. In one-space dimension the model includes:

(1) a wave equation which governs the longitudinal vibrations

(2) a diffusion equation which governs the thermal energy transfer

(3) a diffusion equation which governs the phase–transition process

We analyze this model for the effects of time–scale in each of these phenomena. Specifically, for the case when the mechanical vibrations are *fast* we study the effects of the thermal and phase diffusion speeds. Qualitative features of the numerical solutions are emphasized.

A FUNDAMENTAL THEORY OF THERMOELASTIC DAMPING ORIGINATING IN THE SECOND LAW OF THERMODYNAMICS

Vikram K. Kinra
K. Bryan Milligan
Joseph E. Bishop
Marcus Parche

Center for Mechanics of Composites
Department of Aerospace Engineering
Texas A & M University
College Station, Tx 77843

Passive damping is an important material property from the viewpoint of vibration suppression in a variety of structures: automotive bodies, jet engines, aircraft structures, submarines, and high precision large flexible space structures. Metal-matrix composites are materials of choice for these structures. Therefore, there is an urgent need for a theory for predicting damping in metal-matrix composites.

At the atomistic level, there is a large number of mechanisms of damping; a partial list is: diffusion of atoms and point defects, the Snoek relaxation, the Zener relaxation, dislocations, grain boundary sliding and relaxation, phase transformations, magnetoelastic relaxation, and interaction of electrons with ultrasonic waves (phonons). At the continuum level some of the damping mechanisms are viscous damping, Coulomb friction, and thermoelastic damping. In view of the ubiquitous nature of damping there is a large number of theoretical models for damping. Unfortunately, a majority of these models are semi-empirical in nature: they contain one or more undetermined coefficients (dashpots, for example) whose numerical value is determined a posteriori through a search over a range of values that brings about the best agreement between the model prediction and the experimental data. Since, in some sense, the model outputs the input, it is here that one loses the rigor of testing one's model.

The objective of the present work is to develop a rigorous theory of damping (which does not use any undetermined coefficients); the particular mechanism selected is thermoelastic damping for its obvious application to metal-matrix composites.

The physics of thermoelastic damping is well understood. In 1938, in a classic paper, Zener considered time-harmonic flexural vibrations of a linear, isotropic, homogeneous, thermoelastic beam. He observed that the compressional side heats up while the tensile side cools (Joule-Thomson effect) setting up time-harmonic irreversible heat transfer. He postulated

the existence of thermoelastic damping which he modelled as a series of simple, three-parameter anelastic solids (two springs and a dashpot). It is a testimonial to his enormous insight and genius that in spite of a number of rather restrictive assumptions, the predictions of his model agree remarkably well with a wealth of experimental data produced during the ensuing half-century. Unfortunately, however, Zener's approach cannot be conveniently extended to heterogeneous (composite) materials; hence this paper.

The foundations of our continuum theory of thermoelastic damping are the following universally accepted physical laws:

The First Law of Thermodynamics
The Second Law of Thermodynamics
Newton's Laws of Motion
Fourier Law of Heat Conduction
Hooke's Law for a Linear Thermoelastic Solid

The development of the theory proceeds as follows. For any specific boundary value problem the coupled heat-conduction equation is solved to obtain the temperature field (only time-harmonic deformations are considered); local thermal currents are set up as a consequence. The entropy created due to this irreversible heat exchange is calculated by the use of the Second Law of Thermodynamics. The entropy created leads directly to an expression for the amount of work converted into heat in a cycle; an exact closed-form solution for thermoelastic damping follows. En route to a general theory for thermoelastic damping in structured materials (metal-matrix composites being just one example), the following specific boundary value problems have been solved:

1. The problem solved by Zener in 1938 is revisited: Flexural vibrations of a linear isotropic, homogeneous elastic Bernoulli-Enler beam; an excellent comparison between our exact solution and Zener's approximate solution was noted. An excellent comparison between the theory and the experiments conducted with aluminum beams was also observed.

2. As a canonical problem in the study of heterogeneous media we have considered tow rods of thermoelastically different materials in welded longitudinal contact subjected to uniform uniaxial longitudinal stress.

3. An infinitely long slender rod containing a thermoelastic inclusion subjected to an uniaxial stress.

4. A beam of one material laminated by two symmetric beams of another material (symmetric laminated beam) subjected to uniform uniaxial longitudinal strain.

5. A symmetric laminated beam subjected to pure flexure.

6. A general N-layer material subjected to a uniaxial stress perpendicular to the interfaces with N arbitrarily large.

Modeling of Shape Memory Actuators

John Burns
Ruben Spies
Interdisciplinary Center for Applied Mathematics
Department of Mathematics
Virginia Polytechnic Institute & State University
Blacksburg, VA 24061

In the past few years there has been considerable interest in so-called shape memory actuators. These devices are based on thermomechanical materials that exhibit a shape memory effect. This effect is essentially due to a phase transition and can produce powerful yet compact actuators. In order to design and use such actuators it is essential to control the actuator dynamics which are governed by complex hybrid nonlinear partial differential equations. In this paper we review several models of these materials and present a computational scheme for the full nonlinear system. The problem of identifying the material properties is considered and a new model that captures the dynamic memory of these systems is proposed.

Constitutive Modeling of Phase Transition in Smart Materials

Mehrdad Negahban, Department of Engineering Mechanics and the Center for Materials Research and Analysis, University of Nebraska-Lincoln, Lincoln, NE 68588-0347

Smart, intelligent, and adaptive materials and structures have recently received particular attention due to their active, as opposed to passive, interaction with their working environment. This interaction can range from sensing the material's structural integrity to actively sensing and appropriately responding to different stimuli. Real and pseudo phase transitions are one mechanism which can be used to make a material smart. Shape memory alloys and rheological fluids are two examples of materials which can be made smart by taking advantage of their phase transitions.

Many polymers in a way similar to shape memory alloys remember their previous shape and can be stimulated to return to this shape. Polycarbonate and polystyrene are two materials which after plastic forming below the glass transition temperature will recover their original shape upon heating above their glass transition temperature. Mathematical modeling will allow one to take advantage of this feature in a manner which allows the use of mathematical analysis, and therefore computational simulation of complex problems.

The basic structure of a mathematical model which can account for shape memory is as follows. The material will be considered an elastic-plastic material bellow the transition temperature. In a manner similar to nonlinear visco-plasticity, it will be assumed that the second Piola-Kirchhoff stress, $\mathbf{S}(t)$, is given by a function of the current value of the Cauchy strain, $\mathbf{C}(t)$, the plastic strain, $\mathbf{C}^p(t)$, and the temperature, $T(t)$, in the form

$$\mathbf{S}(t) = \mathcal{S}\{\mathbf{C}(t), \mathbf{C}^p(t), T(t)\},$$

where $\mathbf{C}^p(t)$ is given by a flow rule of the form

$$\dot{\mathbf{C}}^p(t) = \frac{\partial \mathbf{C}^p}{\partial T}\dot{T} + \frac{\partial \mathbf{C}^p}{\partial \mathbf{C}} : \dot{\mathbf{C}} + \frac{\partial \mathbf{C}^p}{\partial t}.$$

The last term in the equation implies that the plastic strain can change without the introduction of any changes in the total strain or temperature (normally considered as material aging). This term is the key feature needed to model shape memory materials.

To better understand the characteristics of the model consider a material which is plastically deformed below the transition temperature and is subsequently heated to above the transition temperature and held at a constant strain and temperature. If the term $\frac{\partial \mathbf{C}^p}{\partial t}$ is negative above the transition temperature, then the plastic strain will gradually reduced. Since the total strain is kept constant, the elastic strain will increase to compensate for the reduction in plastic strain and the stress in the material will increase. This added stress will be available to do work.

It will be assumed that the first two terms in the equation for the rate of plastic strain will go to zero above the transition temperature. The last term in this equation, $\frac{\partial \mathbf{C}^p}{\partial t}$, will be taken as negative above the transition temperature and positive below this temperature. Since the rate of change of plastic strain must go to zero as the plastic strain goes to zero, an equation of the form $\frac{\partial \mathbf{C}^p}{\partial t} = \alpha(T)\mathbf{C}^p(t)$ is a possible selection, where $\alpha(T)$ is positive below the transition temperature and negative above it.

In addition to thermodynamic constraints, some examples will be discussed.

Nonlinear Constitutive Relations for piezoceramic materials

Shiv P. Joshi

Center for Composite Materials
Aerospace Engineering Departments
University of Texas at Arlington, Arlington, TX 76019

ABSTRACT

Piezoelectric material produces electric charges when mechanically deformed and an electric potential causes a mechanical deformation. This property makes it suitable for sensor and transducer applications. The understanding of the electroelastic constitutive behavior is critical to predicting the response of a structure with embedded piezoelectric material. A concise formulation of relevant nonlinear constitutive relations is presented in this paper.

INTRODUCTION

The Curies first showed the presence of piezoelectricity in crystals in 1880. The first practical use of the piezoelectric effect was during World War I when Langevin's sonar emitter was effectively used to detect German submarines. Prior to World War II, researchers at MIT discovered that certain ceramics such as PZT (Lead-Zirconate-Titanate) could be polarized to yield a high piezo response [1]. Piezoceramics consists of a large number of small crystallites sintered together and polarized by an external electric field. Kawai [2] discovered that the polarized homopolymer of vinylidene fluoride (PVDF) developed far greater piezo activity than any other synthetic or natural polymer. Poled PVDF still dominates all other materials in terms of its intensity of piezo activity [1].

Although the behavior of piezoelectric materials in non-structural applications has been investigated extensively, the treatment is often simplistic. The recent interest in "Smart Structures" has put especial emphasis on the rigorous understanding of elctroelastic behavior of piezoceramics as an integral part of a structure.

The nonlinear theory of dielectrics has been studied by Toupin [3], Nelson [4] and Tiersten[5]. The relation between the equations of linear piezoelectricity and the more general electroelastic equations is discussed by Tiersten [5]. Nelson presented a completely deductive derivation of the dynamical equations and constitutive relations for elastic, electric, and electroelastic phenomena based on the fully electrodynamic Lagrangian theory of elastic dielectrics. Penfield and Hans [6] developed a linear piezoelectric theory which does not account for gradient of polarization and electrostatic interference. Mindlin [7] derived a system of two dimensional equations for high frequency motions of crystal plates accounting for coupling of mechanical, electrical and thermal fields. Readers interested in this area may refer to books by Nye [8], Berlincourt et.al. [9], and Landau and Lifshitz [10]. A phenomenological description of the dynamic response of piezoceramics to an external electric field, including domain reorientation processes and the dynamics of dipole moment in each domain, has been developed by Chen et.al. [11,12,13,14].

A concise formulation of linear constitutive equations for piezoelectric materials is presented by Joshi [15]. The linear relations are specialized for piezoceramic materials available for structural applications. This paper extends the formulation presented by Joshi [15] to include some important nonlinear effects encountered by piezoceramics in "smart structures" applications. Some experimental observations will also be included in the full length paper.

PIEZOELECTRIC FIELD EQUATIONS

The physics involved in the piezoelectric theory may be regarded as a coupling between Maxwell's equations of electromagnetism and elastic stress equations of motion. The coupling takes place through the piezoelectric constitutive equations.

Maxwell's equations in vector form are written as,

$$\nabla \times E = -\mu_o \frac{\partial H}{\partial t} \qquad \nabla \times H = \frac{\partial \hat{D}}{\partial t} \qquad \text{(EQ 1)}$$

where E is electric field intensity, H is magnetic field vector and $\hat{D}$ is electric flux density also known as displacement vector. The free-space permeability, μ_o, is used because piezoelectric materials are nonmagnetic. In the quasi-electrostatic approximation [16], which is usually adequate for the study of piezoelectric phenomena, time-derivative terms in the electromagnetic equations may be dropped. Then the electric field may be expressed as,

$$E = -\nabla \phi \qquad \text{(EQ 2)}$$

and the only electromagnetic equation which need to be considered is

$$\nabla \cdot \hat{D} = 0 \qquad \text{(EQ 3)}$$

The elastic stress equation of motion is,

$$\nabla \cdot \sigma = \rho \ddot{u} \qquad \text{(EQ 4)}$$

where σ is stress tensor, ρ is mass density and $\ddot{u}$ is acceleration vector. Coupling among eq. 2-4 is introduced by piezoelectric constitutive equations.

LINEAR CONSTITUTIVE EQUATIONS

We will adopt index notations in the remainder of the paper for convenience. We will employ the thermodynamic Gibbs potential to derive constitutive equations and will consider the σ_{ij} (stress components), E_k (electric field components), and T (absolute temperature) as independent variables.

$$G = U - \sigma_{ij}\varepsilon_{ij} - E_k\hat{D}_k - TS \qquad \text{(EQ 5)}$$

where G is the Gibbs potential, S is the entropy, and U is the internal energy. For adiabatically insulated reversible system, the total differential of internal energy is

$$dU = \sigma_{ij}d\varepsilon_{ij} + E_k d\hat{D}_k + TdS \qquad \text{(EQ 6)}$$

and therefore the total differential of Gibbs potential is

$$dG = -\varepsilon_{ij}d\sigma_{ij} - \hat{D}_k dE_k - SdT \qquad \text{(EQ 7)}$$

Expressing the Gibbs potential in Taylor series and neglecting higher order terms, we obtain,

$$dG = \left(\frac{\partial G}{\partial \sigma_{ij}}\right)_{E,T} d\sigma_{ij} + \left(\frac{\partial G}{\partial E_k}\right)_{\sigma,T} dE_k + \left(\frac{\partial G}{\partial T}\right)_{\sigma,E} dT \qquad \text{(EQ 8)}$$

From eqs. 7 and 8

$$\varepsilon_{ij} = -\left(\frac{\partial G}{\partial \sigma_{ij}}\right)_{E,T} \qquad \hat{D}_k = -\left(\frac{\partial G}{\partial E_k}\right)_{\sigma,T} \qquad S = -\left(\frac{\partial G}{\partial T}\right)_{\sigma,E} \tag{EQ 9}$$

The total differentials of dependent variables ε_{ij}, $\hat{D}_k$, and S is given as a function of independent variables as

$$\begin{aligned} d\varepsilon_{ij} &= \left(\frac{\partial \varepsilon_{ij}}{\partial \sigma_{lm}}\right)_{E,T} d\sigma_{lm} + \left(\frac{\partial \varepsilon_{ij}}{\partial E_n}\right)_{\sigma,T} dE_n + \left(\frac{\partial \varepsilon_{ij}}{\partial T}\right)_{\sigma,E} dT \\ d\hat{D}_k &= \left(\frac{\partial \hat{D}_k}{\partial \sigma_{lm}}\right)_{E,T} d\sigma_{lm} + \left(\frac{\partial \hat{D}_k}{\partial E_n}\right)_{\sigma,T} dE_n + \left(\frac{\partial \hat{D}_k}{\partial T}\right)_{\sigma,E} dT \\ dS &= \left(\frac{\partial S}{\partial \sigma_{lm}}\right)_{E,T} d\sigma_{lm} + \left(\frac{\partial S}{\partial E_n}\right)_{\sigma,T} dE_n + \left(\frac{\partial S}{\partial T}\right)_{\sigma,E} dT \end{aligned} \tag{EQ 10}$$

where

$$s_{ijlm}^{E,T} = \left(\frac{\partial \varepsilon_{ij}}{\partial \sigma_{lm}}\right)_{E,T}, \quad d_{ijn}^{T} = \left(\frac{\partial \varepsilon_{ij}}{\partial E_n}\right)_{\sigma,T} = \left(\frac{\partial \hat{D}_n}{\partial \sigma_{ij}}\right)_{E,T}, \quad \alpha_{ij}^{E} = \left(\frac{\partial \varepsilon_{ij}}{\partial T}\right)_{\sigma,E} = \left(\frac{\partial S}{\partial \sigma_{ij}}\right)_{E,T},$$

$$\varepsilon_{kn}^{\sigma,T} = \left(\frac{\partial \hat{D}_k}{\partial E_n}\right)_{\sigma,T}, \quad p_k^{\sigma} = \left(\frac{\partial \hat{D}_k}{\partial T}\right)_{\sigma,E} = \left(\frac{\partial S}{\partial E_k}\right)_{\sigma,T}, \quad \frac{\rho c^{\sigma,E}}{T_0} = \left(\frac{\partial S}{\partial T}\right)_{\sigma,E}$$

are elastic compliance coefficients, piezoelectric strain constants, coefficients of thermal expansion, dielectric permittivities, pyroelectric coefficients, respectively; and $c^{\sigma,E}$ is the specific heat and ρ is the mass density. Integrating eq. 10, we obtain,

$$\begin{aligned} \varepsilon_{ij} &= s_{ijlm}^{E,T}\sigma_{lm} + d_{ijn}^{T}E_n + \alpha_{ij}^{E}\Delta T \\ \hat{D}_k &= d_{klm}^{T}\sigma_{lm} + \varepsilon_{kn}^{\sigma,T}E_n + p_k^{\sigma}\Delta T \\ \Delta S &= \alpha_{lm}^{E}\sigma_{lm} + p_n^{\sigma}E_n + \frac{c^{\sigma,E}}{T_0}\Delta T \end{aligned} \tag{EQ 11}$$

Piezoceramics are widely used, therefore we will specialize eq 11 for them. The constitutive equations of the polarized piezoceramics are equivalent to the equations for a piezocrystal of the hexagonal 6mm symmetry class. In abbreviated subscript notation these equations may be written as

$$\begin{aligned} \varepsilon_{11} &= s_{11}^{E,T}\sigma_{11} + s_{12}^{E,T}\sigma_{22} + s_{13}^{E,T}\sigma_{33} + d_{31}^{T}E_3 + \alpha_1^{E}\Delta T \\ \varepsilon_{22} &= s_{12}^{E,T}\sigma_{11} + s_{11}^{E,T}\sigma_{22} + s_{13}^{E,T}\sigma_{33} + d_{31}^{T}E_3 + \alpha_1^{E}\Delta T \\ \varepsilon_{33} &= s_{13}^{E,T}\sigma_{11} + s_{13}^{E,T}\sigma_{22} + s_{33}^{E,T}\sigma_{33} + d_{33}^{T}E_3 + \alpha_3^{E}\Delta T \end{aligned}$$

$$\varepsilon_{23} = s_{44}^{E,T}\sigma_{23} + d_{15}^{T}E_2, \qquad \varepsilon_{13} = s_{44}^{E,T}\sigma_{13} + d_{15}^{T}E_1, \qquad \varepsilon_{12} = \left(\frac{s_{11}^{E,T} - s_{12}^{E,T}}{2}\right)\sigma_{12}$$

$$\hat{D}_1 = d_{15}^{T}\sigma_{13} + \varepsilon_{11}^{\sigma,T}E_1 + p_1^{\sigma}\Delta T, \qquad \hat{D}_2 = d_{15}^{T}\sigma_{23} + \varepsilon_{11}^{\sigma,T}E_2 + p_1^{\sigma}\Delta T \tag{EQ 12}$$

$$\hat{D}_3 = d_{31}^{T}(\sigma_{11} + \sigma_{22}) + d_{33}^{T}\sigma_{33} + \varepsilon_{33}^{\sigma,T}E_3 + p_3^{\sigma}\Delta T$$

$$\Delta S = \alpha_1^{E}(\sigma_{11} + \sigma_{22}) + \alpha_3^{E}\sigma_{33} + p_1^{\sigma}(E_1 + E_2) + p_3^{\sigma}E_3 + \frac{c^{\sigma,T}}{T_0}\Delta T$$

In cases where temperature variation is negligible, neglecting temperature terms and writing eq. 12 in compact matrix notation, we obtain

$$\begin{aligned} \{\varepsilon\} &= [S^{E}]\{\sigma\} + [d]\{E\} \\ \{\hat{D}\} &= [d]^{T}\{\sigma\} + [\varepsilon^{\sigma}]\{E\} \end{aligned} \tag{EQ 13}$$

Where,

$$[S^E] = \begin{bmatrix} S_{11} & S_{12} & S_{13} & 0 & 0 & 0 \\ S_{12} & S_{11} & S_{13} & 0 & 0 & 0 \\ S_{13} & S_{13} & S_{33} & 0 & 0 & 0 \\ 0 & 0 & 0 & S_{44} & 0 & 0 \\ 0 & 0 & 0 & 0 & S_{44} & 0 \\ 0 & 0 & 0 & 0 & 0 & \frac{S_{11}-S_{12}}{2} \end{bmatrix} \qquad [d] = \begin{bmatrix} 0 & 0 & d_{31} \\ 0 & 0 & d_{31} \\ 0 & 0 & d_{33} \\ 0 & d_{15} & 0 \\ d_{15} & 0 & 0 \\ 0 & 0 & 0 \end{bmatrix} \qquad [\varepsilon^\sigma] = \begin{bmatrix} \varepsilon_{11} & 0 & 0 \\ 0 & \varepsilon_{11} & 0 \\ 0 & 0 & \varepsilon_{33} \end{bmatrix} \quad \text{(EQ 14)}$$

The alternative constitutive formulations may be obtained by considering other potentials. The final linear constitutive relations in compact matrix notation are presented below. The temperature change is assumed negligible and therefore not included in the relations.

$$\{\sigma\} = [C^E]\{\varepsilon\} - [g]\{E\}$$
$$\{\hat{D}\} = [g]^T\{\varepsilon\} + [\varepsilon^\varepsilon]\{E\}$$
$$\text{or} \qquad \text{(EQ 15)}$$
$$\{\sigma\} = [C^{\hat{D}}]\{\varepsilon\} - [h]\{\hat{D}\}$$
$$\{E\} = -[h]^T\{\varepsilon\} + [\beta^\varepsilon]\{\hat{D}\}$$

Where,

$$[C^E] = [S^E]^{-1} \qquad [g] = [C^E][d] \qquad [\varepsilon^\varepsilon] = [\varepsilon^\sigma] - [g]^T[d]$$
$$[C^{\hat{D}}] = [C^E] + [g][\beta^\varepsilon][g]^T \qquad [h] = [g][\beta^\varepsilon] \qquad [\beta^\varepsilon] = [\varepsilon^\varepsilon]^{-1} \quad \text{(EQ 16)}$$

Eq. 13 and 15 gives alternative forms of linear constitutive relations. The coefficients are related to each other as given by eq. 16.

NONLINEAR CONSTITUTIVE EQUATIONS

The strain, electric displacement and entropy are assumed to depend linearly on the stress, electric field and temperature (eq. 10) in deriving eq. 11. Some higher order effects can be brought about by including second order terms in eq. 10, as follows;

$$d\varepsilon_{ij} = \left(\frac{\partial\varepsilon_{ij}}{\partial\sigma_{lm}}\right)_E d\sigma_{lm} + \left(\frac{\partial\varepsilon_{ij}}{\partial E_n}\right)_\sigma dE_n + \frac{1}{2}\left[\left(\frac{\partial^2\varepsilon_{ij}}{\partial\sigma_{lm}\partial\sigma_{pq}}\right)_E d\sigma_{lm}d\sigma_{pq} + \left(\frac{\partial^2\varepsilon_{ij}}{\partial E_n\partial E_r}\right)_\sigma dE_n dE_r + 2\left(\frac{\partial^2\varepsilon_{ij}}{\partial\sigma_{lm}\partial E_n}\right)d\sigma_{lm}dE_n\right]$$

$$d\hat{D}_k = \left(\frac{\partial\hat{D}_k}{\partial\sigma_{lm}}\right)_E d\sigma_{lm} + \left(\frac{\partial\hat{D}_k}{\partial E_n}\right)_\sigma dE_n + \frac{1}{2}\left[\left(\frac{\partial^2\hat{D}_k}{\partial\sigma_{lm}\partial\sigma_{pq}}\right)_E d\sigma_{lm}d\sigma_{pq} + \left(\frac{\partial^2\hat{D}_k}{\partial E_n\partial E_r}\right)_\sigma dE_n dE_r + 2\left(\frac{\partial^2\hat{D}_k}{\partial\sigma_{lm}\partial E_n}\right)d\sigma_{lm}dE_n\right] \quad \text{(EQ 17)}$$

where,

$$S^E_{ijlmpq} = \left(\frac{\partial^2\varepsilon_{ij}}{\partial\sigma_{lm}\partial\sigma_{pq}}\right)_E \qquad d_{ijnr} = \left(\frac{\partial^2\varepsilon_{ij}}{\partial E_n\partial E_r}\right)_\sigma = \left(\frac{\partial^2\hat{D}_k}{\partial\sigma_{lm}\partial E_n}\right)$$
$$\kappa_{ijlmn} = \left(\frac{\partial^2\varepsilon_{ij}}{\partial\sigma_{lm}\partial E_n}\right) = \left(\frac{\partial^2\hat{D}_k}{\partial\sigma_{lm}\partial\sigma_{pq}}\right)_E \qquad \varepsilon_{knr} = \left(\frac{\partial^2\hat{D}_k}{\partial E_n\partial E_r}\right)_\sigma \quad \text{(EQ 18)}$$

are nonlinear elastic compliance coefficients, electrostriction coefficients, elastostriction coefficients and nonlinear dielectric permittivity coefficients, respectively. Integrating equation 10, we obtain,

$$\varepsilon_{ij} = S^{E}_{ijlm} + d_{ijn}E_n + \frac{1}{2}S^{E}_{ijlmpq}\sigma_{lm}\sigma_{pq} + \frac{1}{2}d_{ijnr}E_nE_r + \kappa_{ijlmn}\sigma_{lm}E_n$$

$$\hat{D}_k = d_{klm}\sigma_{lm} + \varepsilon^{\sigma}_{kn}E_n + \frac{1}{2}\kappa_{klmpq}\sigma_{lm}\sigma_{pq} + \frac{1}{2}\varepsilon_{knr}E_nE_r + d_{klmn}\sigma_{lm}E_n \qquad \text{(EQ 19)}$$

Piezoceramics are brittle materials and elastically behave linearly up to the failure. Electrostriction coefficients are important at high electric field strengths. In cases, where mechanical stresses are applied in addition to electric field (piezoceramic is constrained from freely deforming), the elastostriction coefficients should be included in constitutive relations.

ACKNOWLEDGEMENT

This work is a part of the preliminary studies on damage survivability and damage tolerance of "smart" laminated composites sponsored by the Army Research Office.

REFERENCES

1. Manual, "Kynar Piezo Film", Pennwalt Corporation.
2. Kawai, H., "The Piezoelectricity of Polyrinydene Fluoride", Japan Journal of applied physics, No. 8, 1979, pp. 975-976.
3. Toupin, R.A., "A Dynamical Theory of Elastic Dielectrics", Int. J. Eng. Sci., Vol. 1, 1983, pp 101-126.
4. Nelson, D.F., "Theory of Nonlinear Electroacoustics of Dielectric, Piezoelectric Crystals", J. Acoust. Soc. Am., Vol. 63, June 1978, pp. 1738-1748.
5. Tiersten, H.F., "Electroelastic Interactions and the Piezoelectric Equations", J. Acoust. Soc. Am., Vol. 70, December 1981, pp. 1567-1576.
6. Penfield, P. Jr and Hans, H.A., "Electrodynamics of moving media", Research Monograph, No. 40, Massachusetts Institute of Technology Press, Cambridge, Massachusetts, 1967.
7. Mindlin, R.D., "Equations of High Frequency Vibration of Thermopiezoelectric Crystal Plates", Int. J. Solids Struct., Vol. 10, No. 6, 1974, pp. 625-637.
8. Nye, J.F., "Physical Properties of Crystals", Oxford, Clarendon Press, 1964.
9. Berlincourt, D.A., Curren, D.R., Jaffe, H., "Piezoelectric and Piezomagnetic Materials and Their Function as Transducers, In: Physical Acoustics, Mason W.P. (Ed.), Vol. 1-Part A, Academic Press, New York, 1964.
10. Landau, L.D., Lifshitz, E.M., "Electrodynamics of Continuous Media", Pergamon Press, Oxford-London-New York-Paris, 1960.
11. Chen, P.J., "Characterization of the Three-Dimensional Properties of Poled PZT-65/35 in the Absence of Losses", Acts Mech., Vol. 47, 1983, pp. 95-106.
12. Chen, P.J., "Three-Dimensional Dynamic Electromechanical Constitutive Relations of Ferroelectric Materials", Int. J. Solids Struct., Vol. 16, No. 12, 1980, pp 1059-1067.
13. Chen, P.J., Peercy, P.S., "One-Dimensional Dynamic Electromechanical Constitutive Relations for Ferroelectric Materials, Acta Mech., Vol. 31, No. 3, 1979, pp. 231-241.
14. Chen, P.J., Tucker, T.J., "Determination of the Polar Equilibrium Properties of the Feroelectric Ceramic PZT-65/35", Acta Mech., Vol. 38, No. 3-4, 1981, pp. 209-218.
15. Joshi, S.P., "Constitutive Models for Piezoelectric Materials", Constitutive Laws for Engineering Materials: Recent Advances and Industrial and Infrastructure Applications (Ed: C.S. Desai, E. Krempl, G. Frantziskonis, H. Saadatmanesh), ASME Press, 1991, pp.605-608.
16. Auld, B.A., "Wave Propagation and Resonance in piezoelectric materials", J. Acoust. Soc. Am., Vol. 70, No. 6, December 1981, pp. 1577-1585.

"Intelligent Materials--Keys to the 21st Century

Hiroaki YANAGIDA

Intelligent mechanisms such as self-diagnosis, self-recovery, self-adjustment, capability to tuning installed in materials can save too much complicated circuits(spaghetti syndrome) and too many numbers of control button(cockpit syndrome) in advanced electronics. Examples developed by the speaker are p/n contact chemical sensors and carbon fiber IR sensors. Humidity sensor made by a contact between CuO and ZnO has a self-cleaning mechanism, while porous zinc oxide type sensors, need additional circuits for cleaning. The contact is also used for gas sensors. Sensitivity and selectivity can be tuned by changing the bias voltage between the contact. Carbon fiber IR sensors can defect stationary objects without choppers, while pyroelectric type ones need them. Small heat capacity of the fibers leads to very fast response. Other examples already commerciallized are PTCR-heating elements and humidity sensors made by mixture of ZrO_2 and MgO. They can also reduce complication of circuits and numbers of control button. Electronic devices thus simplified are more friendly and more reliable than those of too much complexity.

Improvement of reliablity in structual materials must be performed by installing self-diagnosis and self-recovery mechanisms, since recyling of materials is the key to clean environment. Recycling program must be designed beforehand the manufacturing. Saving energy/ resource and recycle program are very important keywords to intelligent materials.

Intelligent materials are the keys to the 21st Century where technologies must be developed more friendly to people and environments.

INTELLIGENT MATERIALS FOR FUTURE ELECTRONICS

Kiyoshi TAKAHASHI

Department of Electronics

Tokyo Institute of Technology

ABSTRACT

Up to now, electronic devices have been made using materials with given characteristics. In the future, however, materials should be designed to given the desired characteristics to electronic devices. As may be inferred from the widespread use of semiconductors in electronic devices, semiconductors are at present indispensable to the electronics industry. However, if we rely exclusively upon semiconductors , we run the risk of material shortages. Consequently, there is a need for new, alternative materials. Such new materials are already being referred to by a recently coined term, "INTELLIGENT MATERIALS".

There is little hope for finding such intelligent materials in nature. It may,. however, be possible to develop them through a kind of "GENETIC ENGINEERING IN MATERIAL SCIENCE".

This paper discusses "INTELLIGENT MATERIALS" from the viewpoint of an individual whose experience is in the field of electronics.

APPLICATIONS OF PIEZOELECTRIC CERAMICS IN SMART ACTUATORS AND SYSTEMS

KENJI UCHINO
Materials Research Laboratory
The Pennsylvania State University
University Park, PA 16802-4801

Permanent Address:
Department of Physics
Sophia University, Kioi-cho 7-1
Chiyoda-ku, Tokyo 102
JAPAN

ABSTRACT

In these several years piezoelectric and electrostrictive actuators have become very popular for micro-positioning in optical and precision machinery fields.[1] Aiming at wide commercialization of these actuators, many investigations have been made in the improvement of ceramic materials for actuators, designs of the devices and control and systematization of the actuators. This paper reviews recent applications of piezoelectric/electrostrictive ceramics from a viewpoint of "smart" actuators and systems.

A passively smart material is exemplified by the lead magnesium niobate (PMN) based ceramic, which can exhibit a large electrostriction ($\Delta l/l \sim 10^{-3}$) without any hysteresis and aging effect during an electric field cycle.[2] A composite actuator structure called "moonie" has been developed to amplify the small displacement induced in a multilayer piezoelectric device. Passive damper application is another smart usage of piezoelectrics, where mechanical noise vibration is radically suppressed by the converted electric energy dissipation through Joule heat when a suitable resistance is connected to the piezoelectric plate.[3] Piezoceramic: carbon black: polymer composites are promising useful designs for practical use.

An actively smart material is exemplified by the video tape head positioner made from a lead zirconate titanate (PZT) bimorph with sensor and actuator-divided electrodes.[4]

Monomorphs and shape memory ceramics belong to very smart materials. A monomorph device made of a semiconductive piezoelectric plate generates the Schottky barrier when metal electrodes are coated on the faces, providing non-uniform distribution of the electric field even in a compositionally uniform ceramic. A superimposed effect of piezoelectricity and semiconductivity leads to a bending deformation in a total ceramic plate.[5] The strains associated with phase transitions such as an antiferroelectric-to-ferroelectric transition in lead zirconate stanate-based ceramics reach up to 0.4%, which is much larger than that expected in electrostrictors. Moreover, this field-induced transition exhibits a shape memory effect in appropriate compositions, and such ceramics are useful for the applications to latching relay and a mechanical clamper.[6]

A photostrictive actuator is the best example of intelligent materials including sensing, actuating and drive/control functions in a unique material.[7] In certain ferroelectrics a phenomenon by which a constant electromotive force is generated with exposure of light has been observed. A photostrictive effect is expected as a result of the coupling of the photovoltaic and inverse piezoelectric effects. A remote control miniature walking robot, which is activated with illumination, is currently being fabricated. Two photostrictive PLZT bimorphs were combined together and each plate exhibits a minute photo-induced displacement on the order of 150 μm. Alternative illumination causes a slow moving of the ceramic device.

A flight actuator consisting of a pulse-driven piezoelectric element and a steel ball is a very suggestive mechanism, even if it would not be denoted as a smart system. A 2mm steel ball can be hit up to 20 mm by a 5 μm displacement induced in a multilayer actuator with quick response.[8]

A smart system is typically exemplified by a precision lathe machine. A micro displacement actuator has been manufactured using an electrostrictive multilayer actuator, a magneto-resistive strain sensor and an adaptive control circuitry. The feedback control has suppressed the position deviation of the cutting edge when pushing stress is produced during cutting process. The cutting accuracy in less than ±0.01 μm is now available.

A very smart system contains a reliability test system, which can stop an actuator system safely without causing any serious damages on to the work, e.g. in a lather machine. Acoustic emission measurement of a piezo-actuator under a cyclic electric field is a good candidate for estimating the life time of the actuators.[9]

The bright future of piezoelectric/electrostrictive actuators has been initiated and even greater commercial participation in their continued growth and application is anticipated.

REFERENCES

1. K. Uchino, Piezoelectric/Electrostrictive Actuators, Morikita Pub. Co., Japan (1986).

2. L. E. Cross, S. J. Jang, R. E. Newnham, S. Nomura and K. Uchino, Ferroelectrics 23, 187 (1980).

3. K. Uchino and T. Ishii, J. Ceram. Soc. Jpn. 96, 863 (1988).

4. A. Ohgoshi and S. Nishigaki, Ceramic Data Book '81, Industrial Products Tech. Assoc., Japan, p. 35 (1981).

5. K. Uchino, M. Yoshizaki, K. Kasai, H. Yamamura, N. Sakai and H. Asakura, Jpn. J. Appl. Phys. 26, 1046 (1987).

6. K. Uchino, Proc. MRS Int'l. Mtg. on Adv. Mats. 9, 489 (1989).

7. M. Tanimura and K. Uchino, Sensors and Mater. 1, 47 (1988).

8. S. Sugiyama and K. Uchino, Proc. 6th IEEE Int'l Symp. Appl. Ferroelectrics, p. 637 (1986).

9. T. Hirose and K. Uchino, Ferroelectrics 87, 295 (1988).

THE MACE ACTIVE MEMBER

Warren Hoskins
LMSC, Inc
Sunnyvale, CA

Bob Buchanan
LMSC, Inc.
Palo Alto, CA

David Miller
Space Engineering Research Center
MIT, Cambridge, MA

Javier de Luis
Payload Systems, Inc
Cambridge, MA

ABSTRACT

The MACE program (Mid-deck Active Control Experiment) is a flight experiment conducted in the shirt sleeve environment of the mid-deck of the Shuttle sponsored by NASA LaRC and designed by the Space Engineering Research Center at MIT with PSI as subcontractor and LMSC as a corporate sponsor. The MACE testbed is scheduled to fly in 1994. The objective of MACE is to validate the modeling tools associated with flexible body dynamics and active controls and to exaimine the effects of non zero gravity ground testing. The approach is to design a small flexible testbed (2 Hz) and perform a set of experiments which capture the essential physics of large precision spacecraft structures. The testbed consists of a multi-segmented beam with a torque wheel for attitude control and an active member used for vibration suppression and attached to either end are two axes gimbals for rigid payloads. The first series of experiments include a system identification test for comparison with ground test results and for updating the control system gains corresponding to the revised dynamic parameters. The second test in this series includes a pointing test using the torque wheels for broad band disturbance and the active member coupled with the gimbal to maintain payload point. The third test is a pointing test of one payload while slewing the other payload.

The purpose of this paper is to address the design of the active member. Basically the active member consists of a lexan tube which is about one foot in length and represents one of the four segments. This tube is thicker than the other tubes and is machined flat on all four sides producing an equivalent stiffness to the other four members. A piezo ceramic PZT thin plate is epoxied to all four machined sides along the full length. On one side the PZT is used as the sensor and on the opposite side the PZT is used as the actuator introducing a distributed moment along its length. With the actuator - sensor pair control is provided in both axes. This active member is used to provide active damping over the bandwidth of the disturbance.

This paper covers the derivation of requirements, chacterization of the active member as a component and its effect on the system level performance. The disturbance levels are established by the top level requirement to attenuate the vibration by 40 db while being 20 db above the sensor noise floor level. By integrating the active member into the broad band pointing test simulation the moment requirement is established. It is expected that the active member will provide an order of magnitude attenuation and torque shaping will achieve an additional order of magnitude. Tests results from the component characterization tests are included showing both the moment and motion capability over the control bandwidth. In addition the active member is tested in a psuedo system level testbed (gimbals and payloads are mass simulated) to demonstrate its effectiveness in suppressing vibrations.

Design of Composite Tubes with Embedded Piezoelectric Ceramics for Active Members of Space Structures

by
Candice Snydar, J.B. Cushman, David Wilson

Boeing Defense and Space Group
Structures Technology
P.O. Box 3999
Seattle, WA 98124-2499
Mail Stop 82-97

ABSTRACT

Future space structures will require lightweight intelligent members to perform such tasks as active/ passive vibration control, shape control, precision pointing, and health monitoring. This paper describes work performed on Phillips Lab, Edwards AFB "Advanced Composites with Embedded Sensors and Actuators (ACESA)" contract for techniques to embed piezoelectric ceramics in graphite/ epoxy structures. Different combinations of lead wire types and attachment methods were examined for effectiveness before and after temperature cycling. Methods for insulating the piezoelectric ceramics from the conductive graphite fiber were examined. This paper also describes work focusing on the design of composite tubes with embedded piezoelectric ceramics for active members of space structures. The composite tube was designed to match the stiffness of an aluminum tube it would be replacing. This affected the choice of material, number of layers, and layup for the tube. Steps in fabrication will be described, and the functionality testing of the tubes after cure will be examined. Testing included capacitance readings before and after embedding, x-rays, CT scanning, and structural testing. Actuation authority testing included axial actuation capability/ strain efficiency, force capability, and stroke capability.

Active Vibration Filtering for Optical Delay Line

C. Garnier, and B. Koehler, Aerospatiale, Division Systemes Strategiques et Spatiaux, Establissment de Cannes, France

Stellar interferometry -or optical aperture synthesis- is a very promising technique for astronomical observation owing to the dramatic increase in angular resolution it allows (in the 10^{-3} arcsec range). One of the technical challenge is the recombination of light beams collected by several apertures in order to obtain interference fringes. The Optical Path Lengths of the different beams must be equalized in real time at better than a few nanometers RMS.

This paper describes the concept designed by Aerospatiale for such delay lines. This concept uses accelerometer feedback and magnetic or piezo-electric actuators to control and suppress vibrations in the Optical Path Difference (OPD) between different beams. An on-ground realisation of a delay line filtering stage for the Observatoire de la Cote d'Azur (France) will be presented. This realisation allows a stability of the OPD better than 12.5 nm RMS for all speeds up to 1 mm/s.

Finally a project for Optical Aperture Synthesis in Space will be discussed. New problems rise due to spacecraft equipment induced perturbations and large structure vibrations. Potential solutions using the concept of active vibration filtering will be presented.

CONTROL OF SPACE STRUCTURES USING ACTIVE PIEZOELECTRIC MEMBERS

C. L. Trent
Y. H. Pak
McDonnell Douglas Space Systems Company
Huntington Beach, California 92647

ABSTRACT

A major concern in Large Space Structure (LSS) design is the development of accurate models of the structural dynamics. It is very difficult to obtain accurate LSS model characterization via ground test. Test complications include the size of the structure, loads induced by the one-g environment, air damping and the inherent nonlinearities in the structural joints. In addition, structural characteristics may change during the vehicle lifetime due to on-orbit build-up, hardware failures, and operational or environmentally induced changes. As a result, the actual performance and stability of control systems, designed based on ground test data, may be significantly different from pre-flight predictions. Adaptive control techniques with on-line system identification need to be demonstrated in order to ensure that future platforms meet mission objectives. Control-structure interaction (CSI) is another concern that occurs when the controller bandwidths are sufficiently high that they interact with or overlap the flexible body modes of the structure. In addition, due to the multi-purpose characteristics of these platforms, multiple control systems operate simultaneously, which can cause unwanted interaction among them. Methodologies need to be developed and tested that resolve these CSI issues. This paper reports the results of the first phase of a project that addresses these isses.

The objective of the reported work is to develop the methodology to design and test controllers that will provide stable environments for payloads mounted on space platforms. We use our 9.5-meter long, 19 bay, MDSSC CSI System Demonstration Truss

(CSDT) testbed to verify CSI methodologies and performance of the various controllers. Four aluminum truss members were replaced with Active Damping Struts (ADSs). The ADS consists of a composite-piezoelectric tube bonded to aluminum end-fittings. The composite tube consists of ten tubes bonded end-to-end. One of the tubes at the end is electrically isolated from the rest and functions as a sensor. The rest of the tubes function as the actuator.

A method was developed for optimally placing a specified number of ADSs in the CSDT for bending control. This method allowed new controller gains to be calculated for each new ADS configuration. An optimization algorithm was chosen that maximized the rate of energy dissipation of the CSDT. This algorithm simulates the annealing process of solids which is characterized by a general decrease in energy level with occasional energy increases whose rate of occurrence may be estimated by the Boltzmann probability function (i.e., a decaying exponential function of temperature). Using this analogy, a mechanism is provided for climbing out of local optima in the form of probabilistic acceptance of non-improving solutions. The advantage of this technique is that it does not require an exhaustive search to converge to its final solution.

We designed and implemented an active vibration suppression controller that increased the first mode damping by more than 400% and developed a method for optimally placing the ADSs in the CSDT. This controller utilized non-collocated strain rate feedback. The controller was designed in the discrete domain using MATRIXx, with modal data from a NASTRAN normal modes analysis used to describe the CSDT structural dynamics. The measured first mode damping factor was 0.3%, which yielded an open loop settling time of 27 seconds. The closed loop settling time achieved was 6 seconds. This corresponds to an increase in the first mode damping from 0.3% to 1.33%. This response matched the analytical predictions.

NN

Neural Control of Smart Electromagnetic Structures

Michael Thursby, Kisuck Yoo and Barry Grossman
Department of Electrical and Computer Engineering
Florida Institute of Technology, Melbourne, Fl 32901
Tele (407) 768-8000 X6160 EMail mthursby@zach.fit.edu

Abstract

We are studying a new class of smart structures-smart electromagnetic structures(SEMS). These structures are "smart" in that they integrate sensing elements (e.g.,antennas), processing elements (neural networks) and control elements(diodes) in a manner not previously considered. Smart Electromagnetic Structures(SEMS) have the potential to provide an adaptive electromagnetic(EM) environment to the structure on which they can be mounted. Based on their sensing capabilities they may be able to detect and modify the EM fields around them as well as their far field image. The ability to adapt derives from the closed loop nature of the SEMS, hence the speed of adaptation is determined by the speed of the loop. Factors including bandwidth of the control structure do influence the speed of the system. The speed of the response is primarily determined by the technology of the computational elements. The implementation we are studying includes an Artificial Neural Network(ANN) as the processor. The neural net can respond in no more than three gate delays for each iteration of the loop. We have found that the network takes from three to five iterations of the loop to complete its control task. This results in a total time for system response of less than fifteen gate delays.

Artificial neural networks(ANNs) and their ability to model and control dynamical systems for smart structures, including sensors, actuators, and plants, are directly applicable to the SEMS concept. By incorporating a neural network into the control structure of a single microstrip patch element its electrical characteristics can be changed in response to a received signal. This change can be used to alter the antenna's performance in real time.

The Neural Net Antenna

The micropatch antenna has many advantages including simplicity and size, and a few drawbacks, e.g., narrow bandwidth. The electrical characteristics of the antenna can be adjusted using control elements embedded in the patch itself. We will describe research being carried out in the Autonomous Systems Laboratory(ASL) of Florida Institute of Technology(FIT) into the control of such patch antenna elements using a neural network (NN) in the feed back loop to enhance the operating characteristics of the The neural net can make the required determinations in near real time. The ability of the net to adapt to unknown inputs (generalization) and its fault tolerance makes the neural antenna an ideal candidate for flexible tactical antennas for the future. The combination of a simple neural network with a microstrip patch antenna is shown in Figure 1.

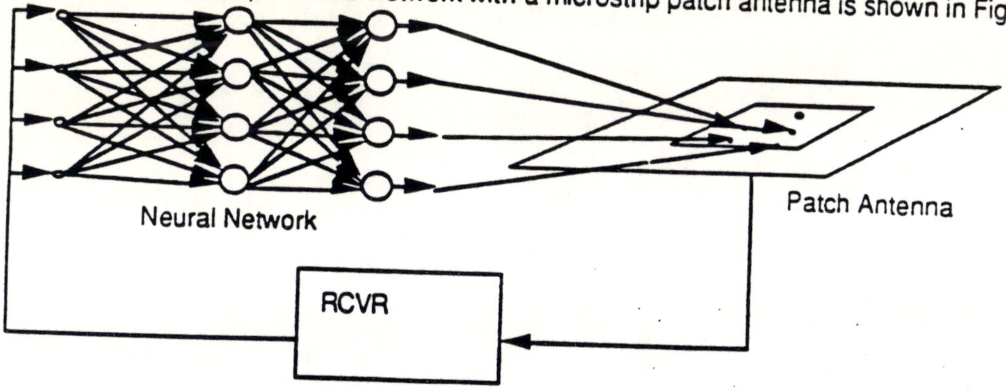

Figure 1. The microwave patch antenna with tuning points and a neural network to drive the points can be considered a smart antenna structure.

The patch neural network antenna system has been developed and this analytical model, as well as experimental models of the antenna are being tested and compared. The model and prototypes are being taught to adapt to the magnitude and phase response of incoming signals.

In order to test the ability of such a system to tune to the frequency of incoming signals a series of experiments were conducted on the trained simulation of the neural antenna system. The ability of the neural antenna to follow the center frequency of incoming signals with time varying frequency characteristics will be presented. Experiments requiring the network to tune the antenna in a stepped frequency environment. The ability of the antenna to follow a continuously varying frequency signal will also be presented. We demonstrate that the patch can be given autonomous adaptive capabilities using neural networks.

Several applications for such an antenna can be postulated. First such an device would improve receiver characteristics in a frequency agile environment. The adaptability of the neural antenna would reduce the manufacturing and siting tolerance requirements normally placed on such conformal antennas. An array of such smart patches could be assembled to create an even more adaptable antenna system.

Acknowledgements

This research was partially supported by U.S. Air Force Contract Nr.F08635-87-C-0460 and Army Research Office Grant DAAL03-89-G-0085.

NN

MULTICOMPUTER NETWORKS FOR SMART STRUCTURES (EXTENDED ABSTRACT)

Scott F. Midkiff[1] and John T. McHenry

The Bradley Department of Electrical Engineering
Virginia Polytechnic Institute and State University
Blacksburg, Virginia 24061-0111

ABSTRACT

Using concurrent processing and high-speed interconnection networks, multicomputers are capable of the high data throughput and real-time computations required for smart structures. However, to realize these capabilities it is essential that the multicomputer network and other components of the system are designed in an integrated manner. This paper discusses the design of multicomputer networks to meet the processing requirements of smart structures. Methods for mapping neural network computational models onto multicomputer networks are presented.

Keywords: *Distributed networks, distributed and embedded processors, neural networks*

I. INTRODUCTION

If multicomputers are to be used in smart structures, the multicomputer network must be designed to provide real-time processing and high data throughput. This paper discusses the design of multicomputer networks for smart structure applications. Requirements for multicomputer networks and opportunities to exploit application-specific features are described. The implementation of neural networks, which have been proposed for processing in smart structures, is discussed as a specific example.

II. BACKGROUND

A. Processing Requirements for Smart Structures

The development and application of smart structures requires the integration of a number of technologies including structures, materials, sensors, control, and actuators. In addition, capabilities for real-time processing and communication are necessary.

The processing functions may be viewed as a three-stage pipeline.

---

[1] Direct correspondence to S. F. Midkiff at the address above; telephone (703) 231-5190; e-mail midkiff@vtvm1.cc.vt.edu.

1. Data is acquired from one or more sensors. Sensor data acquisition may require signal processing to condition the signal into a form suitable for use in the second stage.

2. Sensor data is processed. Control algorithms may compute a new desired system state or diagnosis algorithms may determine if a failure has occurred or is about to occur.

3. Outputs from the processing (second) stage are used to control actuators or indicate failure conditions.

Data communication is needed to acquire sensor data, transfer the data to one or more processing elements, and send outputs to actuators or indicators. In addition, if multiple processing elements are employed, then interprocessor communication is needed to coordinate and share information among tasks executing on different processors.

The computation and communication must be performed in real-time. This implies that the operations are correct if and only if the logical results of the computation are valid, and the time at which the results are produced meets all timing constraints [1]. Real-time computing is not simply fast computing. Fast computation ensures only that the average delay is small, but not that all computations are performed within the allotted time. Thus the predictability of delay for communication and processing operations is of the utmost importance.

B. Multicomputer Networks

Multicomputers, also known as distributed memory computers, are a type of parallel processing system consisting of multiple processing nodes interconnected by a communications network. Each processing node in a multicomputer has memory, computing resources, and interprocessor communication facilities. The computing resources are used to perform the processing assigned to the node. The memory stores both program and data. The communications facilities access the interconnection network and support other input and output operations. A multicomputer contains no shared memory.

Data and other information is shared between tasks executing on different processing nodes via an interconnection network. The interconnection network has one of a variety of topologies, which may include either point-to-point links or shared busses. Communication is particularly important to the overall performance of a multicomputer. Communication performance in a multicomputer depends on four general factors: (1) the speed of links and protocols, (2) the speed of node interfaces and processing, (3) the characteristics of the traffic generated by the application, and (4) the the topology of the interconnection network.

In general, topologies may be either "general-purpose" or "application-specific." General-purpose topologies provide small average and worst-case distances between nodes, and thus minimize the number of links that must be traversed for a variety of different traffic patterns. Application-specific topologies match the communication needs of a particular application or class of applications.

III. MULTICOMPUTER NETWORKS FOR SMART STRUCTURES

To implement a multicomputer system that can meet the requirements of smart structure applications, it is necessary to address processing and communication concerns within an application-specific context including sensors, actuators, processors, and interconnection network.

A. Multicomputer Processing Nodes

In a multicomputer architecture, sensor data acquisition and signal processing, local control functions, and actuator control operations map naturally onto individual processing nodes, as shown in the network of Figure 1. Multicomputer nodes with significant power are currently implemented as single integrated circuits. However, sensor and actuator interfacing issues must be addressed for smart structure applications. It is also conceivable that processing nodes can be embedded in the material of a smart structure, and thus be co-located with sensors and actuators. Alternatively, nodes can be interfaced to sensors and actuators along structure boundaries.

B. Interconnection Networks

The design of interconnection networks for smart structures involves several important issues.

- The topology of the network should support the communication pattern of the application.
- Network links must be fast, reliable, and testable.
- Link and network protocols must provide predictable delays and high data throughput.

Optical communication links offer a number of benefits for smart structures, including high bandwidth, immunity to electromagnetic interference, and the potential for hybrid sensing and communication. Hybrid sensing and communication is attractive for several reasons. First, optical fibers in a material may be used to full capacity by providing both functions; a separate communication network is not required. Secondly, the synergy between structural and network topologies can be fully exploited. Finally, the ability of the many fibers needed for sensing to also be used for communication provides increased robustness in the event of failures.

C. Hierarchical Computations Networks

In addition to structural bases for network topologies, the organization of computations may also be exploited in an application-specific topology. For example, failure detection or control algorithms typically require system-level processing. A three-level hierarchy of nodes can be envisioned. At the lowest level, local control functions are performed at each node. Regional control functions, which may be executed in a distributed manner on a number of nodes or in a centralized fashion on a set of designated nodes, gather state information and provide control for a region of a structure. At the highest level, a global control function, which may again be distributed or centralized, gathers state information and provides control for the entire structure.

IV. EXAMPLE NEURAL NETWORK IMPLEMENTATION

The neural network or connectionist paradigm is a candidate model for computations in smart structures. In one proposed organization, the neural network receives sensor data as input and drives actuators with outputs [2]. The inherent parallelism and distributed state in a neural network allow it to be realized by a multicomputer network [3].

To illustrate a multicomputer implementation of a neural network computation, consider the simple neural network of Figure 2. There are four sensors (S_0, S_1, S_2, S_3) and two actuators (A_0, A_1). The four sensors provide inputs to the input layer cells. Each of the four cells in the input layer sends its output to each of the five cells in the hidden layer. Each hidden layer cell sends its output to each of the two output layer cells that control the two actuators. The cells of the neural network can be mapped to the four-node multicomputer network of Figure 1 as follows.

Node 0: Cells S_0, H_0, H_4, and A_0

Node 1: Cell S_1, H_1, and A_1

Node 2: Cells S_2 and H_2

Node 3: Cells S_3 and H_3

This mapping is one of several that are optimal assuming time division multiplexing of nodes and links, bidirectional links, and synchronous computation of cell outputs and states. The time for one iteration is determined by:

- four computation cycles, one for the four input layer cells, two for the five hidden layer cells (node 0 executes both H_0 and H_1), and one for the two output layer cells, and

- four communication cycles, two to move input layer outputs to hidden layer cells and two to move hidden layer outputs to output layer cells.

V. CURRENT STATUS

Investigation of multicomputer networks for real-time computation and communication in smart structures is currently underway at Virginia Tech. This investigation is a collaborative effort involving researchers in the areas of computer engineering, communications, controls, and fiber-based sensing. A three-node test bed, shown in Figure 3, has been implemented to permit investigation of sensor signal processing, distributed control algorithms for smart structures, communication protocols for real-time operation, and hybrid communication and sensor links [4]. Nodes in the test system are personal computers. A fiber optic MIL-STD-1773 bus connects the three nodes and is also used to implement three microbend sensors. All of the nodes may send and receive data using an 820 nanometer wavelength signal on the bus. For sensing, one of the nodes generates a 1300 nanometer signal that passes through sensors along three different paths, each of which terminates at a different node.

REFERENCES

[1] J. A. Stankovic, "Misconceptions about real-time computing: A serious problem for next-generation systems," *Computer*, vol. 21, no. 10, pp. 10-19, Oct. 1988.

[2] B. Grossman, H. Hou, R. Nassar, A. Ren, and M. Thursby, "Neural network processing of fiberoptic sensors and sensor arrays," *Fiber Optic Smart Structures and Skins III*, E. Udd and R. O. Claus, eds., Proc. SPIE 1370, pp. 205-211, 1990.

[3] J. Ghosh and K. Hwang, "Critical issues in mapping neural networks on message-passing multicomputers," *Proc. Int'l. Symp. Computer Architecture*, 1988, pp. 3-11.

[4] J. T. McHenry, S. F. Midkiff, J. A. Wiencko, and T. W. Reed, "Dual MIL-STD-1773 communications and microbend sensor fiber optic link," *Proc. Fiber Optic Sensor-Based Smart Materials and Structures Workshop*, 1991, to appear.

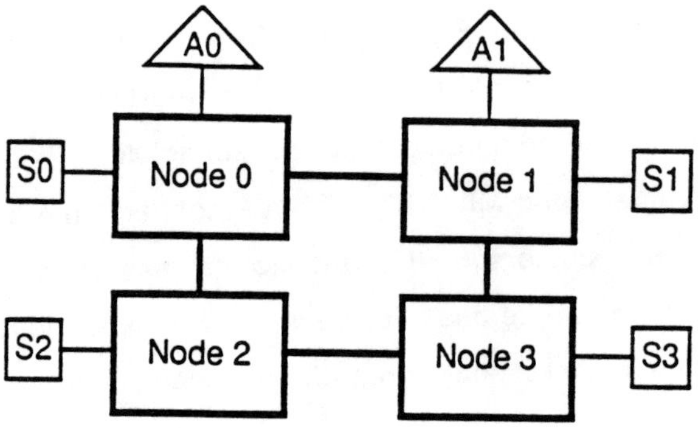

Figure 1. Example multicomputer network with sensors and actuators.

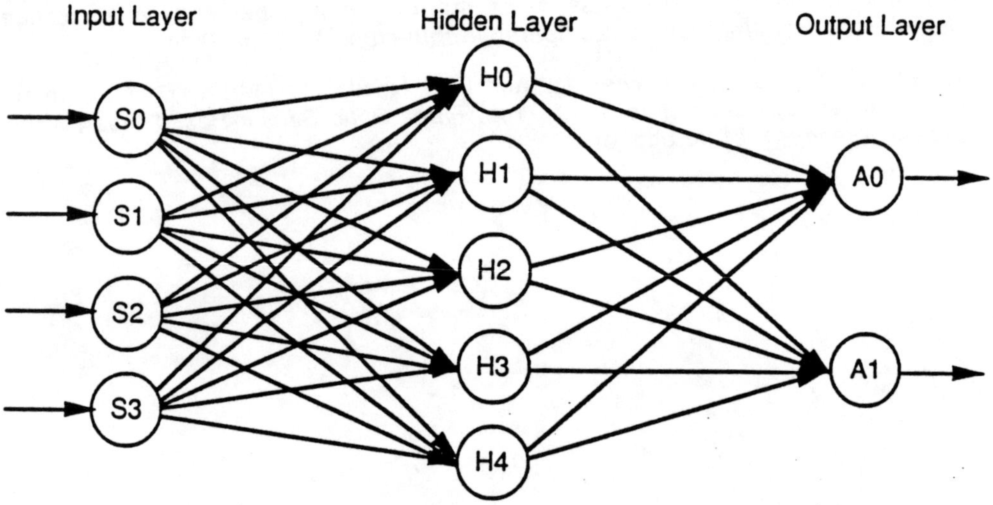

Figure 2. Example neural network.

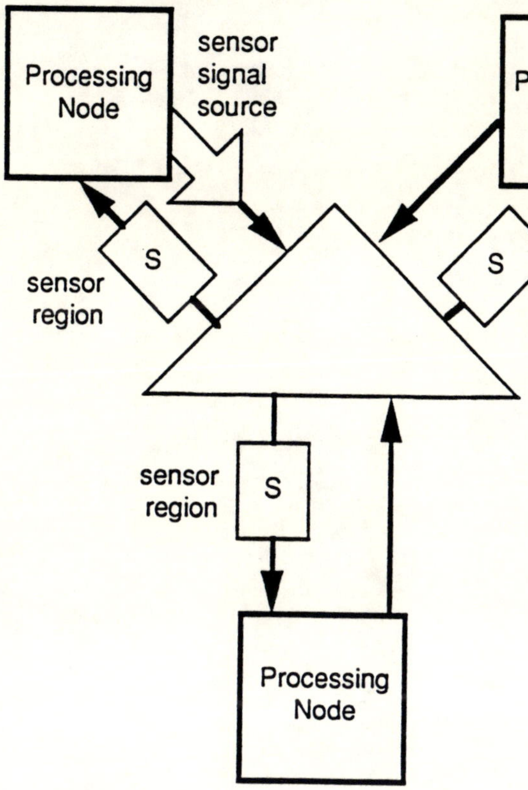

Figure 3. Three-node test system with hybrid sensing and communication.

Neural Network/Knowledge Based Systems for Smart Structures

James M. Mazzu, Scott M. Allen, and Alper K. Caglayan

Charles River Analytics Inc.
55 Wheeler Street
Cambridge, MA 02138
(617) 491-3474
E-Mail: jmm@cra.crasun.com

Abstract

In this paper, we present an approach for the design of intelligent structural monitoring systems. This approach consists of integrating of artificial neural networks (ANNs) and knowledge based expert systems (KBs) in order to achieve maximum benefits from both. We demonstrate our approach using a specific application involving the detection and isolation of in-flight aircraft structural damage, with assessments to determine the structure's residual strength. Our smart structures system uses strain measurements as sensory inputs. These strain distributions are obtained by embedding sensors, such as fiber optics, within composite aircraft structures. In general, ANNs process these multiple sensor measurements in parallel, while the KBs evaluate the results.

The primary development objective is to use the complementary capabilities of neural networks and expert systems within appropriate tasks and to determine integration strategies for creating structural monitoring systems. The smart structures system development takes place within the in-house developed *NueX* Hybrid Environment. In particular, ANN input and output nodes are represented as objects within the knowledge base, thereby supporting the inheritance of structural information. The ANN architectures, including node connection paths and corresponding weights, are stored in an efficient external data structure. The executive controller for the smart structures system is handled by a specialized KB; *NueX* allows the executive KB to pass information to and from the ANNs. Structural information regarding geometry and sensor locations is stored within a structural KB, which also performs structural reasoning on the relationships between sensor locations and critical aircraft components. Failure Detection and Isolation (FDI) is accomplished by both ANNs and KBs. The results from the FDI are evaluated within the Damage Assessor (DA), which also includes ANNs and KBs. The DA is responsible for determining the structural residual strength and the effects of damage on critical components such as hydraulic lines.

In our hybrid smart structures methodology, ANN development is accomplished using a structural class based approach. Typically for large aircraft structures, finite element models are only available for specific critical locations, and it is desirable to minimize the need for further analyses. Therefore, a variety of general structure classes are defined such that the majority of the structure is represented; finite element models need only be available for a representative of each of these general locations. The structures investigated consist of skin sections, skin/spar interface sections, corners, and skin boundaries. For each of these general classes, specialized ANNs are developed to process the sensor measurements which relate to that particular structural class. ANN training data are obtained by subjecting each section's finite element model to the largest range of loading conditions that its particular class may encounter over the structure and over the load spectrum of the aircraft. In doing so, the resulting ANN can be used over any location on the structure that matches its class. One of the ANN tasks involves estimating each sensor's undamaged strain measurement based upon its neighboring strain distribution. If this estimate significantly deviates from the sensor's actual strain reading, structural damage may be present and the expert system is alerted. Additional neural network tasks include serving as a damage dictionary and recognizing damage from temporal strain signatures.

The performance of our smart structures system is evaluated using finite element models and experimental results. The neural networks are trained with finite element data from an advanced high-strain composite wing section (manufactured by Grumman Aerospace Corporation), referred to as the Survivability Element. This structural section consists of one skin panel with three attached spars, therefore incorporating all four structural classes previously defined. The system is initially tested using the Survivability Element model with two damage scenarios. Subsequently, a hardware breadboard demonstration is performed on a composite skin panel incorporating one damage scenario. Finally, high-strain wing subcomponent test results are used to evaluate the system's performance on a large wing center section of which the ANNs have not been specifically trained, yet incorporates the same structural classes.

A smart structures design tool, based upon these hybrid concepts, will assist in the specification of optimal sensor locations, selection and training of ANNs, and the development of a structure-dependent knowledge base. These steps will be performed iteratively by the structural designer using our smart structures design tool, along with CAD and finite element programs. In this manner, engineers will create intelligent structural monitoring systems designed specifically for their applications.

ACTIVE MATERIALS AND ADAPTIVE STRUCTURES CONFERENCE
ARLINGTON, VA , NOVEMBER 5-7 1991

" APPLICATION OF A NEURAL NETWORK TO THE ACTIVE CONTROL OF STRUCTURAL VIBRATIONS "

Dr. Marcello R. Napolitano*
Dr. Roy Nutter**
Dr. Ching I Chen***

WEST VIRGINIA UNIVERSITY
MORGANTOWN , WV 26506-6101

APRIL 1991

---

* Assistant Professor, Department of Mechanical and Aerospace Engineering
** Professor, Chairman, Department of Electrical Engineering
*** Research Associate, Department of Mechanical and Aerospace Engineering

ABSTRACT

In this paper we propose the application of a Neural Network for state estimation purposes in a system for active control of structural vibrations. The implementation of a Neural Network is shown to be a powerful alternative to classic estimation structures, such as Luenberger Observer or Kalman Filter, whose on-line implementation on a large vibrating structure is not very attractive. The need for such estimation structure is due to the fact that most of the control strategies implemented for the problem use as feedback signal modal or physical displacements and/or velocities data directly taken or extracted from the data coming from the sensors located on the vibrating structure. However, for large structures, only a limited number of sensors can be placed, from which it would be extremely difficult to extract all the necessary data to be given to the control algorithm. This study shows the implementation of a Neural Network for state estimation on an aluminum cantilevered beam with random exciting force at the tip. Such Neural Network is designed with an off-line training session using numerical data from a dynamic simulation. The algorithm used for the determination of the weights and the thresholds of the neurons is the well-known Back-Propagation method. The input data of this Neural Network will be the data coming from the sensors located on the vibrating structure; the output data can be either modal or physical displacements and/or velocities. Once the training is completed, the Neural Network can replicate the system dynamics not only for the training conditions but at any other condition. This capability of generalizing its learning make the Neural Network a very practical tool for predicting the dynamics of complex systems, such as a large vibrating structure, from a limited number of training cycles. The results related to the training session and the numerical implementation of the Neural Network will be presented. The differences in the training of the Neural Network for different number of hidden layers, different number of neurons in each hidden layer, different size of the pattern for input data will be discussed.

EXTENDED ABSTRACT

IMPACT LOCATION ESTIMATION BY DISPERSIVE SIGNAL ANALYSIS

Chi-Ming Lu and Shiv P. Joshi

Aerospace Engineering Department

University of Texas at Arlington, Arlington, Tx 76019

Introduction

Monitoring a structure for the impact event is important in many engineering applications. For exmaple, the maintenance of the space platform requires knowledge of the severity and the location of the impact event in a real time. The impact detection techniques are also required in smart structures. There is a vast amount of research literature which has concentrated on nondispersive waves. For example, in seismology [1] the origin of an earthquake can be determined by noting the arrival times of the P, S and Rayleigh waves since each of these travels at a constant speed. These waves do not change their shapes as they propagate. The reason is that their phase speeds do not depend on frequency. Consequently, it is easy to keep track of the waves in space and time. A more complicated range of problems arise when the signal is dispersive. The ability to identify the waves as they propagate is difficult because the phase speed is frequency dependent. The pulse will change its shape as it propagates .

Recompressing the dispersive signal by time varying pulse-compression filter to give sharp arrival times is given by Booer et.al.[2] and Brazier-Smith et.al.[3]. The scheme is based on the conversion of the frequency transform of the signal to a wavelength transform as originally introduced for imaging faults in coal seams [2]. This method can not tell the absolute positions but if there exist multiple reflections then the distant between boundary can be told.

Whiston [4] and Jordan and Whiston[5] presented a different approach. First an estimuate of a position is obtained by the transient duration. This estimate is refined successively until the reconstructed force has no significant negative portions. This approach is suitable for an impact-type original pulse but not for a general disturbance. This approach is adopted in one of the methods presented in this report.

The phases difference in signal recorded at two points to locate the impact source is used by Doyle [6]. The phase angles difference between signal recorded at two points can be obtain by FFT. The distance of the origin of the signal is obtained from the phase difference information. This method is suitable for small structures.

The goals of this research are to find the distance of the origin of the signal from the recording site and the magnitude of the pulse at the origin. The dispersive signal from measurement site and the known dispersive relations for the medium are the only two known conditions. Three approachs are presented for achieving the above mentioned objectives. These three methods are all base on the Fast Fourier Transform and its inverse [7]. The first method uses a moving time window technique to calculate the origin of the signal. The second method [6] uses the phase difference of signals recorded at two locations to estimate the origin. The third method [4] makes the first guess of distance of the origin by estimating the transient time between the maximum and minimum frequency components arriving at the recording location and reconstructs the signal by transformation. The iteration procedure used in this method is based on the fact that the impact pulse at the origin is all positive. The last two methods are based on the formulation presented by Whiston [4] and Doyle [6].

Constructing Dispersive Waves

The solution of a wave problem is represented as

$$\begin{aligned} u(x,t) &= \sum_n F_n[G_1(k_{1n}x) + G_2(k_{2n}x) + \cdots] \exp(i\omega_n t) \\ &= \sum_n F_n G(k_{mn}x) \exp(i\omega_n t) \end{aligned}$$

where G is the transfer function of the problem [3]. Fn is forward FFT spectrum of the time input F(t). Consider the dispersive case

$$Gn = \exp(-ik_n x) \quad (1)$$

$$k_n = \frac{\sqrt{\omega_0 \omega_n}}{c_0}$$

$$\begin{aligned} u(x,t) &= \sum_n F_n \exp(-ik_n x) \exp(i\omega_n t) \\ &= \sum_n F_n \exp[i\omega_n(t - t_n)] \end{aligned}$$

$$t_n = \frac{x}{c_n} = \frac{x}{\frac{\omega_n}{k_n}}$$

when

$$F_n = \begin{cases} 0 & \text{if } t < t_n \\ F_n & \text{otherwise} \end{cases}$$

According to the above equation, the wave equation can be constructed as shown in Figure 1.

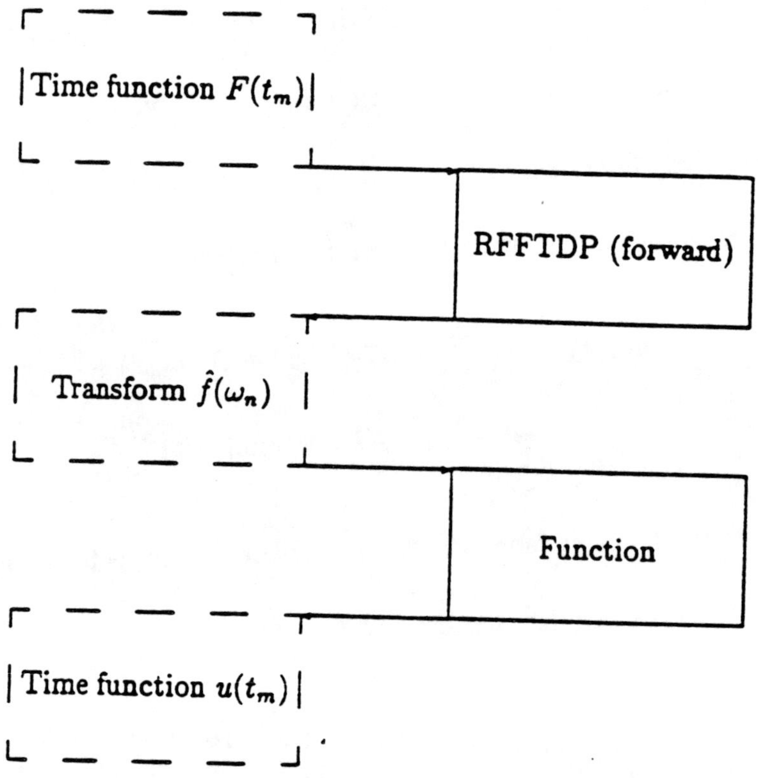

Figure 1   Flow diagram for the wave construction program

Methods For Locating The Origin of Dispersive Signals

Three methods for locating the origin of dispersive signals are brifely discussed in this section.

Moving Window Approach

The solution of a dispersive wave can be written as:

$$u(x,t) = \sum F_n \exp[i\omega_n(t - t_n)] \tag{2}$$

when

$$F_n = \begin{cases} 0 & \text{if } t < t_n \\ F_n & \text{otherwise} \end{cases}$$

The superposition of simple sine waves is used to simulate u(x,t) to clearly show some features of the method.

$$v = \sum_n \sin 2\omega_n \ll t - t_n \gg \tag{3}$$

$$t_n = 0.001(n-1) * 100(sec.)$$

$$\omega_n = 2\pi * 0.9765625[5 + (n-1)50]$$

when

$$\ll t - t_n \gg = \begin{cases} 0 & \text{if } t < t_n \\ t - t_n & \text{otherwise} \end{cases}$$

Figure 2 shows a dispersive signal for one second time duration. Higher frequency sine waves are superimposed on lower frequency sine waves at every 0.1 sec. time interval. This simulates the arrival time difference of 0.1 sec. between frequency components.

The information about the relation between frequencies and arrival times can not be obtained by using the power spectrum analysis of the dispersive signal as shown in Figure 3. However, if the wave is divided into small windows, the power spectrum of each small window will differ from other small windows. The frequency content of small windows and their relative positions can be utilized in obtaining the origin of the dispersive signal. The results obtained from this method be will discussed in the full length paper.

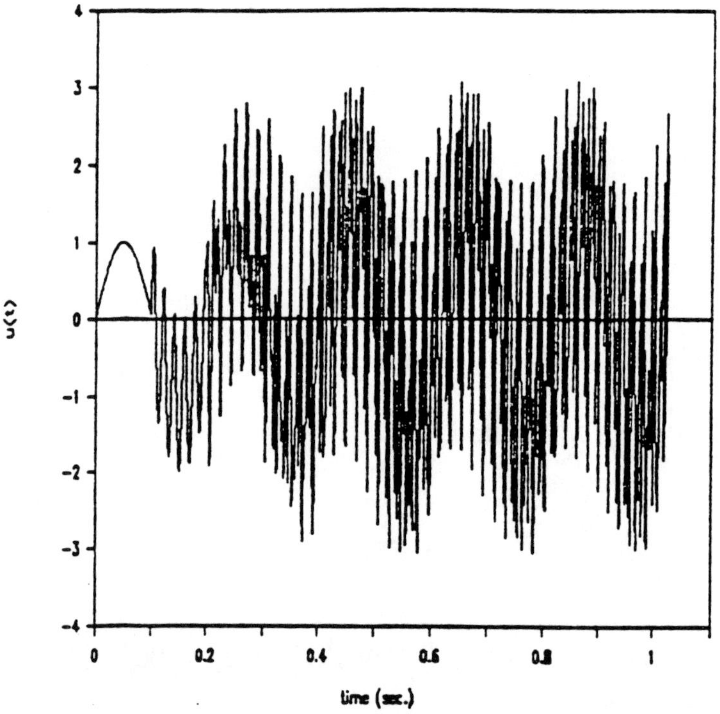

Figure 3 Dispersionlike signal $v = \sum_n \sin 2\omega_n \langle t - t_n \rangle$

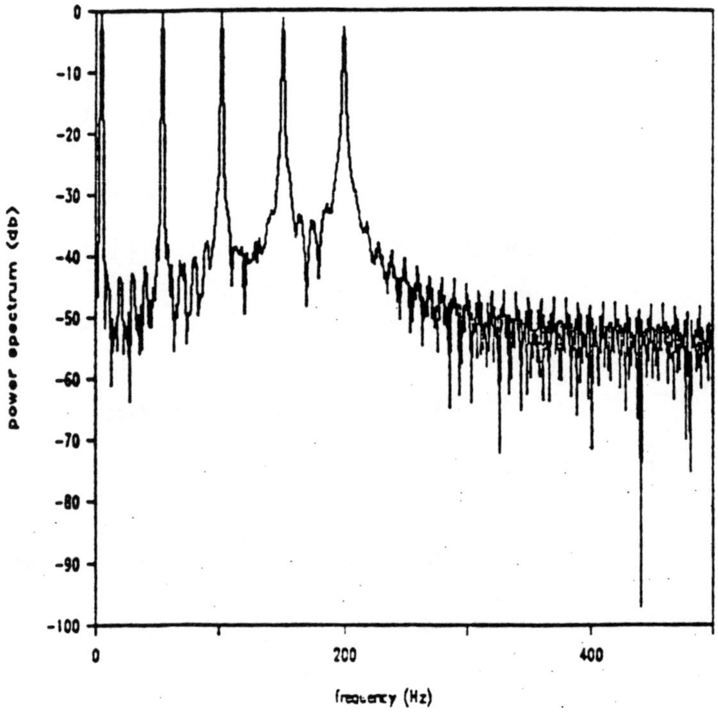

Figure 4 Power spectrum of the dispersionlike signal

Phase Difference Approach

The phase ϕ_n at two locations can be obtained by taking FFT of the recorded signals. The first guess is chosen to be the point between two recording points. After two or three iteration, the accurate position of the impact point can be obtained. The results obtained from the phase difference method are presented in Table 1

position	B	C
computation	0.203 m	0.457 m
real data	0.2 m	0.46 m

Table 1 Results of the phase difference approach

Reconstruction Approach

The results obtained from the reconstruction approach are presented in Table2. The computational procedure and the convergence scheme for the iterative procedure will be included in the full length paper.

position	B	C
computation	0.20007 m	0.46005 m
real data	0.2 m	0.46 m

Table 2 Results of the reconstruction approach

References

1. Ewing, W.M., Jardetzky, W.S. and Press, F., Elastic Waves in Layered Media, 1957, McGraw-Hill, New York.

2. Booer, A.K., Chambers, J. and Mason, I.M., "Fast Numerical Algorithm for the Recompression of Dispersed Time Signal," Electronics Letters,13, Aug. 1977, pp. 453-455.

3. Brazier-Smith, P.R., Butler, D. and Halstead, J.R., "The Determination of Propagation Path Lengths of Dispersive Flexural Waves Through Structures," Journal of Sound and Vibration, 75 (3), 1981, pp. 453-457.

4. Whiston, G.S., "Remote Impact Analysis by Use of Propagated Acceleration Signals, I: Theoretical Methods," Journal of Sound and Vibration, 97 (1), 1984, pp. 35-51.

5. Jordan, R.W. and Whiston, G.S., "Remote Impact Analysis by Use of Propagated Acceleration Signals, II: Comparison between Theory and Experiment," Journal of Sound and Vibration, 97 (1), 1984, pp. 53-63.

6. Doyle, J.F., "An Experimental Method for Determining the Location and Time of Initiation of an Unknown Dispersing Pulse," Experimental Mechanics, 27 Sept. 1987, pp. 229-233.

7. GEÇKİNLİ, N.C. and Yavuz, D.,Discrete Fourier Transformation Its Application to Power Spectra Estimation,Chap 2 and 3, 1983, Elsevier Scientific Publishing Company, The Netherlands.

8. Doyle, J.F., Wave Propagation in Structures, Chap. 2 and 4, Springer-Verlag, New York.

Modeling and Identification of the JPL Phase B Testbed*

John Spanos and Andy Kissil

Jet Propulsion Laboratory, California Institute of Technology

Pasadena, California 91109

Abstract

One of the most challenging problems in the study of large flexible space structures is the development of accurate mathematical models. The desired models are linear, time-invariant, and are to be used for structural redesign, actuator-sensor placement, control system design, and end-to-end performance assessment. Two types of models have been widely used in practical applications: the finite element model, and the experimentally obtained parametric input-output model.

The finite element model is consistent with linear elasticity theory and is created from knowledge of the geometry (i.e., size and connectivity of structural members) and the physical properties (i.e., mass density, modulus of elasticity, etc.) of the materials that make up the structure. An important feature of this model is that it can be assembled in a computer data base before the structure is built thereby allowing the designer to perform analysis and detect flaws in the early stages of the design process. However, as recent studies indicate [1][2], a major disadvantage of the finite element model is that it is not sufficiently accurate to be used for high bandwidth control system design.

The *nonparametric* input-output model is a set of actuator-sensor measurements taken from the physical structure after it is built (i.e., time responses, frequency responses, etc). On the other hand, the *parametric* input-output model is obtained by approximating the nonparametric model with a linear system. This approximation process is referred to as *linear system identification*. Numerous identification techniques have appeared in the literature [3] and, depending on the application, some tackle the approximation problem in the time domain while others deal with it in the frequency domain. Loosely stated, the objective is to find the linear system which, when excited with the same set of inputs as the structure, will produce outputs closely approximating those measured. Alternatively, the goal may be to find a linear system whose frequency response is a good approximation to the measured frequency response. It should be noted that the selection of input excitations as well as the locations of actuators and sensors used in the measurement process are as important here as the identification of parametric models from their nonparametric counterparts.

As pointed out in [1] and [2], input-output models identified in the frequency domain are more accurate representations of the structure than finite element models. This is primarily due to the fact that the finite element model is strongly dependent on the modeling assumptions used by the structural analyst while input-output models are closer to reality as they are generated by fitting actual measurements. In order to obtain a more accurate finite element model it is necessary to "tune" the various physical parameters of the model (i.e., cross-sectional area of members, modulus of elasticity, etc.) such that its modal

* Abstract submitted for presentation at the 1991 ADPA/AIAA/ASME/SPIE Conference on Active Materials and Adaptive Structures (5/1/91).

parameters (i.e., eigenvalues, eigenvectors) are in close agreement with those of the input-output model. The updated finite element model can then be used to make structural modifications and to more effectivelly place the control actuators and sensors on the structure.

This paper addresses the modeling and identification of an experimental flexible structure designed at the Jet Propulsion Laboratory known as the the "JPL Phase B testbed" [4]. Finite element models as well as input-output models (parametric and nonparameteric) were derived and compared. A single shaker and 56 accelerometers attached to the structure were used to obtain single-input, multi-output (SIMO) frequency response measurements via band limited white noise excitation, windowing, and spectral averaging. An input-output transfer function model was parameterized so as to match the second order modal form of the finite element model. A nonlinear least squares curve-fitting problem was formulated where the unknown model parameters were the modal frequencies, modal dampings, and the eigenvector elements at the accelerometer locations. The goal was to find the transfer function whose frequency response approximates the measured response such that the l_2 norm of the error is minimized. Frequency weighting was also introduced in the formulation in order to shape the error in the frequency band of interest. A SIMO frequency domain curve fit algorithm was used to obtain the optimal parameters and, subsequently, the finite element model was tuned to match the input-output model. An important factor in the success of the SIMO curve fitter was the excellent initial estimates of the modal parameters obtained from an initialization procedure involving optimal SISO transfer function curve fits [5]. Details of the technical approach and the results obtained will be reported in the full version of the paper.

References

[1] Spanos, J. T., and O'Neal, M. C., "Nanometer Level Optical Control on the JPL Phase B Testbed," ADPA/AIAA/ASME/SPIE Conference on Active Materials and Adaptive Structures, Alexandria, VA, Nov. 5-7, 1991.

[2] Dailey, R. L., and Lukich, M. S., "Recent Results in Identification and Control of a Flexible Truss Structure," Proceedings of the American Control Conference, Atlanta, GA, June 15-17, 1988.

[3] Denman, et. al., "Identification of Large Space Structures on Orbit," AFRPL TR-86-054-054, Airforce Rocket Propulsion Laboratory, Sep., 1986.

[4] O'Neal, M. C., and Eldred, D., "The JPL Phase B Testbed Facility," ADPA/AIAA/ASME/SPIE Conference on Active Materials and Adaptive Structures, Alexandria, VA, Nov. 5-7, 1991.

[5] Spanos, J. T., "Algorithms for l_2 and l_∞ Transfer Function Curve Fitting," Guidance, Navigation, and Control Conference, New Orleans, LA., Aug. 1991.

Computational Methods for Identification and Control in Smart Structures

H.T. Banks
Department of Mathematics
University of Southern California
Los Angeles, CA 90089

D.J. Inman
Department of Mechanical & Aerospace Engineering
State University of New York at Buffalo
Buffalo, NY 14260

Y. Wang
Department of Mathematics
University of Southern California
Los Angeles, CA 90089

We consider models for structures with distributed actuators/sensors typical of those arising in smart material structures. These models involve unbounded input/output operators of a special class. A rigorous computational methodology for both the identification and feedback control problems will be presented. We illustrate these ideas using experimental data from a structure with piezoceramic actuators/sensors.

Placement of a Limited Number of Sensors for Modal Identification of a Space Station Photovoltaic Array

Daniel C. Kammer[1]
Assistant Professor
Department of Engineering Mechanics
(608) 262-5724

and

Leehter Yao
Graduate Student
Department of Electrical and Computer Engineering

University of Wisconsin
Madison, WI 53706

Abstract

An iterative method, called Effective Independence, is used to place a small number of sensors on a representative Space Station photovoltaic array for identification of a set of dynamically important mode shapes. The work is important in the area of on-orbit modal identification where a very limited number of sensors will be available. The method ranks sensor locations based upon their contribution to the linear independence of the target modal partitions. Linear independence must be maintained to perform test-analysis correlation to determine the accuracy of the analytical model with respect to the test data. The derived sensor configuration tends to maintain the spatial independence of the target modes and the determinant of the Fisher Information matrix. Three other sensor configurations based upon kinetic energy and engineering judgement are also considered. Out of all the sensor sets considered, the Effective Independence configuration produced superior values for the Fisher Information matrix determinant and condition number, and the target mode observability. The Eigensystem Realization Algorithm was used to extract mode shapes, frequencies, and damping from simulated response data for each of the sensor configurations. Seven cases were studied possessing varying levels of noise and damping. A test-analysis cross-orthogonality computation was used to determine the accuracy of the extracted test mode shapes. In all cases considered, the Effective Independence sensor configuration identified more target modes than the other three sensor sets.

[1] Senior Member AIAA

Submitted to the AIAA *Journal of Guidance, Control, and Dynamics*

ABSTRACT

Aircraft Structural Integrity and "Smart" Structural Health Monitoring

M.F. Nahan, B. Westerman
Boeing Defense & Space Group
Military Airplanes Division
Seattle, Washington 98124

A "Smart" Structural Health Monitoring System (SHMS) could short-circuit current aircraft maintenance and yield performance and costs benefits. However system integration issues need to be addressed. This paper discusses the benefits and issues of SHMS integration with the current system approach to ensuring aircraft structural integrity.

SHMS would consist of a network of advanced photonic sensors embedded into an airframe. An illustration of the SHMS concept is shown in figure 1. SHMS would monitor fatigue cracks, corrosion, impact events, disbond and temperature in order to ensure flight safety at a minimum cost to aircraft availability and maintenance. Complete sensor coverage would be afforded to specific critical parts. These parts would be identified in the aircraft damage tolerance and durability analysis or during vehicle service in response to incured damage. SHMS would allow for a two-fold expansion of sensor coverage over the life of the aircraft to enable service critical part coverage. Lighter coverage would be applied to structure that is simply difficult and costly to access and inspect. Sparse sensor coverage would be afforded to monitoring aircraft usage and flight load spectra. Sensors would be conformal, unobtrusive and easy to install in the field. Sensors would be locally networked into distributed processors for data assessment, capture and interrogation of sensor and data integrity. Pertinent data would be passed to the aircraft central processor for further assessment and action management. Non-critical data would be discarded. Ground support would interface with the central processor for a data dump and to receive damage verification orders.

SHMS offers significant benefits in aircraft autonomy, availability, and costs. Improved aircraft availability would result from the elimination of todays' precautionary manual and tear-down

inspections. Improved aircraft autonomy would result from real time structural monitoring, enabling immediate condition awareness for inflight assessment of structural integrity. Improved costs would result from applying sensor networks to monitor structure that is highly integrated with other systems. Cost and performance would also improve from greater repair flexibility. Structural damage, developed in service, could be instrumented and efficiently incorporated by SHMS. This flexibility could enable continued flight in spite of potential flaw growth. Damage repair could similarly be instrumented and incorporated by SHMS for the continuous assessment of the repair performance.

Today Aircraft structural integrity is achieved through a disciplined durability and damage tolerance design and test validation procedure. It is maintained by an on-going structural maintenance plan. The Air Force Aircraft Structural Integrity Program (ASIP) is an umbrella effort that promotes these activities. The aircraft structural design provides the required strength, stiffness and safety at the least cost over the economic life of the aircraft. Minimum flaw size detection, flaw growth rates, load path plurality, component accessibility, and replacement costs are all considered in selecting a safe and least cost design approach. Based on the expected operational usage and accepted flaw growth rates, a maintenance plan is implemented that continuously checks for minimum-sized flaws during scheduled inspections. Individual aircraft usage and flight load spectra is monitored using a few strain gages and accelerometers. This data is used to modify the inspection schedule based on revised flaw growth rate estimates.

The future adoption of SHMS by ASIP would have an impact on both the structural design and the approach to maintenance. Damage tolerance design would address a new set of initial flaw size assumptions based on the minimum detectable size using SHMS sensors. It would also become the responsibility of the designer to position SHMS sensors and to define guidelines for data interpretation. The required duration of damage tolerance could be reduced because damage would be immediately detected. Currently structure must wait for a periodic manual inspection that may not always be afforded. Thus, less design conservatism could be employed if a sufficient understanding of flaw growth is available. One negative issue, component economic life design (replacement costs) would be affected by the added costs of sensor installation and network integration. The elimination of precautionary inspections, from todays maintenance planning, would reduce the burden on base and depot resources. When damage is identified by SHMS, smaller scale equipment at the base level could suffice in performing damage verification. The aircraft base would have SHMS and verification data with which to judge repair potential. This knowledge would enable increased use of base level repair. The accumulation of structural health data could become a problem that requires too much maintenance to be considered practical. To avoid this problem, data correctness must be easily verified, damage verification and repair records must be cross referenced to SHMS data and insignificant data must be discarded.

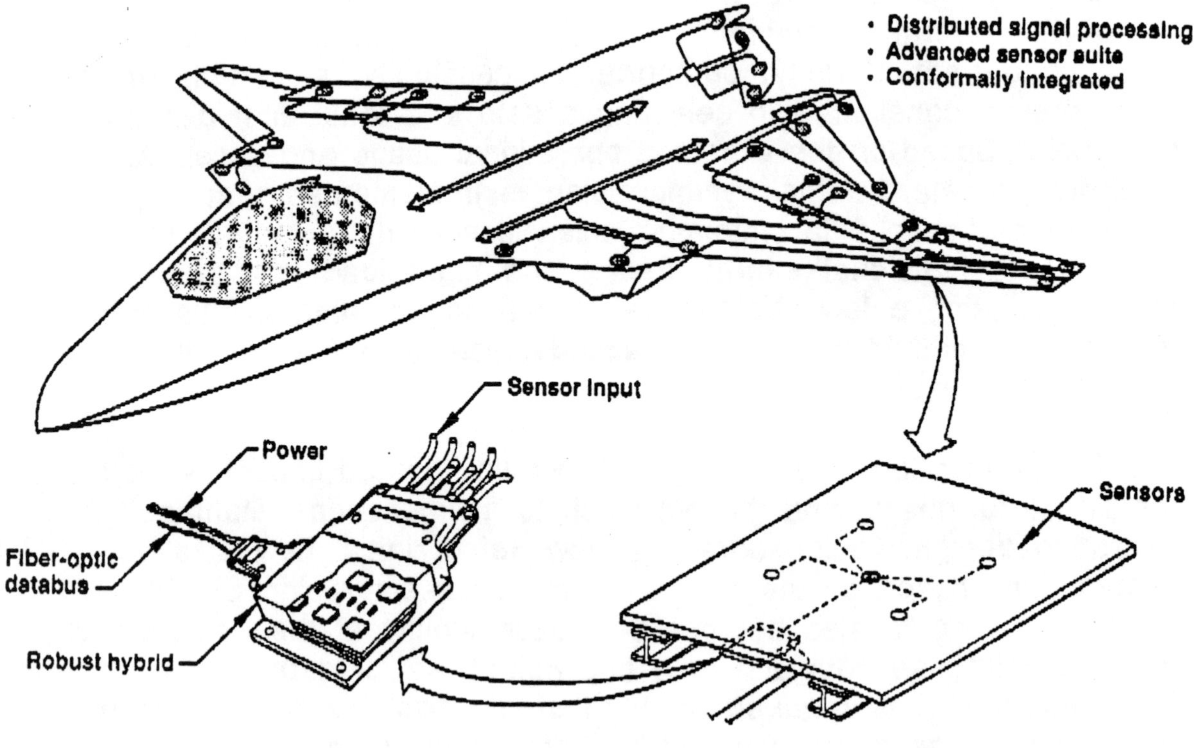

Figure 1: "Smart" Structural Health Monitoring System Concept

Development of an Intelligent Rotor

Inderjit Chopra
Center for Rotorcraft Education & Research
University of Maryland, College Park, MD 20742

(Abstract for: *Conference on Active Materials and Adaptive Structres*)

The principles of smart structures technology are applied to the rotary-wing field in order to build an intelligent rotor with reduced vibration. A higher harmonic control (HHC) system based on piezoelectric technology is being developed to suppress vibration. This system will use piezoelectric crystals for both sensing and actuation and will require the development of an active feedback control system. An analytical model of the piezoelectric HHC system is being formulated and incorporated into a comprehensive rotorcraft code (UMARC). This model will then be validated against numerous experimental data. To then prove the feasibility of this concept, a six foot diameter Froude-scaled bearingless intelligent rotor model is being built in house using modern composite technology. The model rotor will be systematically tested at the University of Maryland in a vibration laboratory, a 10 foot diameter vacuum chamber, a hover stand, and finally in the Glenn L. Martin Wind Tunnel. The fabrication, testing, and analysis procedures proposed for this research effort are outlined below. This work will be an important step towards demonstrating the feasibility of the intelligent rotor concept, a concept which may revolutionize future rotor designs.

Recently, there has been an increased emphasis on research activities in the area of smart structures. Much of this work has focused on the application of piezoelectric technology to space related activities, such as the control of large space structures (Refs. 1-2). Some of the initial work in applying this emerging technology to the rotor has been pioneered at Maryland by Barrett (Ref. 3). Using a simple feedback system, it was demonstrated that the forced flap vibration of a rotating blade in a vacuum could be significantly reduced. The objective of the proposed research will be to expand on this initial work by developing a higher harmonic control system for vibration reduction based on piezoelectric technology and incorporating an active feedback control system. The use of HHC systems to suppress helicopter vibration is not new and in fact has been widely investigated (Refs. 4-5). The most common system involves blade pitch control through excitation of the swash plate with servo-actuators. This action results in the generation of new unsteady aerodynamic forces which in turn suppress vibration. Although this method has been proven to be quite effective, a number of significant drawbacks do exist. The power requirements needed to drive the servo-actuators can be substantial, particularly at extreme flight conditions where vibration becomes most pronounced (Refs. 6-7). In addition, current systems are limited to blade root control for both the multi-blade and individual blade control systems. However, with the application of piezoelectric technology, these seemingly inherent limitations may be overcome.

For the fabrication of the model blades, rigid foam is cut to the desired airfoil shape, joined with a glass-epoxy spar, and covered with a fiber-glass skin. The complete blade assembly is then installed in a numerically machined aluminum mold having a NACA 0012 airfoil section and cured in an autoclave. Tantalum masses are placed in the foam at various locations in order to achieve the desired cg locations. Specially cut piezoelectric crystals are embedded in certain directions along the blade spar. These crystals are used as both sensors and actuators. To achieve appreciable signals or actuation forces, a large number of crystal banks are embedded in the blade at several locations. Also, the structural couplings associated with various composite layups, such as bending-twisting and extension-twisting, will be exploited in order to magnify the actuation strains. Such couplings are achievable with symmetric and antisymmetric ply layups, respectively (Ref. 8).

After the fabrication is completed, static and dynamic tests will be performed on the individual flexbeams, torque tubes, and blades in order to determine their stiffness and inertial properties. Considerable effort will be expended to ensure that the properties of each blade assembly are as nearly identical as possible. Where ever possible, comparison with known properties will be made in order to assess the accuracy of the experimental procedures. These experimentally determined properties will also be used as input data for theoretical analyses.

A 10 foot diameter vacuum chamber facility, located at the University of Maryland, will be used to determine the dynamic characteristics of the model blades in a rotating environment. Signals acquired in the rotating frame will be transmitted to the fixed frame via a 100 channel slip ring. This type of facility is useful since aerodynamic forces are not present. Blade natural modes will be identified by exciting piezoelectric crystals with a function-generator and measuring the response with piezoelectric sensors. A FFT analyzer and various other instrumentation will be used for data acquisition and analysis (Ref. 9). In addition to obtaining vibration characteristics, the vacuum chamber will be used to aid in developing an efficient piezoelectric HHC system which is capable of suppressing vibration. Initially, a very simple feedback system, which uses piezoelectrics for both excitation and actuation, will be employed to suppress flap vibration. Then, a more advanced distributed control system will be developed which will use crystals, placed at predetermined locations along the blade, to identify and suppress vibration. The system proposed for this research will use a selected number of crystals to excite the blades at higher harmonics of the rotational speed. Additional crystals will then be used as sensors in order to measure the level of vibration present at various blade stations. Finally, the remaining crystals will be used as actuators to absorb the vibration. Thus, the vibration will be suppressed at its source through the use of an active feedback control system. A hover stand, which is located immediately adjacent to the vacuum chamber, will be used to evaluate the feedback system in an aerodynamic environment and may be used to investigate the stability of the rotor in hover.

The vacuum chamber/hover stand can be used for preliminary design of the piezoelectric system and to verify its operation. However, the final design or fine tuning of the HHC system must be performed in a wind tunnel under forward flight conditions. It is under such conditions that vibratory excitation of the blades occurs due to the unsteady aerodynamic forces associated with forward flight. The rotor model will be tested in the Glenn L. Martin Wind Tunnel using our bearingless rotor rig (Ref. 10). This tunnel is capable of a maximum speed of 230 mph in the 8 by 11 foot test section. The goal of this stage of the proposed research will be to examine the functioning of a higher harmonic control system based on piezoelectric technology in a realistic aerodynamic environment. Some crystals will be used to sense vibration while others will be used to excite the blade at various stations. This will result in a time varying radial distribution of blade twist or effective camber thereby creating new unsteady aerodynamic forces which may suppress vibration. Since excitation is not limited to root control, this may yield significant improvements in performance and efficiency over existing HHC systems.

Along with the experimental work outlined above, a parallel and complimentary analytical analysis is undertaken. A comprehensive rotorcraft code (UMARC) has recently been developed at the University of Maryland (Ref. 11). Within UMARC, rotor blades are represented as elastic beams undergoing flap and lag bending, elastic twist and axial extension. The resulting equations of motion are discretized by using a finite element method in space and time. A number of blade aerodynamic options are available including simple quasisteady aerodynamics, unsteady attached flow, and unsteady separated flow. Wake analyses range from simple inflow models all the way to a complex free wake analysis. The code has the capability to perform a dynamic analysis of bearingless rotors in both hover and forward flight. NASA Ames is currently upgrading the code in order to simulate higher harmonic control of helicopter vibration. The powerful capabilities of this comprehensive code will be used throughout the course of the proposed research. In addition, an analytical model of the piezoelectric system will be incorporated into UMARC and validated against the experimental results.

Finally, a study is being made to examine the feasibility of implementing the intelligent rotor concept on a full scale rotor. This is being performed in collaboration with Ames Research Center scientists, to investigated the implementation of a piezoelectric HHC system on a full scale rotor for testing in the Ames 40 by 80 foot tunnel. This project may be incorporated into the active rotor control program currently existing at NASA Ames which involves HHC/IBC testing of several different rotor systems. It is hoped that the research effort proposed here will provide an important step towards determining whether such systems are practical and indeed possible for routine use in the future.

References

1. Crawley, E.F., and de Luis, J., "Use of Piezoelectric Actuators as Elements of Intelligent Structures," AIAA Journal, Vol. 25, (10), October 1987.

2. Anderson, E. H., and Crawley, E. F., "Detailed Models of Piezoceramic Actuation of Beams," 30th AIAA/ASME/ASCE/AHS/ASC Structures, Structural Dynamics and Material Conference, Mobile, Alabama, April, 1989.

3. Barrett, R., *Intelligent Rotor Blade and Structures Development Using Directionaly Attached Piezoelectric Crystals*. MS. Thesis, University of Maryland.

4. Shaw, J., *Higher Harmonic Blade Pitch Control: A System for Helicopter Vibration Reduction*. Ph.D. Thesis, Massachusetts Institute of Technology, May 1980.

5. Ham, N. D., " A Simple System for Helicopter Individual-Blade-Control Using Modal Decomposition," Vertica, Vol. 4, (1), 1980.

6. Nguyen, K., *Higher Harmonic Control Analysis for Vibration Reduction of Helicopter Rotor Systems*. Ph.D. Thesis, University of Maryland.

7. Miao, W., Kottapali, S. B. R. and Frye, H. M., "Flight Demonstration of Higher Harmonic Control (HHC) on S-76," 42nd Annual Forum of the American Helicopter Society, Washington, D.C., June 1986.

8. Smith, E. C. and Chopra, I., "Formulation and Evaluation of an Analytical Model for Composite Box-Beams," Proceedings of the 31st AIAA SDM Conference, Long Beach, California, April, 1990.

9. Chandra, R. and Chopra, I., "Influence of Elastic Couplings on Vibration Characteristics of Thin-Walled Box Beams under Rotation," Accepted for Publication in the AIAA Journal of Aircraft.

10. Wang, J. M., et. al., "Theoretical and Experimental Investigation of the Aeroelastic Stability of an Advanced Bearingless Rotor in Hover and Forward Flight," AHS National Specialists Meeting on Rotorcraft Dynamics, Arlington, Texas, Nov. 1989.

11. Bir, G. S., Chopra, I. and Nguyen, K., "Development of University of Maryland Advanced Rotorcraft Code (UMARC)," Proceedings of the 46th Annual Forum of AHS, Washington, D.C., May 1990.

<u>References</u>

1. Crawley, E.F., and de Luis, J., "Use of Piezoelectric Actuators as Elements of Intelligent Structures," AIAA Journal, Vol. 25, (10), October 1987.

2. Anderson, E. H., and Crawley, E. F., "Detailed Models of Piezoceramic Actuation of Beams," 30th AIAA/ASME/ASCE/AHS/ASC Structures, Structural Dynamics and Material Conference, Mobile, Alabama, April, 1989.

3. Barrett, R., *Intelligent Rotor Blade and Structures Development Using Directionaly Attached Piezoelectric Crystals.* MS. Thesis, University of Maryland.

4. Shaw, J., *Higher Harmonic Blade Pitch Control: A System for Helicopter Vibration Reduction.* Ph.D. Thesis, Massachusetts Institute of Technology, May 1980.

5. Ham, N. D., " A Simple System for Helicopter Individual-Blade-Control Using Modal Decomposition," Vertica, Vol. 4, (1), 1980.

6. Nguyen, K., *Higher Harmonic Control Analysis for Vibration Reduction of Helicopter Rotor Systems.* Ph.D. Thesis, University of Maryland.

7. Miao, W., Kottapali, S. B. R. and Frye, H. M., "Flight Demonstration of Higher Harmonic Control (HHC) on S-76," 42nd Annual Forum of the American Helicopter Society, Washington, D.C., June 1986.

8. Smith, E. C. and Chopra, I., "Formulation and Evaluation of an Analytical Model for Composite Box-Beams," Proceedings of the 31st AIAA SDM Conference, Long Beach, California, April, 1990.

9. Chandra, R. and Chopra, I., "Influence of Elastic Couplings on Vibration Characteristics of Thin-Walled Box Beams under Rotation," Accepted for Publication in the AIAA Journal of Aircraft.

10. Wang, J. M., et. al., "Theoretical and Experimental Investigation of the Aeroelastic Stability of an Advanced Bearingless Rotor in Hover and Forward Flight," AHS National Specialists Meeting on Rotorcraft Dynamics, Arlington, Texas, Nov. 1989.

11. Bir, G. S., Chopra, I. and Nguyen, K., "Development of University of Maryland Advanced Rotorcraft Code (UMARC)," Proceedings of the 46th Annual Forum of AHS, Washington, D.C., May 1990.

DESIGN, MODELING, ANALYSIS, AND TESTS OF SENSORS AND ACTUATORS UTILIZED IN A MISSION ADAPTIVE WING

Charlie D. Turner
Nichols Research Corporation
4040 So. Memorial Parkway
Huntsville, Alabama 35802

Phone : (205) 883 - 1140
Telefax : (205) 880 - 0367

Abstract

With the development of the Space Shuttle in the seventies plans were formulated to deploy, erect, or fabricate on-orbit large space systems. Studies during this time frame conducted by NASA, DOD, and Industry indicate that in order to meet future users needs, large antennas and platforms would be required in either low earth orbit or in geo-synchronous orbit. Specific applications were identified in studies which examine future civilian and military needs. An example is the need for continuous monitoring/mapping of rainfall and soil moisture, which will require microwave antennas 40 meters or larger in order to obtain the 10 km resolution needed to measure the smaller storms. About half of earth's rain each year falls from short duration storms about 10 km across. Such applications call for large/flexible structures that not only have to be stabilized, but also required precise control of both geometry and pointing.

The technical requirements involved in control of large space structures are many times more complex than have been experienced in the development of smaller rigid spacecraft since large space structures inherently possess low structure rigidity, high modal density, and low damping. Because of these characteristics environmental effects or interaction of structural components with the spacecraft control system can occur which reduces performance or restricts operations. These effects also occurs in smaller components if precision pointing and/or surface shapes/orientations at near-optical wavelengths are critical performance factors. These instruments require extremely accurate surfaces to enable their operation at near-optical wavelengths indicated by an achievable root-mean-square (RMS) surface accuracy. Achievement of the necessary RMS surface distortion of 0.0001 inch will require active surface control. It was estimated that up to five thousand control points would be required on the larger space structures, therefore requiring a distributed control system/neural network.

Much of the growing technology base is due to the requirements that have been established by the need for precision control of large space structures, but the highly distributed sensor/actuator/control concepts have also found applications within the surface ship, submersible, building, optics, aircraft, and helicopter communities. In order for the next generation of advanced structures to meet more demanding performance requirements, advances in critical technology areas are needed. These areas include dynamics and control of structures; system identification and health monitoring; computation control hardware and software; and adaptive structures; design, modeling, and analysis of advanced structures.

This paper presents the experimental test results and supporting analytical analysis of a combined set of sensors, actuators, and data processing/control system that provides for health monitoring and static/dynamic control for a mission adaptive wing. The mission adaptive wing employes both conventional aerodynamic control surfaces and internal actuators to provide for static and dynamic control/load distribution. A hybrid neural network concept provides for an interface between the conventional aerodynamic control system and the internal sensor/actuator system. The primary focus of the paper is the dual distributed sensors that are employed for both health monitoring and static shape control along with the segmented distributed actuators that are used for both dynamic and static control. A structural beam with segmented actuators and distributed sensors forms the basic internal control unit. The mission adaptive wing aeroelastic wind tunnel model which is assembled from a set of these sensors/actuators/structural beams is presented in detail. This wind tunnel model provides the capabilities for studying structural, sensor, actuator, and processing/control failures during tests.

A Compliant Wing Section for Adaptive Control Surfaces

B. J. Maclean
B. F. Carpenter
J. L. Draper
M. S. Misra

Materials and Structures / Research & Technology
Martin Marietta Defense and Space Communications
Denver, Colorado 80201

ABSTRACT

Martin Marietta is utilizing shape memory alloy (SMA) wires as embedded actuator elements in compliant wing sections to develop adaptive control surfaces for aircraft applications. Shape memory alloys utilize a reversible crystalline phase transformation to recover their original heat-treated shape when heated above a critical transformation temperature range (recoverable strain can be as high as 8%) or, if constrained during heating, to generate high recovery stresses (in excess of 100 ksi in some alloys). Utilizing a combination of these two effects, SMA wire "tendons" can be used as embedded actuator elements to control the level of facesheet strain in adaptive structural components which utilize sandwich panel construction. As facesheet strain is varied, the degree of curvature and magnitude of tip deflection of a panel section can be controlled. For example, a wing section with chord-to-thickness ratio of 12:1 can demonstrate tip deflection on the order of 50% chord with less than 4% facesheet strain. Electrical resistance heating of the SMA wires is used to control facesheet contraction (and, therefore, the amount of panel deformation) and a closed-loop strain/displacement sensor feedback loop has been successfully demonstrated to provide that control. Significant possible applications include mission adaptive aircraft wings for extended range and expanded flight envelope, and compliant control fins for submarines and torpedoes to reduce noise, turbulence, and system weight. This paper reports the design, fabrication, and testing of two compliant wing test sections. A "flex-biased" wing section consisted of a 1" wide, 12" long honeycomb winglet wedge section, 1" thick at the root. This particular design utilizes a conventional glass/epoxy laminate on one side and an elastomeric thermoplastic facesheet (elastic modulus <30 ksi) with embedded SMA actuator elements on the opposite side of a common aluminum honeycomb core. The SMA actuator elements utilized in this winglet were Cu doped Ni-Ti 0.020" diameter wires. The wires were "conditioned" to exhibit a "two-way" memory effect by isothermally straining 12" wire bundles at 70°C over a range of 4% stain at 1 Hz for a total of 1000 cycles. This process produced a spontaneous elongation of 2.1% upon cooling down form the cycling temperature, with zero applied stress.. An "antagonistic" second wing section had the same dimensions as the first but with two opposing active facesheets (with embedded SMA actuator elements) bonded on either side of a sandwich panel with a composite laminate located in the middle. Another important point to be made regarding the design of the winglet is its ability to balance externally applied loads. With 12 each 0.020" diameter wires contracting with a potential of 65 ksi operational recovery stress (= 245 lbs total contraction force potential), against a facesheet stiffness of 150 lbs/in, 4% wire recovery corresponds to only 72 lbs of wire tension. This leaves a wire operating margin of 173 lbs. That means that with a lever arm of 12:1, the 1.0" wide winglet is capable of lifting almost 15 lbs at its tip. However, to generate a "balanced" or symmetric force response for externally applied loads, i.e. establish the capability to balance loads equally whether applied from above or below, the "antagonistic" control approach can be used. The control of such winglets can be accomplished using a closed-loop, displacement sensor feedback approach. A proportional/integral/differential (PID) law provides the ability to supply electrical power to the SMA wires based on a sensed displacement error. As error is reduced, the integral term takes over and accounts for ambient heat losses while the differential term controls the rate of approach to the command set point. Performance test results show the relationship between power requirements and bandwidth response, as well as the ability to maintain a commanded winglet "shape" despite a variety of adverse static and ambient loading conditions.

THE JPL PHASE B TESTBED FACILITY

by
Michael O'Neal and Daniel Eldred

To develop enabling technologies for future large space and lunar missions, the Jet Propulsion Laboratory (JPL) is participating in NASA's Control-Structure Interaction (CSI) program to explore, validate and demonstrate emerging control technologies and design methodologies. Towards this end JPL has developed the Phase B Testbed as part of an evolutionary chain of ground-based testbeds. The goal of the program is to reject disturbances in flexible structures and maintain ultra precise alignments with extremely small jitter with several orders of magnitude better performance than conventional technologies. This paper describes the Testbed structure and optical components, the sensors and actuators, the real-time control computer environment, and the classes of experiments which the Testbed supports.

The configuration resembles part of a stellar interferometer and consists of two 4 foot long horizontal box truss structures cantilevered from the top of a ten foot tall vertical box truss column, which is rigidly attached to the ground. In the current configuration, modal surveys show that there are 7 modes under 25 Hz and 20 under 100 Hz, with between 0.1% and 0.8% critical damping. The structure is designed to be easily reconfigured, and accordingly each structural element can be removed and replaced by either an active strut or a passive damper without disturbing the rest of the structure. Proof mass actuators and other disturbance sources can be located at any node of the structure. This, coupled with the modular nature of the design, ensures expandability and versatility; for example, the system modes can be tuned or modified by adding mass at different locations. Sensors, including accelerometers, eddy current devices and strain gauges, can be located anywhere on the structure. In addition, the active struts incorporate displacement and load sensors.

An optical motion compensation system rests on one of the two horizontal arms. This system, which is used to actively compensate for changes in optical pathlength, incorporates a cat's-eye retroreflector on a moveable assembly and is identical to one being used in an existing ground-based interferometer. The optical path can be changed easily to achieve a desired level of coupling with structural dynamics. Control of the optical pathlength is effected by a combination of a voice-coil actuator which reacts between the fixed and the moving assemblies, and a small mirror which is positioned via a reactionless two-stage piezoelectric actuator. The pathlength itself is determined using a laser interferometer with a resolution of better than a nanometer. Control of the pathlength is achieved using a combination of 68000-based single board computers and array processors running VX-Works real-time operating system, and are integrated into a network-based Unix development environment.

The Phase B Testbed is designed to support experiments employing a variety of control approaches including pathlength compensation, active structure control and disturbance isolation. Ultimately, these approaches will be combined in a "layered" control strategy in which all approaches are applied simultaneously and synergistically. At the same time, issues such as optimal active and passive strut location and reconciliation of experimental and analytical models can be investigated. The Phase B Testbed is currently fully functional, and experiments using it are proceeding as planned.

Extended Summary of Paper for ADPA/AIAA/ASME/SPIE Conf. on Active Materials and Adaptive Structures, Nov 5-7, 1991

Adaptive Structures Technology Effort at the Phillips Laboratory

Alok Das
Structures & Controls Division, Directorate of Space & Missiles Tech.
Phillips Laboratory (AFSC), Edwards AFB, California

Introduction

Future Department of Defense (DoD) and NASA space systems, such as, Space Based Interceptors, Space Surveillance Systems, Precision Segmented Reflector, and Optical Interferometers, will require structures with very precise pointing and vibration control requirements. Some of these systems, like the Optical Interferometers, are tens of meters in size and require the relative position of specific nodes of the support structure be maintained within a few microns or less during the observation period[1]. On the other hand, although the space defense and surveillance systems are much smaller in physical size, their mission imposes very stringent pointing and vibration suppression requirements on selected subsystems, often in the presence of severe on-board or environmental / threat induced disturbances. In addition, most of these space systems will be required to operate successfully over a 10-20 year life with minimal ground support. This will require the development of autonomous on-board guidance, control and health monitoring system capable of continuous monitoring of the spacecraft health, the threat environment and its impact on the satellite, detect significant changes in key performance parameters, and autonomously reconfigure itself to compensate for these changes. Degradation in system performance could be small or large due either to gradual aging or catastrophic failure of structural members and control components such as sensors/actuators. In either case, the control system must sense the changes and autonomously compensate to recover the required system performance. Finally, the weight penalty imposed by the health monitoring and control system is of major concern for most space systems. This is particularly true for the near term SDI systems, where the system weight plays a crucial role in determining the feasibility of the concept.

The performance requirements of these precision space systems have motivated a new approach for their structural and control system design. This approach, referred to as "adaptive structures", utlizes recent breakthroughs in a number of fields to provide an integrated structure / controller capable of sensing its environment and dynamic characteristics and autonomously adapting to meet the mission requirements.

A number of recently developed technologies contribute to enable the adaptive structures approach. These include advances in high modulus composite materials, resulting in lightweight, yet stiff, thermally stable structures; sensors and actuators, including distributed fiber optic sensors, piezoceramic sensors/actuators embedded inside the layers of the composite structure, and shape memory alloys; and intelligent control concepts, including highly distributed control utilizing integral sensors and actuators, and use of neural networks for autonomous system identification, rapid failure detection, and control system reconfiguration (Figure 1).

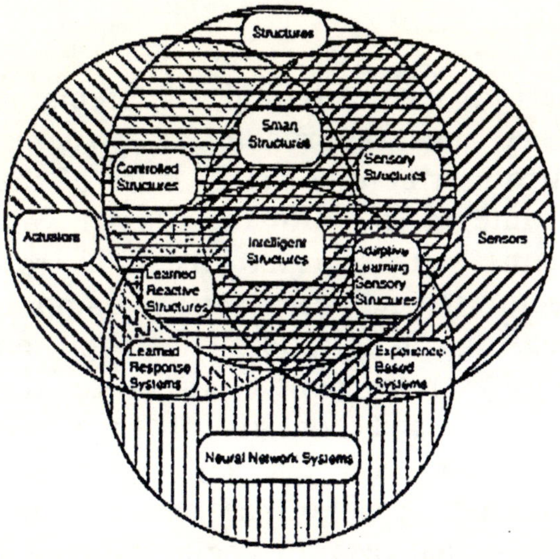

Figure 1. Components of adaptive structures

This paper provides an overview of adaptive structures technology efforts on-going at the Phillips Laboratory.

Efficient Feedback and Active Member Location Using Discrete Control/Structure Design Optimization Techniques

by

Roy Ikegami
David G. Wilson
K. Scott Hunziker

Boeing Defense and Space Group
Structures Technology
P.O. Box 3999
Seattle, WA 98124-2499
Mail Stop 82-97

ABSTRACT

The future performance required of precision space structures has spawned a new approach to structural design where feedback control principles and state-of-the-art sensors and actuators are integrated into the design of highly advanced structural systems. This paper discusses research in the area of smart structures used for active damping of large space structures, focusing on efficient feedback and active member location using discrete control/structure design optimization techniques. The testbed consists of an eighteen bay prototype erectable space truss with ten active elements that are placed in various locations. Each of the active elements are essentially smart components that use piezoelectric actuators embedded in an advanced composite to provide axial control authority. The axial piezoelectric strut design consists of a hollow square tube cross-section with eight hard piezoelectric ceramics embedded along the longitudinal direction, for each face of the square tube. by applying a voltage across the piezoelectric ceramics a mechanical strain is realized. For an electrical field of 640 V/mm, 400 microinch of stroke was achieved. In the constrained configuration this resulted in approximately 15 lbs of actuation force. A piezoelectric sensor which eliminates cross-axis sensitivity is used to provide uniaxial strain measurements. Due to the piezoelectrics low mass, small size, fast response and their capability of measuring strain rate directly, these sensors can be used as effective point sensors in most vibration control algorithms. Traditionally, strain rate has been achieved by differentiating the time domain signal output of a position sensor such as a strain gage. The differentiation is a noisy process, causing very low accuracy and additional low pass filtering which degrades closed loop performance. Therefore, a sensor such as the piezoelectric sensor, that directly measures strain rate, is highly desirable.

To achieve effective distributed control a multi-input multi-output linear control system is proposed to control a finite number of modes simultaneously which does not induce spillover into the uncontrolled modes. The goal in controlling the elastic modes is to increase damping and/or stiffness of the structure so as to achieve the desired time behavior. The behavior depends on the number and location of sensors and actuators. To determine physically realizable optimization criterion the weighting of the state and control terms used in standard Riccati design are replaced with criterion used for actuator/sensor positioning and feedback determination. The criterion chosen is the dissipation energy of the system. As the dissipation energy depends on the initial conditions of the flexible structure, realistic conditions have to be selected. Several conditions involving impulsive point loads, generated from a single axis shaker at a specific location, are investigated. The disturbances are transformed to corresponding initial conditions of the state vector.

The control/structure optimization method used, designs a discrete time linear control system for a given configuration of active struts. By alternating the location of the active members new linear controllers are designed. A technique called simulated annealing is used to determine the optimal or near-optimal discrete locations of the 10 active members and feedback gains through the maximization of the dissipation energy, which is extracted by action of the feedback system. Several of the best designs will be implemented on the actual eighteen bay truss and the experimental data will be correlated with the analytical predictions.

Abstract

for the

ADPA/AIAA/ASME/SPIE Conference

on

Active Materials and Adaptive Structures

C-SIDE: Control-Structure Interaction Demonstration Experiment

James B. Mohl *   Hugh W. Davis †

1 July 1991

Objectives

Ball Electro-Optics/ Cryogenics Division (BECD) is currently fabricating the Control-Structure Interaction Demonstration Experiment (C-SIDE) to illustrate capabilities in actively controlled structures. The C-SIDE is part of an internal project to develop common technology needed for detecting, measuring, evaluating and counteracting the deformations of large flexible space structures. This activity includes the system level study of:

- structural analysis and modelling,
- materials and fabrication techniques,
- sensing devices,
- signal processing,
- system identification,
- actuating devices,
- control system methodologies,
- computational technology, and
- verification testing methods.

One objective of the project is to demonstrate a method of solution to a broad class of structural control problems while utilizing currently available hardware. The C-SIDE is a testbed that incorporates figure control for a flexible primary structure, a passive secondary reaction structure, remote and unobtrusive position sensing, non-contacting linear

*Principal Analyst, Control and Dynamics, Ball Space Systems Division

†Principal Systems Engineer, Pointing and Tracking Control Products, Ball Electro-Optics/ Cryogenics Division

force actuators, and a single digital control processor. The control algorithm is based on a reduced-order model (ROM) of the system dynamics with a residual mode filter (RMF) to compensate for any interaction with unmodelled dynamics. The various components of the testbed can be individually modified or upgraded when follow-on research opportunities are pursued.

A second objective is to foster cooperative research between BECD and the University of Colorado - Boulder, Aerospace Engineering Sciences Department. A consulting arrangement with Professor Mark Balas of the University's Center for Space Structures and Control is already in place.

Executive Overview

In the initial demonstration, we are assembling an antenna-like structure which is cantilevered from a massive central body. The concept is illustrated in Figure 1. The "antenna" is a thin, flexible facesheet one meter in height, three meters in length and 1.6 millimeters thick. It exhibits modes of vibration starting below 0.3 Hz. A truss made of graphite/PEEK thermoplastic composite material is placed behind the facesheet to act as a reaction structure; its first cantilever vibration mode is below 2 Hz. A figure control system is imposed to maintain physical flatness of the facesheet to within plus or minus one millimeter. The control system is composed of a Remote Attitude Measurement Sensor (RAMS), up to ten linear force actuators and a single digital processor.

The RAMS sensor provides an unobtrusive position measurement and is ideally suited to monitor structural dynamic behavior. Two single axis sensor heads are positioned to view most of the facesheet backside and a portion of the reaction structure. The single-axis RAMS is sufficient for this case since only translations normal to the facesheet are of interest. A minimum of ten position readings are available at an update rate in excess of 50 Hz. The resolution of the readings vary with distance along the facesheet. The worst condition is at the free end where motions of 0.015 mm are resolvable. Targets on the facesheet are set at the actuator locations. Targets on the reaction structure are also provided for system identification.

A non-contacting actuator is used to avoid creating bending moments on the thin facesheet. The selected device is shown in Figure 2. The lightweight winding is attached to the facesheet with a yoke. The iron armature is attached to the reaction structure. The force capability and travel are oversized for better demonstration capability. Smaller, lighter devices would be used in actual space applications. Configurations without the captured winding are possible using the same force generation principle.

The control algorithm uses a simple, reduced-order model (ROM) of the flexible system dynamics with provision for a residual mode filter (RMF) to eliminate potential destabilizing structural interactions. For the purposes of demonstration, the control actions will be applied so that disruptive interaction with unmodelled flexible dynamics does occur. The interaction is subsequently suppressed by the RMF to restore desirable performance.

Presentation Agenda

The C-SIDE is currently being integrated and operations are expected to commence this fall. A more comprehensive system description, hardware photographs, laboratory results and initial lessons learned will comprise the presentation at the conference.

Authors

Presenter:

James B. Mohl
Mail Station CO-10B
Ball Space Systems Division
P.O.Box 1062
Boulder, CO 80306
(303) 939-5064
FAX: (303) 939-5914

Co-Author:

Hugh W. Davis
Mail Station FM-2
Ball Electro-Optics/ Cryogenics Division
P.O.Box 1062
Boulder, CO 80306
(303) 939-4022

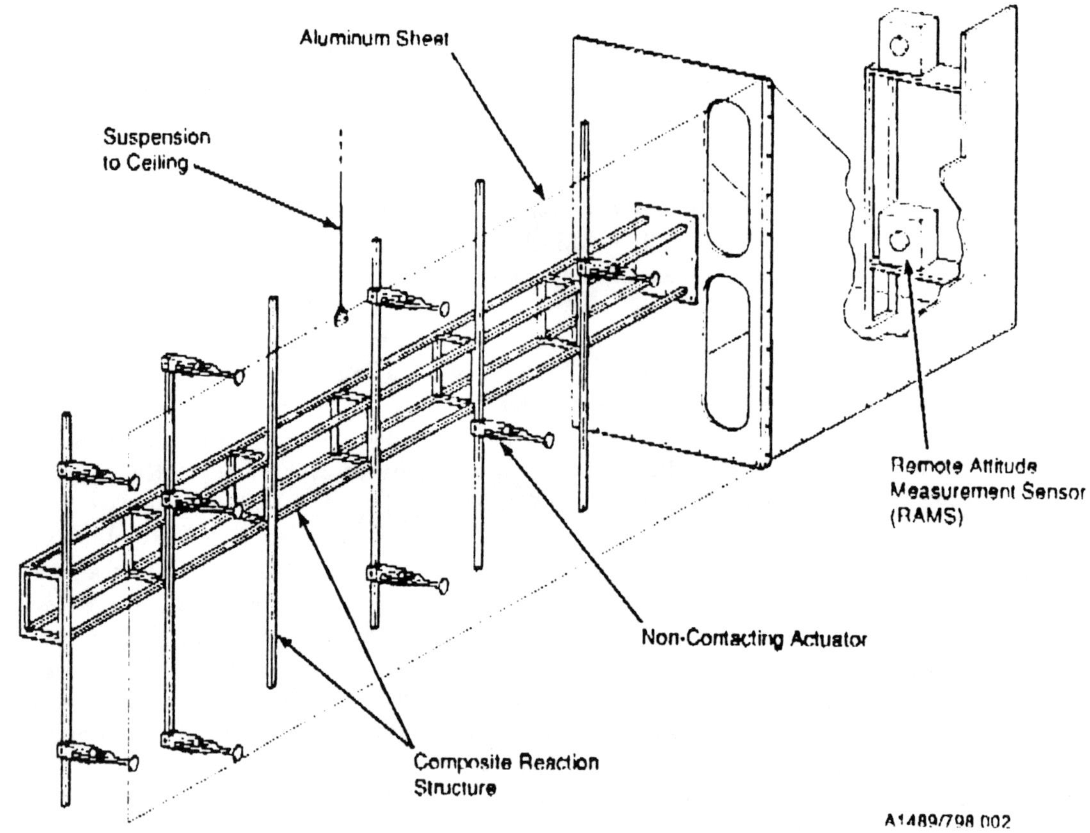

Figure 1: The C-SIDE, patterned after a space radar system, provides flatness control for a thin, flexible plate.

Figure 2: The actuator is a standard BEI design; modifications are made in-house to increase lateral clearances.

ACTIVE MATERIALS AND ADAPTIVE STRUCTURES
ADPA, AIAA, ASME, SPIE CONFERENCE
Alexandria, Virginia, Nov 5-7, 1991

Analysis of Multiple Frequency Interference in Photorefractive Media

David E. Cox and Sharon S. Welch
NASA Langley Research Center
Hampton, Virginia, 23665-5225

This paper describes the use of a simulation to predict the dynamic behavior of a photorefractive crystal exposed to interfering light waves at multiple frequencies. Properties of the crystal, such as response time and diffraction efficiency, are determined for various operating conditions. These properties are evaluated in terms of their effect on the sampling rate and linearity of a dynamic sensor which uses the crystal as a holographic recording medium. The simulation is based on the response of a photorefractive material to intensity patterns and the diffraction of waves due to volume refractive index gratings. Experiments are being conducted to validate the simulation and to determine specific characteristics of the photorefractive crystals. A description of these experiments and in progress results will also be presented.

A noncontact distributed sensor based on photorefractive crystals has been proposed for measuring the surface distortions of large reflectors.[1] The sensor produces an image of the reflector surface where the intensity of each point in the image is related to the displacement of a corresponding point on the surface. The use of photorefractive crystals allows these images to be generated dynamically. Photorefractive crystals are nonlinear optical materials in which the refractive index of the crystal changes in response to light. Obtaining distributed position information from these crystals relies on holographic reconstruction from interferograms recorded at multiple frequencies. Multiple frequency holography or contour holography is an established method of determining distributed position information, and photorefractive crystals have been used to record real-time phase holograms. Combining these technologies to create a dynamic sensor requires accurate models of holographic recording and reconstruction based on the physics of photorefractive materials.

The formation of gratings in a photorefractive crystal occurs due to a nonlinear interaction between light and the medium. The presence of an optical power distribution in the crystal causes the diffusion of charge carriers away from areas of intense light. These charge carriers become trapped by acceptor sites in the crystal lattice, creating a charge distribution. The nonequilibrium charge distribution generates an electric field in the medium. This electric field acts as a feedback mechanism which limits further development of the charge distribution. The electric field also causes refractive index variations to occur through the electro-optic effect. The refractive index variations are phase gratings which will diffract incident light. Charge transport equations[2] can be used to describe this process and to predict the transient response as well as saturation characteristics of the resulting index modulation.

Gratings in photorefractive crystals typically occur over a large volume, therefore, coupled wave equations[3] must be used to predict diffraction effects. Coupled wave equations describe diffraction as a continuous coupling of power between a signal beam and reference beam due to sinusoidal variations in the medium. For multifrequency operation this description must be extended to include nonsinusoidal gratings and high order modes. When light waves of multiple frequencies interfere, a stationary sinusoidal intensity pattern forms for each frequency present. The resulting index modulation is the incoherent superposition of these sinusoidal intensity patterns. Since the gratings are closely spaced, the object beam from one grating can serve as a reference beam for another grating. Because of this coupling between gratings the diffraction from multiple sinusoidal gratings cannot be independently superimposed.

The diffraction of light from the gratings is present during both recording and readout. A full dynamic model of the medium must combine the dynamics of grating formation with the variation in the intensity distribution due to diffraction. The simulation provides such a model by combining space-charge dynamics with coupled wave analysis.

REFERENCES

1. S. S. Welch and D. E. Cox, "Characteristics of a Dynamic Holographic Sensor for Shape Control of a Large Reflector," Proceedings of the

SPIE Symposium on Optical Engineering and Aerospace Sensing, Paper No. 1480-01, Orlando Florida, April 1991.

2. N. V. Kukhtarev, V. B. Markov, S. G. Odulov, M. S. Soskin, V. L. Vinetskii, "Holographic storage in electro-optic crystals," Ferroelectrics, Vol. 22, pp. 949-960, 1979.

3. H. Kogelnik, "Coupled Wave Theory for Thick Hologram Gratings," Bell System Technical Journal, Vol. 18, pp. 2909-2947, 1969.

High Temperature and Ultrasonic Wave Optical Fiber Sensor Instrumentation for Aerospace Applications

K. A. Murphy, A. M. Vengsarkar, R. G. May and R. O. Claus
Fiber & Electro-Optics Research Center
Bradley Department of Electrical Engineering
Virginia Tech
Blacksburg, VA 24061-0111

This paper reviews major smart structure research areas addressed by the Fiber & Electro-Optics Research Center in cooperation with the NASA-Langley Research Center from 1979 through 1991.

Two technical areas are described in particular. The first is the development and application of sapphire optical fibers and fiber sensor systems which utilize those fibers to allow environmental measurements at high temperatures. Specific developments have been methods for cladding and coating sapphire rod waveguides, techniques to allow the splicing of sapphire and silica fibers, and ways to implement interferometric fiber sensors using the fabricated sapphire fiber elements.

The second major area is the quantitative detection of ultrasonic waves using fiber sensors. Our group published some of the first work in this area more than a decade ago, but recent developments indicate the ability to measure quantitative acoustic field components at frequencies up to several megahertz with minimum detectable surface displacement amplitudes of less than a nanometer. Such waves have been detected on and in materials over extended temperature ranges and in simulated harsh environments.

[This work has been supported by a continuing research grant program from the NASA-Langley Research Center. The authors acknowledge many discussions with R. Rogowski, J. Heyman, E. Madaras and J. Cantrell.]

Paper Abstract for Conference on Active Materials and Adaptive Structures

Fiber-Optic Interferometric Sensors for Ultrasonic NDE of composite materials

Kexing Liu and Raymond M. Measures

University of Toronto Institute for Aerospace Studies (UTIAS), 4925 Dufferin Street, Downsview, Ontario M3H 5T6, CANADA

Composite materials are finding a widespread use in the aerospace and other industries. Because of their special properties, it is desirable to establish new non-destructive evaluation (NDE) techniques for in-service (in-situ) monitoring of composite structures. These advanced NDE techniques should be capable of sensing damage arising within a structure in real-time. The use of a dielectric sensor is preferable for embedding in composite materials. The fact that optical fibers are dielectric material, free from electromagnetic interference, small in size and light in weight and can survive high temperature and high pressure has led to the investigation of embedding optical fibers into solid materials, particularly the advanced composite materials for nondestructive testing and structure integrity monitoring[1].

Optical fibers as intrinsic sensing elements have been studied for the detection of ultrasonic strain waves within composite materials since more than a decade ago [2]. More recently, an ultrasonic detection system based on fiber Michelson interferometry for composite damage monitoring was reported [3]. The system employed an active homodyne demodulation technique to maintain linearity and maximum sensitivity. The fibers were embedded in both Graphite/epoxy and Kevlar/epoxy composite specimens. Acoustic emission signals were detected for both tensile loading and out of plane loading. The system provided single-ended sensing with real-time monitoring capabilities, and had a broadband response of 100kHz to 2MHz.

The sensors were embedded and surface adhered to the composite materials. The result for the detection of acoustic emission (AE) and laser generated ultrasound in a laboratory environment will be presented. The potential application of such sensors for the monitoring of structure integrity and composite cure will be discussed. Other interferometric type fiber sensors studied for ultrasonic detection will also be reported.

The use of interferometers based on ordinary single-mode fibers for the detection of low frequency strain [4] and ultrasonic waves [3] within composite materials has the obvious advantages of simpler configuration, higher sensitivity and lower cost compared with those based on polarization preserving fibers. However, the problems associated with the embedded ultrasonic strain wave sensors based on ordinary single-mode fibers such as localization and sensitivity variation due to the state of polarization (SOP) have drawn little attention in the smart structures sensing community [5][6]. In this paper, the results of a sensitivity analysis will be given for the detection of elastic strain waves with embedded ordinary single-mode fibers. Problems of localization and visibility variation due to SOP and birefringence effects will be discussed.

REFERENCES

1. Measures, R.M., "Smart Structures with Nerves of Glass", *Prog. Aerospace Sci.* Vol.26, pp.289-351, (1989).

2. Claus, R.O., and Cantrell Jr., J.H., "Detection of Ultrasonic Waves in Solids by an Optical Fiber Interferometer", *Proc. IEEE Ultrasonic Symposium*, pp.719-721, (1980).

3. Liu, K., Ferguson, S.M., and Measures, R.M., "Fiber-Optic Interferometric Sensor for the Detection of Acoustic Emission within Composite Materials", Optics Letters, Vol.15, No.22,pp.1255-1257, (1990).

4. Valis, T., Tapanes, E., Liu, K., and Measures, R.M., "Passive Quadrature Demodulated Localized Michelson Fiber Optic Strain Sensor Embedded in Composite Materials," *J. of Lightwave Technology*, Vol.9, pp.535-544, (1991).

5. Kersey, A.D., Marrone, M.J., Dandridge, A., and Tveten, A.B., "Optimization and Stabilization of Visibility in Interferometric Fiber-Optic Sensors Using Input-Polarization Control", *J. of Lightwave Technology*, Vol.6, pp.1599-1609, (1988).

6. Liu, K., and Measures, R.M., "Detection of High Frequency Elastic Strain Waves with Ordinary Single-Mode Optical Fibers", to be published, (1991).

FIBER OPTIC SENSORS AND ARCHETECTURES FOR LARGE STRUCTURES

Eric Udd
McDonnell Douglas Electronic Systems Company
1801 E. St. Andrew Place
Santa Ana, California 92705
714 566-5687
FAX: 714 566-4100

ABSTRACT

This paper provides an overview and design considerations for fiber optic sensors and overall system archetectures that are particularly well suited for application to large structures. Issues that will be addressed include embedding fiber optic sensors into large composite structures, performance criteria and reliability and maintainability.

ABSTRACT

DISTRIBUTED FIBER OPTIC SENSORS

J. P. Kurmer and A. A. Boiarski
Battelle Memorial Institute

A Summary of Battelle's efforts in the area of distributed fiber optic sensor technology will be presented. Battelle has been actively involved in fiber optics for over ten years with emphasis in theoretical analysis, materials development, and system level hardware development for specific applications. In addition, Battelle has helped develop several discrete fiber optic sensors used by the medical industry to monitor, in situ, the pH, CO_2, and O_2 levels in various liquids. These point type sensors will also be briefly discussed.

For specific applications or special environments, Fiber optic sensors have several advantages over their electrical counterparts. A particular environment of interest is the electrical power industry where conductor temperature monitoring would provide information for safe reliable operation of a power plant. For example, measurement of stator or rotor winding temperature could determine winding integrity. Techniques presently exist to measure average temperature of a stator coil or local temperature at specific points; however, no method exists for determining peak conditions by measuring temperature along the winding length. For the past several years Battelle has been developing a distributed fiber optic temperature sensor suitable for such applications.

To provide continuous high-spatial resolution monitoring of temperature at many locations within a power system, a high resolution, distributed fiber optic temperature sensor (DFOTS) was developed. The key advantage to the DFOTS approach is the fact that a single fiber can be used to obtain measurements at many points along its length. Further, optical fiber sensing does not perturb the electrical environment because of the nonmetallic nature of this sensing technology.

Optical time domain reflectometry was used as the sensing technique to measure the backscattered light pulse intensity in the fiber produced by temperature variations along the fiber optic sensor. Scattering in optical fibers is caused principally by the Rayleigh effect which results from inhomogeneities in the core glass that are formed in the production process. This type of scattering is at the same wavelength as the incident light, and core scattering is largely independent of temperature changes. There is, however, a small contribution to the scattered power from the Raman effect which originates from molecular and crystalline effects within the core glass. A key attribute of the Raman effect is that it results in temperature-dependent scattering at a different wavelength than the incident light. Raman scattering has received the most attention as a method of providing a DFOTS system and several companies produce commercial devices.

Battelle, however, has developed an approach which relies on the Rayleigh effect in the fiber optic cladding material. The backscattered light is caused by scattering in the clad material and is coupled back into the core and contributes to the overall backscattered light intensity. The present DFOTS approach uses changes in this backscattered light. A proprietary, ultraviolet light curable material was developed which produces the change in the local backscattered light intensity with a change in local fiber temperature. The amount of change is proportional to the temperature value resulting in a distributed temperature sensor. A key advantage of this approach is that Rayleigh scattering is used which is 100 to 1000 times more intense than the Raman scattering intensity from a typical fiber. This increased intensity provides the ability for improved spatial resolution in the backscattered signal measurements.

The system so designed has a temperature accuracy of ±5 C with a spatial resolution of 10 cm over a 40 m length. A second generation system is currently being developed which will extend the active sensing length to over 200 meters. The results of these investigations will be reported.

Battelle is also currently developing a distributed fiber optic acoustic sensor which will enable detection of liquid or gas leaks in underground piping systems over several hundred meter distances. The system is built around a Sagnac interferometer. Light from a laser diode or high-radiance LED is split by a distributed directional coupler into oppositely directed paths through a single-mode optical fiber. The fiber used is ordinary single-mode fiber with a protective jacket of low acoustic loss plastic material. The fiber passes through the pipe twice, having been folded back on itself. In one direction the fiber is covered with an additional jacket consisting of a thick coating of acoustically-lossy material to absorb acoustic energy. In this way, at any point along the fiber, only the uncoated fiber is sensitive to the acoustic field. A non-reciprocal phase modulator is placed next to the distributed coupler, the purpose of which is to induce a large non-reciprocal phase dither for signal processing purposes, since for optimum sensitivity one would normally have to introduce a quadrature optical bias. The output of the photodetector will contain major harmonics at even multiples of the phase modulator frequency. About or between these harmonics will be the "noise" envelop caused by the acoustic (leak) signal. Spectral analysis of these "noise" sidebands will indicate the position of the acoustic source (leak). Calculations have indicated that the acoustic sensitivity of the distributed fiber optic sensor should be comparable to the commercially available B&K 8103 hydrophone. Experimentation is currently underway to verify device performance the results of which will be reported at the conference.

Optical Fiber Sensor-Based Smart Materials and Structures

R. O. Claus, K. A. Murphy, A. M. Vengsarkar and R. G. May
Fiber & Electro-Optics Research Center
Bradley Department of Electrical Engineering
Virginia Tech
Blacksburg, VA 24061-0111

.s paper provides an overview of the application of optical fiber sensors to smart materials and smart structures during the past ten years. It specifically describes the uses of fiber sensors for the evaluation of materials 1) during composite cure monitoring and as part of structural fabrication evaluation, 2) during the normal use lifetime of structures, and 3) during the degradation phase of structural materials. Emphasis is placed on the attachment of fibers to materials, the embedding of fibers within advanced composites, and the implementation of sensors for mechanical measurements including strain, temperature, vibration and acoustic wave propagation. Optical fiber sensor methods are described in general, advantages indicated and examples of demonstrated systems presented.

It is anticipated that this would be one of several overview papers in a special session devoted to optical fiber sensor applications. Part of the purpose of the papers in this session would be to describe the general background of this part of the technical area of smart materials and structures, thus providing a foundation for more specific related technical discussions which would follow as part of other sessions.

Development of a Fiber Fabry-Perot (FFP) Strain Gauge with High Reflectivity Mirrors

Dayle Hogg*†, Doug Janzen*, Gary Zuliani*, Tomas Valis†, and Raymond M. Measures*†

The emerging field of Smart Structures will require a new generation of strain gauges suitable for mounting to or embedding in composite material structures. These sensors are going to be expected to perform without degradation over the serviceable lifetime of the structure in a harsh (high temperature, strain, EMR) environment. Fiber optic based sensors have been recognised as one possible replacement for traditional resistive foil strain gauges. This paper will review the choice of the Fiber Fabry-Perot (FFP) as the preferred fiber optic strain gauge and describe an optimal fabrication method, demodulation technique, and apparent strain considerations.

The FFP strain gauge was selected over other fiber optic sensors (modalmetric, polarimetric, Bragg) for its high sensitivity. Strain measurements are often required with 3mm spatial resolution and a 1με strain resolution, such as a rosette to determine the 2D strain tensor at a point. An intrinsic, back-reflected FFP strain gauge has been fabricated using two, high-reflectivity metallic mirrors to define the sensing region (i.e. gauge length). The front (i.e. semi-silvered) mirror was accomplished by depositing a metallic mirror over the core of a single mode fiber and fusion splicing it with a second similar optical fiber. A second endface mirror was formed by coating the entire endface of the sensing region fiber with a metal film. This approach yielded a strong sensor with a cross-sectional area defined by the optical fiber, typically 80μm to 125μm in diameter. No reinforcing members or tubes were required, thus facilitating its use as an embedded sensor.

High reflectivity metallic mirrors were chosen over low reflectivity dielectric or air gap approaches in order to maximise back-reflected energy and simplify construction. By maximising the back-reflected light, the power budget constraints on the source and detector pair can be minimised, and the signal to noise ratio improved. A high reflectivity front mirror with an associated high loss allows the use of a fully reflective second mirror, while still maintaining the optical signal balance. Additional lengths of fiber with index matching fluid to couple out excessive transmitted light are no longer required.

The performance of any sensing system can be measured by the linearity of the sensor's output as a function of the measurand. A linear strain output for the FFP strain gauge has been achieved by wavelength modulation of the source, and subsequent digital phase tracking. A description of the system will be given, and results shown.

In addition to a linear output, the absolute strain on a sensor is also required. Most interferometric demodulation techniques are incremental by nature and must be arbitrarily reset to zero whenever a power failure or temporary disconnect of the sensor from its demodulation optics/electronics occurs. Interruptions are expected to occur over the lifetime of a structure and a means of recovering the signal is required. In order to obtain absolute strain, a short coherence interferometric sensing system is currently being developed and initial results will be reported.

Apparent strain is defined as a change in sensor output resulting from a change in temperature of the host structure, but misinterpreted as a change in strain. While the problem of thermally induced apparent strain has been extensively dealt with by manufactures of resistive foil strain gauges, it has largely been ignored by designers of fiber optic strain gauges. A brief review of apparent strain in resistive foil gauges will be given with the corresponding FFP strain gauge analogy. A comparison of apparent strain with commercially available optical fibers bonded to various materials will be given. Problems with temperature compensation in embedded sensors will also be introduced.

* - University of Toronto Institute for Aerospace Studies, 4925 Dufferin St., Downsview, Ontario, CANADA, M3H 5T6.

† - FiberMetrics Corporation, 4925 Dufferin St., Downsview, Ontario, CANADA, M3H 5T6.

The Use of Adaptive Structures in Reducing Drag of Underwater Vehicles

K.J. Moore, M. Noori, J. Wilson and J.V. Dugan, Jr.

Cortana Corporation, 520 North Washington Street, Suite 200, Falls Church, VA 22046

The possibility of using vortical flow to reduce drag has been known for a number of years. Suitably positioned transverse slots can result in entrained vortices and a significant reduction in end drag over the closure of a bluff body. In addition, experiments at NASA Langley have shown that ordered periodic vorticity injected into a turbulent boundary layer near a wall results in about 25% drag reduction over nominal flat plate values.

A related concept in fluid flow dynamics has been investigated with a view to improving the hydrodynamic performance of underwater vehicles. A traveling wave with specified and controllable phase velocity, amplitude, wavelength and shape is used to generate and trap vortices in the troughs of a flexible wall. The resulting secondary flow condition cannot be classified as conventional laminar, transitional or turbulent flow, but rather constitutes a new regime described as "controlled vortical flow". Analytical studies have shown that an appropriate tailoring of traveling wave to flow parameters results in ordered vortical flow and an associated reduction in drag to a level substantially below flat plate values. Further computational verification of the traveling wave concept will likely be costly, and may not engender the same level of confidence as an experimental demonstration. Thus the need for a practical investigation has been identified.

Experimental demonstration of the traveling wave concept is a multidisciplinary problem involving aspects of hydrodynamics, materials, and control systems technologies. In order to demonstrate controlled vortical flow experimentally, it will be necessary to design and construct an active wall device to generate a traveling wave. A number of design and materials aspects of active wall structures are discussed in the present paper.

An active wall is defined as one which uses energy from an external source for deformation, in contrast to a passive wall which uses energy from the flow to deform. Auto-oscillatory, or floating, walls are intermediate between active and passive systems. Oscillations are injected into the flexible wall in a continuous manner (active), but the wall is then allowed to interact with the flow (passive), and the observed oscillation results from both active and passive effects.

The principal objectives of the proposed experimental investigation of the traveling wave concept are as follows:

- To demonstrate vortex entrapment using flow visualization techniques;
- To determine the effect of controlled vortical flow on drag;
- To assess the trade-off between the energy required for wall activation and the energy gained through drag reduction;
- To examine the possibility of tailoring wave to flow parameters for maximum drag reduction.

One of the major design goals for an active wall test plate is that the wave amplitude, together with two of the three related parameters - wavelength, wave speed and frequency - should be independently controllable. The potential to investigate various waveforms, eg. sine wave, cat's eye, cycloid, would be an added advantage. Hence a fully-controlled, adaptive structure is deemed necessary for experimental investigation of the traveling wave. An operational device designed to function at fixed, predetermined parameter values might use a lower-cost auto-oscillatory system.

Computational fluid dynamics (CFD) studies of traveling wave behavior have shown that the amplitude-to-wavelength ratio, a/λ, and ratio of wave velocity to freestream velocity, c/U_∞, are critical parameters influencing vortex entrapment and establishment of a controlled vortical flow. Parameter ranges of interest are

$$0.10 \leq a/\lambda \leq 0.25 \qquad (1)$$

$$c/U_\infty \geq 0.5 \qquad (2)$$

Target ranges for active wall parameters have been defined on the basis of the above values, together with information on wavelength, freestream velocity, and Reynolds number (based on wavelength) for practical applications. The need to reduce complexity and minimize costs has also been taken into account.

Wavelengths should be in the range

$$\lambda = 5 - 20 \text{ cm} \qquad (3)$$

The active plate length L, should be in the range

$$L \sim 0.5 - 2 \text{ m} \qquad (4)$$

such that the active length incorporates at least 10 wavelengths, i.e. $L \geq 10\lambda$.

Amplitudes of oscillation should be in the range

$$a = 0.5 - 5 \text{ cm} \qquad (5)$$

such that (1) is satisfied.

Values of frequency, f, and wave speed, c, should be within the ranges specified below:

$$f = 10 - 100 \text{ Hz} \qquad (6)$$

$$c = 2 - 5 \text{ m.s}^{-1} \qquad (7)$$

A number of active wall devices have been reported in the technical literature. The majority of these are mechanically-driven wave machines utilizing a series of cams positioned on a common camshaft with a successive phase difference between adjacent cams. The cams are attached to a deformable rubber surface by a set of ribs. The systems described are generally fixed amplitude and wavelength devices, although frequency and wave speed can be varied by altering the speed of rotation of the camshaft. In general a/λ values are ≤ 0.05, but one article reports investigation of a/λ values in the range specified above (equation 1). Interestingly, vortex entrapment was demonstrated in this case, but no attempt was made to measure drag, and only limited parameter optimization was possible with the mechanically-driven active wall device.

An extremely sophisticated active wall system using electronically-driven piezoelectric ceramic activators has been built by Park and coworkers at Southwest Research Institute in San Antonio. Waveform, wave speed, frequency, and amplitude can all be selected independently. A lead zirconate titanate (PZT) ceramic was chosen as the actuator material on account of its high piezoelectric strain constant (274 pm/V). However, peak wave amplitudes are limited to 15 microns, i.e. three orders of magnitude less than present requirements.

The possibility of using an adaptive structure to generate a traveling wave has been investigated further by the present authors, in the light of recent developments in actuator materials technology. In order to achieve shape control, actuators which will induce forces/strains in response to external stimuli (electric, thermal, magnetic) must be incorporated in the flexible wall. The potential advantages and limitations of piezoelectric, electrostrictor, magnetostrictor, ferrofluid, and shape memory alloy actuators will be discussed.

Acknowledgment.

This project was funded by the Defense Advanced Research Projects Agency (DARPA) as part of Cortana's Advanced Submarine Technology and Integration (ASTI) Program under Contract No. MDA972-88-C-0064.

ELECTRICALLY CONTROLLED POLYMERIC MUSCLES AS ACTIVE MATERIALS USED IN ADAPTIVE STRUCTURES

D. Segalman, W. Witkowski and D. Adolf

Sandia National Laboratories

and

M. Shahinpoor

University of New Mexico

Abstract

Discussed are some applications of certain classes of ionizable polymeric solutions containing polyacids and polybases that are capable of undergoing substantial expansions and contractions if subjected to electrolytic ionization that changes the pH of the solutions. Further discussed are conceptual designs of a number of smart devices made with such polymeric muscles that can be electrically activated, thus, creating the possibility of direct and active computer control of their expansion and contraction capabilities. Finally presented are the experimental arrangement and results on the relationships between the voltage applied, its frequencies and the forces and deformations generated by such polymeric muscles.

Introduction

The possibility that certain co–polymers may be chemically contracted and expanded like a pseudo–muscle was first discussed by Kuhn and Katchalsky [1, 2].

As originally reported by Kuhn, Horgitay, Katchalsky and Eisenberg [3] a three dimensional network, consisting of polyacrylic acid, can be obtained by heating a foil of polyacrylic acid containing a polyvalent alcohol such as glycerol or polyvinyl alcohol. The resulting three–dimensional networks are insoluble in water but swell enormously in water on addition of alkali and contract enormously on addition of acids. Linear reversible dilations and contractions of the order of more than 400 per cent has been observed. Furthermore, as reported recently by Li and Tanaka [4], the structural deformation (swelling or collapsing) of these gels is homogeneous in the sense that, for example, for a long cylindrical gel, the relative changes of the length and the diameter are the same.

Polymethacrylic acid cross–linked by divinyl benzene copolymerized in methanol exhibit similar properties, as shown by Kuhn, Horgitay, Katchalsky and Eisenberg [3]. Chemically stimulated pseudo–muscular actuation has also been discussed recently by De Rossi, Chiarelli, Buzzigoli, Domenici and Lazzeri [5] and Caldwell and Taylor [6].

The possibility of using these polymeric gel muscles or actuators for mechanochemical engines and turbines was originally discussed by Steinberg, Oplatka, and Katchalsky [7], and Sussman and Katchalsky [8].

As originally mentioned by Hamlen, Kent and Shafer [9], the same effect can be obtained electrolytically, provided a conductor is included within the polymer. Now if a voltage is applied it causes the solution to become either acidic or alkaline depending on the direction of current or the sign of the voltage. For example, the polymeric fibers may be filled with platinum by alternatively treating them with solutions of platinic chloride and sodium borohydride. Depending on the sign of the voltage applied to the gel in a 1% sodium chloride solution, the solution can be made alkaline by evolution of hydrogen thus forcing the gel to expand. Otherwise, the solution becomes acidic and the gel contracts. Thus, a reversible expansion and contraction of the fibre is obtained with the application of an electric field. It is shown in the present paper that the expansion and contraction forces may depend linearly on the strength of the voltage applied based on a simple theory of motion of charged counterions in an electric field. It is further shown that direct motion control of these polymeric muscles with position and velocity feedback is feasible and in fact one may present a set of control algorithms for this purpose. Presented are also a number of applications of such electrically controlled polymeric muscles as active materials, sensors and actuators for use in adaptive structures. Such applications have been discussed by Eric Cross [10], Crawly, de Luis, Hagood and Anderson [11] Tzou and Tseng [12] in the context of piezoelectric sensors and actuators, Gandhi, Thompson, Choi and Shakir [13], Choi, Gandhi and Thompson [14] and Choi, Sprecher and Conrad [15] in the context of electro–rheological fluid sensors and actuators, Hanagud, Wan and Obal [16] and Tadjbakhsh and Su [17] in the context of optimal placement of generic sensors and actuators in adaptive and intelligent structures.

It is further shown in the present paper that polymeric muscles as sensors and actuators offer certain advantages over piezoelectric and electro–rheological fluid sensors and actuators in as much as the control voltage, the frequency characteristics, and active control of adaptive structures are concerned.

REFERENCES

1. W. Kuhn, "Reversible Dehung and Kontraction bei Anderung der Ionisation lines Netwerks Polyvalenter Fadenmolekulionen", Experentia, Vol,. 5, pp. 318–319, (1949).
2. A. Katchalsky, "Rapid Swelling and De Swelling of Reversibel Gels of Polymeric Acids By Ionization," Experientia, Vol. 5, pp.319–320, (1949).
3. W. Kuhn, B. Horgitay, A Katchalsky and H. Eisenberg, "Reversible Dilation and Contraction By Changing The State of Ionization of High–Polymer Acid Networks," Nature, Vol. 165, No. 4196, pp. 514–516, (1950).
4. Y. Li and t. Tanaka, "Kinetics of Swelling and Shrinking of Gels", J. Chem. Phys., vol. 92, no. 2, pp. 1365–1371, (1990).
5. D.E.De Rossi, P. Chiarelli, G. Buzzigoli, C. Domenici and L. Lazzeri, " Contractile Behavior of Electrically Activated Mechanochemical Polymer Actuators", Trans. Am. Soc. Artif. Intern. Organs, vol XXXII, pp. 157–162, (1986)

6. D. G. Caldwell and P. M. Taylor, "Chemically Stimulated Pseudo–Muscular Actuation," Int. J. Engng Sci., vol. 8, pp. 797–808, (1990).

7. I. Z. Steinberg, A. Oplatka and A. Katchalsky, "Mehcano Chemical Engines", Nature, vol. 210, no. 5036, pp. 568–571, (1966).

8. M. V. Sussman and A. Katchalsky, "Mechano Chemical Turbine: A New Power Cycle," Science, vol. 167, pp. 45–47, (1970).

9. R.P. Hamlen, C.E. Kent and S.N. Shafer "Electrolytically Activated Contratile Polymer," Nature, Vol. 206, No. 4989, pp. 1148–1149, (1965).

10. L. Eric Cross, "Piezoelectric and Electrostrictive Sensors and Acutators for Adaptive Structures" edited by B. K. Wado, California Institute of Technology, pp. 9–17, (1989).

11. E. F. Crawley, J. de Luis, N.W. Hagood and E. H. Anderson, "Development of Peizoelectric Technology For Applications In Control of Intelligent Structures, " Proc. 1988 Am. Cont. Conf., vol. 3, pp. 1890–1896, (1988).

12. H. S. Tzou and C. I. Tseng, "Electromechanical Dynamics of Peizoelectric/Elastic Structures Applied To Micro–Actuation and Distributed Structural Identification/Controls," ASME, Computers in Engineering, vol. 1, ASME, pp. 473–480, (1990).

13. M. V. Gandhi, B. S. Choi and S. Shakir, "Electro–Rheological–Fluid–Based Articulating Robotic Systems," Trans ASME, J. of Mechanisms, Transmissions, and Automation in Design," vol. 111, pp. 328–336, (1989).

14. S. B. Choi, M. V. Gandhi and B. S. Thompson, "An Active Vibration Tuning Methodology For Smart Flexible Structures Incorporating Electro–Rheological Fluids: A Proof–of–Concept Investigation," Proc. 1989 Am Cont. Conf., vol. 1, pp. 694–699, (1989).

15. Y. Choi, A. F. Sprecher and H. Conrad, "Vibration Characteristics of A Composite Beam Containing An Electro–Rheological Fluid," J. Intelligent Mater. Syst. and Struct., vol. 1, pp. 91–104, (1990).

16. S. Hanagud, C. C. Won and M. W. Obal,"Optimal Placement of Piezocramic Sensors and Actuators," proc. 1988 Am. Cont. Conf., vol. 3, pp. 1884–1889, (1988).

17. I. G. Tadjbakhsh and Y. Su, "Optimal, Coupled–Modal Control of Distributed Parameter Systems With Discrete Actuators," Trans. ASME, J. Appl. Mech., vol. 56, pp. 941–946, (1989).

Submitted to Conference on Active Materials and Adaptive Structures

ON-LINE ADAPTIVE STIFFNESS CHANGES TO TAILOR MODAL ENERGY CONTENT IN STRUCTURES

by
R.A. Osegueda and D.C. Nemir
Civil Engineering Department
University of Texas at El Paso
El Paso, TX 79968
(915) 747-5470

When a structure exhibits a high degree of flexibility, the control of motion within that structure becomes important. Structures can have a significant flexibility due to size (eg: a suspension bridge), geometry (eg: a skyscraper where the height to base area ratio is large), or material (eg: a space station where structural members are necessarily lightweight). Motion in such structures can arise due to periodic influences such as wind, or due to nonperiodic influences such as earthquakes (civil structures) or attitude adjustments (space structures). For lightly damped structures in particular, disturbances are dissipated slowly and even small amplitude disturbances can eventually lead to unstable large oscillations, disrupting performance and causing structural damage. As such, structural motion control has been and continues to be an active area of research.

This paper presents a case study on using semi-active control of system stiffness to steer the energy in a structure to a desired modal make-up. The approach falls within the domain of "smart structures" since the structure itself is automatically tailoring local system properties to achieve a specified global system objective. Such a technique can be used to limit structural motion in specified directions or to enhance passive energy removal within the structure itself. In this paper the theory is reviewed and feasibility is demonstrated on a simulation model.

The proposed modal energy transference approach for structural motion control can be explained by using the idealized lumped spring-mass-dashpot of Figure 1. This structure has baseline idealized springs, dashpots and lumped masses of magnitudes k_i, c_i and m_i. Additional springs and dashpots Δk_i, and Δc_i can be actively connected or disconnected in parallel with the baseline. For initial studies, the connection and release of these auxiliary springs is constrained to occur only when the strain energy in the spring is zero.

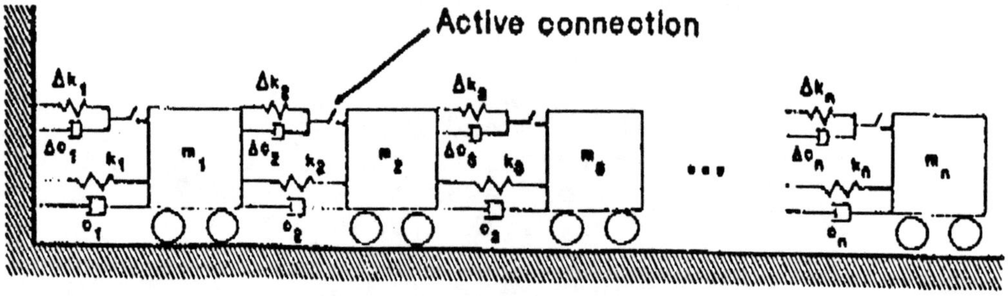

Figure 1 - A Spring-Mass-Damper System with Controllable Stiffness

The idea in our technique is to determine the modal energy makeup at all candidate control instances (when some springs are relaxed) and then determine the modal energy make-up that would accrue from the new basis systems obtained by every combination of clamped/released relaxed springs. The control action that gives a basis in which the modal energy makeup is steered closer to the desired make-up is then chosen. The technique is semi-active since no external out-of-phase excitation is required. Consequently, the technique will never result in modal spillover.

A simulation case study is given on a 20 degree of freedom planar truss. Studies are presented on accelerating the viscous damping inherent in the structure when the structure undergoes free and forced vibration. Studies are also made on the preferential transference of energy in transverse modes into axial modes. These studies demonstrate the feasibility of the approach. Practical implementation issues are discussed.

Frequency-Shaped Passive Damping using Resistively-Shunted Piezoceramics

George A. Lesieutre
Ass't Prof of Aerospace Engineering
The Pennsylvania State University

Christopher L. Davis
M.S. Student
The Pennsylvania State University

The use of piezoelectric materials in combination with resistive and resonant shunting circuits to achieve passive vibration energy dissipation and resonant response reduction has been suggested and demonstrated by several researchers. This paper extends earlier work by addressing the use of multiple resistively-shunted piezoceramic elements to achieve specified frequency-dependent damping. Different kinds of typical material damping behavior can be approximated, including and hysteretic and fractional derivative damping. However, a designer need not be constrained to duplicate such "classical" material damping behavior.

The task of developing a collection of resistively-shunted piezoelectric elements to achieve a desired variation of effective material damping with frequency is shown to be similar to the task faced by workers in internal friction in obtaining a relaxation spectrum from an experimentally-determined dynamic response function. However, depending on the specifics of the structure to be damped as well as the dimensions of the piezo-elements relative to the shortest wavelengths of interest, element placement has a significant impact on achievable structural modal damping (as it does for constrained-layer viscoelastic damping treatments).

The paper presents two different strategies for resistively-shunted piezoelectric damping design, one based on a modified modal strain energy method and the other, on an energy dissipation approach. Experimental results obtained using a cantilevered aluminum beam with surface-mounted ceramic patches are described. Implications for, and progress in, engineering structural composite materials with shunted piezoelectric constituents are also discussed.

Case Studies in Passive Piezoceramic and Viscoelastic Damping

A.H. von Flotow, N.W. Hagood
MIT
C. Johnson, D. Kienholz
CSA

Abstract

Recent research into active control of structural dynamics has refocused interest upon passive damping treatments. The desire to predict close-loop behaviour *a priori* has placed greater emphasis upon damping techniques which can be thoroughly characterized and analyzed. This is particularly important for the precision structures required in some space optical applications. The damping inherent in most structures is often highly nonlinear, detail-dependent and difficult to predict. Damping augmentation schemes can provide predictably high levels of damping.

A classic approach to structural damping augmentation involves the use of viscoelastic materials and geometries (constrained layers are typical), which ensure that the material is strongly strained during important deformations [Nashif]. This approach has found much application in plate-like structures, and has recently been extended to truss structures [Pacoss]. Recently, discrete viscous dampers have been developed with damping coefficients compatible with truss structure stiffness and eigenfrequencies [D-strut] and are being used in space and in the CSI research community. Even more recently [Forward, Hagood], passively-shunted piezoelectric materials have been shown to be useful for structural damping applications.

Table 1 offers a heuristic comparison of these three approaches.

This paper attempts an honest comparison of these three approaches to damping augmentation of structures in two case studies. The first case study involves a truss structure; the ASTREX structure, primarily a truss. Passive damping goals for this structure are intimately related to its active control and to its envisioned use; high speed stop-to-stop slew maneuvers. The paper reports on passive damping treatments designed for shortened transient ring-down, an for robust feedback control of structural dynamics.

The second case study is to focus upon a plate structure. Details have not yet been determined.

Table 1 offers a heuristic comparison of these three approaches.

Table 1

	Viscoelastic	Viscous (D-strut)	Passive Piezo
rengths	- experience base - low modulus	- experience base - high damping coeff. - temp. independant	- simultaneously can be actuated - high modulus - high damping coeff. - temp. independant
aknesses	- temp. sensitive - low modulus	- trusses only	- brittle - little design experience - high modulus

DESIGN AND DEVELOPMENT OF PASSIVE AND ACTIVE DAMPING CONCEPTS FOR ADAPTIVE SPACE STRUCTURES

D. L. Edberg
A. S. Bicos
McDonnell Douglas Space Systems Company
Huntington Beach, California 92647

ABSTRACT

Many types of spacecraft are expected to need shape control and structural vibration suppression systems to meet their performance requirements. Slewing and maneuvering, as well as normal spacecraft operating conditions and on-board systems can induce unacceptable structural vibrations requiring many minutes to naturally dissipate. Imaging systems will require precise spatial positioning of their components, and such motion can result in the need to introduce some means of vibration dissipation to the structure. This dissipation can take the form of some passive damping mechanism, shape controls, an active vibration control system, or a combination of passive and active systems. Because the operational lifetimes of these spacecraft will be long (~10's of years), changes in the systems will occur which necessitate that the vibration control subsystems adapt to these changes so that spacecraft performance is maintained throughout the lifetime.

This investigation is concerned with the development of various vibration suppression techniques with the goal of integrating some of these into structural struts in a generic truss structure. These techniques are intended to increase or enhance the damping available compared to that obtained from an unmodified member. Passive methods use viscoelastic dissipation or the piezoelectric electromechanical conversion to dissipate the electrical voltage. Active methods involve the use of feedback control systems to modulate the length of the actuator.

A piezoelectric material behaves as a capacitance in parallel with a voltage source. Voltage is generated by the piezoelectric effect, which transforms mechanical strain into an electric field. If the piezoelectric material is built into a structural member such as a strut, its inherent capacitance may be used to enhance the passive damping of a structure.

The terminals of the PZT may be externally shunted with resistor (R) or resistor/inductor (RL) networks. In conjunction with the capacitance C of the piezoelectric, this creates RC or RLC networks which may be "tuned" to frequencies near those of the vibrations, similar to a tuned mechanical damper. The shunt networks serve to dissipate the electrical voltage piezoelectrically generated by the structure's motion, which increases the dissipation an order of magnitude compared to an unshunted (open-circuit) piezoelectric.

One difficulty with the attachment of inductance is the large mass and size of the inductive members. Because the vibratory frequencies are often less than 100 Hz, inductances on the order of Henries are required. To fabricate such a large inductance, there is a need for a heavy iron core and low resistance thick copper windings, which in turn means significant non-structural mass. Our work has vastly reduced the necessary mass of the inductance by developing an "active inductor" which is an electronic circuit which simulates the behavior of an inductance with operational amplifiers and some passive components. Not only is the mass reduction achieved, but it is easy to change the electrical properties of the inductance by changing the value of a passive component. This is useful for exactly matching the shunt network to the frequency of vibration and can be retuned to adapt to changes in the structure during its lifetime.

We will also present the results of a design effort using a composite that includes an embedded Lead Zirconate Titanate (PZT) piezoelectric ceramic material inside a structural material, a viscoelastic material, and a PZT constraining layer. Viscoelastics may serve as passive energy dissipators, while piezoelectrics can act as actuators and sensors for both passive and active vibration control systems. Using the piezoelectric effect, the PZT layers act together to increase the shear strain in the VEM layer so as to increase the damping of the total laminate over that of a classical constrained-layer damping treatment. Analytical and numerical (NASTRAN) models have been developed for designing this damper into structural elements, such as truss struts.

The Acousto-Electromagnetic Waveguide Coupler

Tomas Valis, Andreas von Flotow, Nesbitt W. Hagood
Space Engineering Research Center
Massachusetts Institute of Technology
Cambridge, MA02139

May 14, 1991

Piezoelectric materials have been integrated into structural systems to provide either passive or active damping. In the case of shunted-piezoelectric damping, electrical/mechanical impedance concepts are used to optimize the coupling between mechanical and electrical parts of the network. Another area of technology where impedance-matched coupling plays an essential role is in the design of traveling electromagnetic-wave (microwave and optical) devices. A standard and well understood device for transferring power between electromagnetic waveguides is the so-called '2×2' coupler. The associated mathematics is often referred to as 'coupled-mode theory.' By considering a structural element as an acoustic waveguide (i.e., a beam) and coupling it to an electromagnetic waveguide (i.e., a L-C ladder), an analogous device may be envisioned. It is dubbed as an 'acousto-electromagnetic waveguide coupler.' It provides for the coherent and reciprocal conversion of guided acoustic and electromagnetic power. Such a coupler allows for the construction of hybrid structural-acoustic and guided-electromagnetic systems that take advantage of the unique capabilities of their respective components. Potential applications include impedance-matched structural joints, structural acoustic-wave suppression, and recirculating acousto-electromagnetic loops as would be required for efficient linear ultrasonic-motors.

This paper derives first-order coupled-mode theory for the case of longitudinal and flexural acoustic-waveguides coupled to a L-C ladder using finely segmented piezoelectrics. It is shown that in both cases, odd and even normal-modes arise, and that power beats back and forth between guides when the guides are phasespeed matched. The normalized coupling-length is shown to be proportional to the inverse of the effective electromechanical-coupling coefficient. The effective electromechanical-coupling coefficient is expressed in terms of the open- and closed-circuit stiffness ratios of a structure containing a piezoelectric layer laminated to a passive layer.

A proposed experimental geometry is considered, and the corresponding coupling length is derived. Sizing the various parameters indicates that the device is practical. Construction and testing of the coupler is expected to prove the concept and provide further insights regarding its performance, and applicability.

ABSTRACT

PERFORMANCE OF HIGH FORCE, HIGH STRAIN LINEAR ACTUATORS DRIVEN BY TERFENOL-D, A MAGNETOSTRICTIVE ALLOY WITH ADAPTIVE CHARACTERISTICS.

Recent advances in the production of "Giant" magnetostrictive rare earth iron alloys have opened new opportunities in the design of linear actuators. The nearly single crystal alloys of terbium (Tb, dysprosium (Dy) and iron (Fe) known as Terfenol-D offer the highest strain ($\Delta L/L > 1x10^{-3}$) and energy density ($1.4-2.5x10^{4} J/m^{3}$) of any commercial material. Terfenol-D piezomagnetic constants will be reviewed, with emphasis on characteristics such as strain, modulus, elastic compliance and magnetic permeability which change or adapt in response to external variables such as stress.

Terfenol-D has the formula $Tb_x Dy_{1-x} Fe_y$ with $y \leq 2$. Common stoichiometry for active antivibration is $Tb_{.3} Dy_{.7} Fe_{1.9-1.95}$. Stoichiometry can often be varied to optimize characteristics for a given application.

Performance data for prototype actuators will be presented along with a discussion of magnetostrictive actuator design elements and illustrations of how Terfenol-D characteristics affect actuator performance.

These small actuators (up to 12.5cm (5.0inches)) long and 3.8cm (1.5inch) diameter) can provide clamped forces over 1700 N (380 lbs.) strokes over 100μm (0.004inch) peak to peak, and a frequency range to almost 3kHz with wide bandwidth. This actuator performance fills the gap between low force, high stroke, low frequency solenoid type actuators and high force, low stroke, high frequency piezoelectric devices. The combination of stroke, force and frequency capabilities of these actuators make them ideal for many active sound and vibration control applications.

Terfenol-D actuators can be self sensing. Relationships have been developed for voltage and current as functions of force, frequency and mechanical stress. These relationships may be applicable to advanced control theories for Terfenol-D actuators.

Mel J. Goodfriend and Kevin M. Shoop
ETREMA Products, Inc.
Subsidiary of Edge Technologies, Inc.
Ames, IA 50010

Carl G. Miller
MES, Inc.
Severna Park, MD 21146

MAGNETOMECHANICAL TRANSDUCTION MATERIALS

A. E. CLARK
Naval Surface Warfare Center
Silver Spring, MD 20903-5000

Research on the fundamental magnetic and magnetoelastic properties of two classes of materials: (1) rare-earth iron alloys and (2) high permeability amorphous magnetic materials led to remarkable advances in both high-force, high-power output magnetomechanical devices, as well as ultrasensitive mechanical strain detectors. For many years, it was recognized that huge magnetostrictions were available in rare earth alloys, particularly the Tb-based Fe alloys. A proper balance of magnetic anisotropy and magnetostriction, plus a proper choice of crystal axes led to a material which can switch a large quantity of energy between the internal (magnetic) and external (mechanical) states with the application of a small triggering magnetic field. These rare-earth iron alloys are the most powerful active transducer materials. Power densities 2000 times those of conventional magnetostrictive materials and 10-20 times those of typical piezoceramics are available in the lanthanide compounds containing Tb and Sm. Within the last two years, new materials, based upon highly magnetostrictive $TbFe_2$ (Terfenol) have been formulated which reach high magnetostrictions at low applied magnetic fields. This is critical for many power transducers. These high force materials are particularly valuable for large energy transfer applications such as active structure stiffening, high-power low-frequency sonar, and active vibration control of heavy machinery. The second class of new magnetomechanical materials are the magnetostrictive amorphous iron alloys. Because of the lack of crystal structure, grain structure, and crystal anisotropy in amorphous materials, the magnetization process proceeds relatively smoothly via magnetization rotation. Here, Fe-based high magnetization amorphous magnetic ribbons were heat treated under magnetic fields transverse to the ribbon axis to yield extremely large strain dependent permeabilities and vanishingly small magnetic anisotropies. Magnetomechanical coupling factors were increased from 0.26 for the untreated ribbon to 0.95 for the heat treated ribbon. Almost perfect transduction has been achieved. The strain gage figure of merit in amorphous field-annealed materials is 5 orders of magnitude larger than that found in conventional strain gages.

This presentation will be divided into two sections: (I) High Power Lanthanide Transduction Materials and (II) Ultrasensitive Magnetomechanical Detector Materials. In the first part, we will present the most recent "jump" magnetostrictive materials, which make possible rapid strain changes with only a triggering magnetic field. In the second part we will describe the huge strain gage Figure of Merit and the immense dynamic range of the high coupling magnetoelastic materials.

Exploitation of Chaos In Amorphous Magnetoelastic Materials

W.L. Ditto and M.L. Spano
Naval Surface Warfare Center, Silver Spring, Maryland 20903-5000

Abstract

A new class of adaptive materials (transversely annealed amorphous magnetic $Fe_{81}B_{13.5}Si_{3.5}C_2$) has been developed that exhibit large changes of Young's modulus (factor of 10 or more) with small (~1 Oe) magnetic field changes. This effect is exploited to drive a vertically aligned ribbon of the material, clamped at the bottom and free at the end, by buckling the ribbon gravitationally through changes in Young's modulus. Broad band *chaotic* oscillations,are observed in the vibrating ribbon. A recently developed theory of *exploiting* chaos for active control is implemented in real time to control the chaotically oscillating ribbon. The method requires only small time dependent perturbations of a *single* system parameter (in our case a magnetic field) and no *a priori* knowledge of the system. Only a brief history of the motion of the system is initially required. Additionally, we demonstrate the potential of the method for smart control of vibrating systems by stabilizing the chaotically vibrating ribbon about unstable periodic motions of period one, two, and four, exploiting the chaos to switch between these orbits. We have attained control of these chaotic oscillations with mean time between failures greater than 4 days. The method is found to be very robust experimentally and far from being a control that requires experimentally unattainable precision. We believe this method can be widely implemented in a variety of systems including adaptive structures and materials. The advantages of the method include: 1) no a priori knowledge of the dynamics is required; 2) the computations required to implement the control are minimal; 3) the required changes in the control parameter are quite small; 4) different periodic orbits in a chaotically oscillating system can be stabilized for the same system in the same parameter range; 5) the method of control is quite general.

MAGNETOSTRICTIVE LINEAR AND ROTARY MOTOR DEVELOPMENT

J.P.Teter and J.B. Restorff, Naval Surface Warfare Center, Silver Spring, MD 20903
J.M. Vranish and D.P. Naik, NASA Goddard Space Flight Center, Greenbelt, MD 20771

Abstract---Highly magnetostrictive materials such as the pseudo-binary compound $Tb_{.3}Dy_{.7}Fe_2$, commercially known as TERFENOL-D, have been used to date in a wide variety of devices such as sonar transducers, high power actuators and simple linear motors. Improvements in the processing parameters have increased the available magnetostriction , at moderate magnetic field input strengths, to over 2000 ppm. This makes possible a new class of magneto-mechanical devices operating at unequaled energy density levels. Two such devices will be presented, a linear translation motor and a rotary torque motor, both based on an improved inch-worm concept. These motors are ideally matched for applications requiring large and precise force levels; e.g., control of space structures, robotics, advanced machine tool control, avionics, etc.

The linear motor is a self contained unit capable of linear motion along suitable support rails or inside an arbitrary (but uniform) cross section tubular structure. The maximum axial force would be 20 MPa and the maximum traverse speed would be 50 cm/second. The rotary motor is a direct drive, high torque micro-radian stepper, capable of precision movements. The unit is designed to be self braking, at the maximum drive torque level, in the power off state. One prototype of this design yielded a record, for its size, of 12.2 Newton-meters of torque directly off its shaft at 0.5 RPM. The step resolution was 800 micro-radians for a complete full power cycle.

Gearing and mechanically "soft " structures such as pivot joints and active friction brakes are absent from both designs, thereby allowing the full stroke of the magnetostrictive element(s) to be utilized to produce the required motion. Each motor incorporates mechanically prestressed Terfenol-D rods, electrical coils to energize the rods, and a precision machined support structure to translate the stroke while protecting the rods from shear and bending loads. Permanent magnet static pre-biasing can also be included to preset the position of the rods, in the case of the rotary motor, or to prevent frequency doubling and reduce the power consumption, in the case of the linear motor. Results of mathematical modeling techniques will be presented , to include magnetic, structural and both linear and non-linear dynamic calculations. Test results made on prototype hardware will also be presented.

REAL-TIME SENSING WITHIN COMPOSITE MATERIALS

by

Brian E. Spencer, Ph.D.
Technical Director
Spyrotech, Inc.
4930 Superior Street
Lincoln, NE 68523

ABSTRACT

Feed-back systems are available for monitoring and controlling curing of composite materials. The majority of these devices use piezoelectric elements. This technology has been extended to include monitoring structural components. Optical fibers are now being built to monitor pressure, temperature, strain, acceleration and acoustic emission in composite structures. The fibers are small enough to allow implanting within the composite laminate without altering the laminate integrity. Another concept consists of placing fine copper wire in the laminate and using it to measure strain as in a typical strain gage. Although this strain gage will provide average strain over the entire embedded length. These embedded real-time sensors provide the ability to sense out-of-tolerance loads or environmental conditions or eminent failure which could allow time to correct the situation. Other applications allow measuring power transmitted in a drive shaft, heat flow across a critical component and dynamic strains in moving or rotating equipment. This paper discusses the current state-of-the-art in real-time sensing in composite structures and its application.

Active Materials and Adaptive Structures - Meeting , Virginia, Nov 1991

Embedded Optical Fiber Sensors for Acoustic Wave Detection and Cure Monitoring within Composites

K. Liu, B. Park, M. Ohn, A. Davis and R. M. Measures

University of Toronto Institute for Aerospace Studies
4925 Dufferin St., Downsview, Ontario, M3H-5T6, Canada
and
The Ontario Laser and Lightwave Research Centre

Abstract

We describe the use of sensitive interferometric fiber optic strain sensors embedded within composite materials to detect acoustic energy associated with the formation of internal damage.We shall report the *real-time* correlation of *acoustic emission* signals, detected by such embedded interferometric fiber optic sensors, to specific delamination events and cracks within Kevlar/epoxy specimens. We will also present our findings on the potential use of *optoacoustics*, undertaken with embedded interferometric fiber optic sensors, to monitor the state of cure of thermoset composites during their fabrication.

THERMAL-PLASTIC METAL COATINGS ON OPTICAL FIBER SENSORS FOR DAMAGE DETECTION

by

J. S. Sirkis and A. Dasgupta
Department of Mechanical Engineering
University of Maryland
College Park, Maryland 20742

Optical fiber sensors coated with linear work hardening elastic-plastic materials are analytically explored to determine the effects which the coating material properties and thickness have on the sensor performance. The optical fiber system is subjected to both an axial load and an arbitrary thermal gradient. This non-linear analysis reveals a mechanism for designing coatings which provide a "memory" to the fiber optical sensor by forcing the sensor system to undergo permanent deformations in response to predefined excursions in the strain field. A sensor with an elastic-plastic metal coating has obvious potential as a damage sensor working on the same principal as the fiber breakage sensor, except that in this case the fiber sensor is available for post damage measurements. This damage sensor concept can be used with most every intrinsic optical fiber sensor type, and finds great advantage in its simplicity of operation.

This paper presents only a theoretical analysis of the ductile metal coated optical fiber sensor. However, indirect experimental verification of the concept is provided by Inada and Shiota in their paper discussing the development of aluminum hermetic coatings. Inada and Shiota reported that the mismatch between the optical fiber and the metal coating thermal coefficients of expansions lead to significant microbending losses after the coating was cooled to room temperature. They greatly reduced the microbend losses by applying an axial load of sufficient magnitude to permanently deform the coated fiber system. These experiments verify that plastically deforming metal coated microbend sensors are feasible. Since the microbend senor is an intrinsic device, the feasibility conclusion extends to all other intrinsic optical fiber sensors.

The analytical description of the stress and strain fields developed in a metal coated optical fiber subject to axial loading and arbitrary radial thermal gradients is used to investigate the amount of permanent strain which is coupled from the ductile coating to the optical fiber. The optical fiber is modeled as linear elastic, while the metal coating is modeled as linear work hardening obeying a von Mises yield criterion. The optical fiber is assumed to be operating as an interferometric strain sensor, so that the elastic-plastic analysis can be coupled with a three-dimensional phase-strain theory to predict the sensor output. Detectable damage is arbitrarily defined to occur when the optical phase shift of the light propagating in the sensor corresponds to that of an uncoated optical fiber experiencing a 500μ axial strain. This axial strain level represents one quarter of the .02% offset condition typically used to define the yield point on a uniaxial stress-strain curve.

The sensitivity of the sensor response to the coating Young's modulus, Poissons's ratio, yield strength, thermal coefficient of expansion, and thickness are explored for a fixed temperature and a varying axial load, and then a fixed axial load and a varying temperature. As one might expect, the yield strength and coating thickness dominate all other parameters in the sensor performance under both thermal and mechanical loading. As the yield strength increases, larger stresses are required to produce permanent deformations. As the coating thickness goes to zero, the elastic recovery stresses in the fiber dominate the elastic-plastic stresses in the coating so that permanent deformation in the fiber never occurs. Under only mechanical loading, the axial force must approach infinity as the coating thickness does the same in order to produce permanent deformation in the fiber. The coating Young's modulus has an order of magnitude lower influence when compared to the yield strength, and Poison's ratio plays almost no part in determining the amount of permanent deformation occurring the fiber. The mismatch between the thermal coefficients of expansion plays a critical role in determining the performance of the metal coated fiber sensor experiencing thermal loading. Interestingly, the temperature levels required to cause damage level permanent deformations must exceed roughly 500°F for any combination of the other coating parameters. For many coating parameter combinations the threshold temperature is much higher. The implication here is that the thermal and mechanical behavior of this type of damage sensors is automatically decoupled. One should note that all material properties are consider independent of temperature in this analysis.

HEALTH MONITORING SYSTEM FOR AIRCRAFT

Gail A. Hickman and Joseph J. Gerardi
Innovative Dynamics
Ithaca, NY 14850

Abstract

Work is currently in progress to develop an advanced Health Monitoring System (HMS) for application to large airframes. HMS is designed to continuously monitor, over the life of the aircraft, the dynamic properties of the structure. This system is based on the concept of smart structures which integrates sensory systems into the structure to serve as health monitors, analogous to a central nervous system. By monitoring the resultant structural vibration signature, HMS determines structural abnormalities using a network of distributed sensor modules and signal processors. Using an active sensing technique and pattern recognition software, the system is taught to interpret the sensor signals and identify structural damage in real time. The memory of the system is formed through a learning process in which a systematic series of experiments are presented to the system. Conventional minimum distance classification algorithms have been used to obtain a high recognition rate in classification of rivet line failures and fatigue cracks. Current research efforts are aimed at developing a modular prototype system for demonstration on a representative aging aircraft testbed. Application of this technology for detecting structural defects during ground checks has the potential of extending the lifespan of aircraft. The payoff of such a system is to be able to replace aircraft by need or cause rather than by time or statistics.

Active Materials and Adaptive Structures Conference

Aspects of the Microstructural Mechanism of Active Damping in Shape Memory Effect NiTi for use in Vibration Isolation and Cavitation-Erosion Applications.

A.Peter Jardine
Dept. of Materials Science,
S.U.N.Y at Stony Brook,
Stony Brook NY 11794-2275

Shape Memory Effect (SME) NiTi wire has been engineered as part of a vibration-isolation system consisting of a platform consisting of a Copper block suspended by helices providing a resonant frequency of 1.5 Hz for the system. The helices were either stainless steel springs or SME NiTi helices of approximately the same spring constant. Vibrations on the platform were measured by reflection of a collimated light beam from an external, rigidly mounted laser off of a mirror mounted onto the copper block and reflected onto split-cell photodiode, which is mechanically coupled to the frame holding the laser. Vibrations of the copper block with respect to the external frame causes the laser beam to move across the split cell photodiode surface, resulting in different currents from each cell. These currents were then converted to a difference voltage which was measured against time.

Results are presented comparing the vibration characteristics of the system when subjected to identical stimuli using either stainless steel springs or NiTi SME helices. The damping characteristics are shown to differ significantly, with the NiTi helices introducing significant internal damping to the system. Further, the damping is shown to be dependent on the temperature of the NiTi helices. These results are discussed with respect to the relative amounts of martensitic material with temperature.

The high internal damping of NiTi in its pseudo-elastic state may have application in cavitation-erosion. Recent studies by this author and others has demonstrated that cavitation-erosion of NiTi coatings or bulk NiTi is exceptional. Studies were undertaken to ascertain whether this property is a consequence of either the general intermetallic properties of NiTi or by active stress-dissipation of the cavitation-generated shock wave by a microstructural mechanism related to the shape memory effect.

In cavitation, an oscillating pressure field causes the formation and implosion of air bubbles. As a surface easily nucleates bubbles, the implosion of the bubbles also occurs at the surface with stresses approaching several MPa, which are large enough to ablate material, and are also high enough to generate stress-induced Martensite or Austenite, depending on whether the applied stress is tensile or compressive. The implication is that the stress wave may be accommodated by the stress-induced transformation, which can dissipate the energy as heat on retransformation to the materials unstressed state.

Experiments were performed to test this mechanism. Using thin-films of Shape Memory Effect NiTi, the corresponding resistance change of the film to an externally applied 20 KHz cavitating piezoceramic oscillator was performed in distilled water at temperatures in between the M_f and A_f temperature of the NiTi. The results will be discussed with respect to the depth of the material which see the cavitation-induced stresses and the associated problems of heat transfer.

Implications of these results to acoustic damping will be discussed.

MATERIALS CHARACTERIZATION FOR MICROMECHANICS USING DEEP ETCH LITHOGRAPHY

John B. Warren
Brookhaven National Laboratory
Upton, N. Y., 11973

Deep etch lithography is a synchrotron-based lithography process developed in Germany that is used for the manufacture of microstructural components with sub-micron dimensions.* The process steps of deep etch lithography are similar to the X-ray lithography methods used in sub-micron integrated circuit fabrication - both methods require specialized X-ray masks and X-ray sensitive resists such as polymethylmethacrylate (PMMA). Deep etch lithography, however, requires resist layers hundreds of times thicker than those required for the fabrication of integrated circuits. Once these thick PMMA layers are patterned and developed, the remaining PMMA microstructure is used as a sacrificial substrate for subsequent deposition of materials such as electroformed nickel. After deposition of the metal, the PMMA is dissolved, leaving the completed metal microcomponent. Only highly-collimated synchrotron radiation is suitable for deep etch lithography, as the pattern on the original X-ray mask must be transferred without distortion to locations that can be as much as 500 microns below the resist surface.

Work by other investigators has concentrated on electroformed nickel microcomponents because of ease of fabrication. In this application, deep etch lithography is extended to utilize more versatile deposition methods such as plasma enhanced chemical vapor deposition (PECVD). This method permits the fabrication of very fine-grained or amorphous microstructures with widely varying mechanical properties. Of particular interest are high elastic modulus materials such as tungsten and diamond-like-carbon alloys.

These materials are formed into microcomponents that are small enough to be subjected to in-situ mechanical testing on a PZT deformation stage designed to fit in the vacuum chamber of a scanning electron microscope. By observing the degree of deformation or the fracture mode and comparing these values with the predictions of finite element analysis, accurate determination of such properties as the elastic modulus, yield strength and notch toughness for a specific PECVD alloy can be obtained. Such data will be useful in building a material data base for future micromechanical design efforts.

* Requirements on Resist Layers in Deep-etch Synchrotron Radiation Lithography. J. Mohr, W. Ehrfeld & D. Munchmeyer, J. Vac. Sci. Technol. B 6 (6) Nov/Dec 1988. p. 2264-2267.

"Investigation of Shape Memory Properties of Electrodeposited Indium-Thallium Alloys"

by

C. H. Sonu, T. J. O'Keefe, S. V. Rao, and L. R. Koval
Departments of Metallurgical Engineering, Electrical Engineering,
Mechanical Engineering, and Graduate Center for Materials Research
University of Missouri-Rolla
Rolla, Missouri 65401

Abstract

Materials which exhibit the shape memory effect are finding increased usage as both sensors and activators in smart or adaptive structures. The essential requirement for metal alloys to be classified as shape memory alloys is a martensitic transformation upon cooling, with the appropriate phase change. Advantage is then taken of the dimensional change which accompanies the alteration in structure to perform some corrective action in a structure.

The shape memory related phenomenon was first reported in Au-Cd alloys and is a relatively recent development. The alloys most commonly used in commercial applications are Nitinol (Ni-Ti) and Cu-Zn-Al; however, others are now being considered as well, particularly in certain microscale applications.

The alloys are usually fabricated using conventional metallurgical melt-cast-shape techniques. Other more elaborate technologies such as sputtering and rapid solidification have also been investigated for making thin film shape memory alloys, and currently there is considerable activity in this area of research.

Recent studies in our laboratories have demonstrated that it is possible to produce alloys using electrodeposition techniques which exhibit the shape memory effect. This unique processing method offers a number of potentially attractive features, which in time might be incorporated into advanced responsive control systems. The alloys can be deposited in place, at ambient temperature, in thin films or layers, and in a variety of structures and compositions. The electrodeposition technique would also have some limitations, particularly if only aqueous electrolytes are considered, because many of the alloy systems of most interest (such as Nitinol) cannot be deposited from solutions of this type. But since the concept of electrodeposition has now been demonstrated, other advances may occur more rapidly in the future.

It is well documented that the phase-composition-temperature relationship for alloys made electrolytically can vary substantially from similar alloys made thermally. The latter are in the stable, equilibrium state, whereas the electrodeposited alloys may deviate strongly, from the expected state and can remain in this non-equilibrium state for extended time periods. Since many physicochemical properties are directly related to their structure, the opportunity exists to generate a new series of materials which might be effectively used in smart structures or other microelectronic applications.

In this research, the objective was to produce alloys electrolytically which had compositions in a range where the shape memory effect was known to occur. The indium-thallium system was chosen to show the feasibility of the concept. There were several reasons for this choice, but primarily it was due to the fact that both components can be deposited from aqueous solutions, have nearly identical electrode potentials and have a transformation temperature in a practical and useful range from 0 to 70°C. Also, thermal In-Tl alloys have been extensively studied and their behavior well documented in the literature.

The research focused on the production of In-Tl alloys, determining their properties and comparing them with equilibrium alloys of a similar composition made thermally. An acidic sulfate electrolyte was used, with a platinum anode and glassy carbon cathode substrate. Initially the polarization behavior of the system was evaluated using cyclic voltammetry.

After rather extensive testing it was found that satisfactory deposits, in terms of both physical and chemical properties, could be obtained using pulsed current. Parameters evaluated included peak current density, pulse frequency, duty cycle, and temperature. Alloys with Tl contents from 15 to 37% were deposited. A comprehensive statistical screening design was also conducted to determine the influence of various operating parameters on alloy composition.

Qualitative shape memory tests were made on both thermally and electrolytically produced In-Tl alloys. Similar responses were obtained, but the latter were more sluggish in response.

The determination of the phase structures of the alloys at both ambient and elevated temperatures was made using x-ray diffraction techniques. The relationship between phases and compositions showed the structures and transformation temperatures of the alloys made by the two different techniques varied substantially. The FCT (face centered tetragonal) phase was found to be more prevalent for electrodeposited alloys for both higher Tl content and temperature than comparable thermal alloys.

The ability to produce these new types of shape memory materials by electrolysis appears to offer some exciting possibilities for incorporation into adaptive structures. However, considerably more research is required to more clearly define and characterize these materials in order to optimize their potential applications in actual control systems. Future plans call for optimizing the properties of the In-Tl electrodeposited alloys and investigate their use as possible sensors in the active control of a simple cantilever beam. Other binary metal systems will also be evaluated to determine their potential applicability as shape memory materials. In addition, other variations to the processing parameters used for electrodeposition will be made to determine their influence on the structure and properties of the alloys produced.

THE ENERGY TRANSFER EFFECTIVENESS OF A PIEZOELECTRIC ON SILICON BIMORPH MICROMOTOR

Jan G. Smits, Wai-shing Choi, Tom K. Cooney, Department of ECS, Boston University, Boston

Piezoelectric on silicon bimorph micromotors could offer an atractive alternative to electrostatic micromotors in the sense that they are much less sensitive to disturbances of their environment. A major disadvantage of the electrostatic motor is its requirement of an electric field, which precludes the operation of the motor in a dirty environment or in a conductive medium, such as water.

A piezoelectric symmetric or heterogeneous bimorph can be configured as a clamped cantilever beam and is subjected to an applied voltage and three mechanical boundary conditions at the free end: a mechanical moment, M, a perpendicular force, F, and a uniform body force,p.

In deriving the effectiveness of the heterogeneous bimorph, a work cycle is assumed for the bimorph. The work cycle depends on the nature of the load. For a constant load, such as gravity, the workcycle differs from that when a spring type load is applied. The energies stored in the bimorph and the work done are derived for each combination of boundary conditions. The performed work is related to the electrical energy taken in by the bimorph. The effectiveness λ for constant load is defined as the ratio of the energy transferred by the bimorph to the energy put in by the generator. For a symmetric bimorph loaded with a moment type constant load we find:

$$\lambda_M = \frac{3k_{31}^2}{2 + 4k_{31}^2}$$

For the spring type load the results are more complicated because the effectiveness depends on the springconstant of the loading spring. When it is optimized we find for the effectiveness under a moment:

$$\mu_{M_{max}} = \frac{3k_{31}^2}{4(1 - \frac{k_{31}^2}{4})(2 - \frac{3k_{31}^2}{4(1-\frac{k_{31}^2}{4})} + 2\sqrt{1 - \frac{3k_{31}^2}{4(1-\frac{k_{31}^2}{4})}})}$$

For a heterogeneous bimorph, such as a piezoelectric on silicon cantilever beam the expressions are more complex because the symmetry is broken, and the composing materials may have differing elastic moduli and thicknesses. We find for instance for the case that the bimorph is loaded by a constant force F,

$$\lambda_F = \frac{-18k_{31}{}^2nt^2(1+t)^2}{-4(n^3 + 5n^2t + 4nt^2 + 6n^2t^2 + 6nt^3 + 4n^2t^3 + 5nt^4 + t^5) + k_{31}{}^2(4n^2t - 23nt^2 - 54nt^3 - 23nt^4 + 4t^5)}$$

where n is the ratio of the elastic moduli of the non-piezoelectric and the piezoelectric material, and t is the thickness ratio of these materials. k_{31} is the coupling factor of the piezoelectric materials.

Similar expressions are derived for loads being moments and pressures. Optimum values of n and t will be presented as well as optimizations for spring type loads.

In our laboratory we have experimentally fabricated ZnO on SiN heterogeneous bimorphs with dimensions of $3000 \times 400 \times 3\mu$. The fabrication was accomplished by the oxidation of a Si wafer, removal of the top oxide, sputter deposition of a sacrificial layer of ZnO and subsequent shaping of this material by means of photolithography, on which a film of Si_3N_4 was sputtered. On top of that AuCr electrodes were evaporated,and another layer of ZnO was sputtered. The layer was patterned with a photolithography step and a set of AuCr topelectrodes was evaporated on top of the ZnO. Using photoresist as a protective mask the ZnO was covered and the SiN film was opened in HF. After that the ZnO sacrificial layer was etched in a mixture of acetic acid and phosporic acid. Preliminary investigations show that the bimorph has an amplitude deflection of around 2 microns per Volt. The constituent equations of heterogenous bimorphs predict deflections in the order of $40\mu/V$. The discepancy between experiment and theory can not yet be explained.

ROBUSTNESS ISSUES IN MODEL ADAPTIVE CONTROLLERS

S.Hanagud*G.L.NageshBabu†and S.G. Savanur †
School of Aerospace Engineering
Georgia Institute of Technology
Atlanta,Ga-30332

INTRODUCTION

During the past several years there has been a considerable amount of research activity to use bonded or embedded piezoceramic (or PVDF) sensors and actuators to control vibration (or jitter) in light weight structures. Bonded piezoceramic sensors and detection circuits can be designed such that the rate of deformation of a beam structure will result in a signal that is proportional to the difference of the slope rate at the two ends of the transducer[2] Similar results can be derived for voltage time histories as a function of the deformation rates of other types of structures. The detected signal can be conditioned by operations such as filtering, phase shift and amplification. The conditioned signals are used as inputs to bonded or embedded piezoceramic actuators to transmit energy to the structure. The objective of the operations of sensing, conditioning and feedback to selected actuators is to design a vibration or jitter control of the structure.

An early application of piezoceramic transducer to control vibration has been attributed to olsen[1]. Other reported early applications of piezoceramic transducers to vibration control, that followed Olsen's work are due to Forward[2],Forward and Liu [3],Forward and Swiggert[4], Forward, Swigert and Obal[5], Hanagud and Obal[6,7,8] , Crawley and deLuis[9], and Fanson[10].

Since then, there has been an explosion in the research activity on using piezoceramic transducers, PVDF films, shape memory alloys, electrostrictive transducers, voice coil actuators and electro-rheological fluids to control vibrations. Jet propulsion laboratory has demonstrated the use of piezoceramic transducer in the control of vibrations in a space truss like structure. Hanagud and his colleagues have used artificial intelligence techniques to control vibrations in

*Professor
†Graduate Student

a time varying adaptive structure[11,12] and health monitoring of structures[13] In a recent work Hanagud[14] and Fanson [15] have discussed robust vibration control by the use of piezoceramic sensors and actuators. Robustness has been addressed to account for unmodeled dynamics[14,15] and imperfect sensors[14]

Accomodating defects like imperfect sensors and flaws by using robust controllers may involve excessive control forces and eleborate controllers. Another option to accomodate such problem will be the detection of the flaw or imperfection and an incorporation of corrections needed for this condition by modifying the plant only when necessary and changing the controller. This can be called as a model adaptive controller$_{12}$. The purpose of this paper is to develop such Model Adaptive controllers for specific structures with smart sensors and actuators and compare the system with a system that is designed for robustness only.

A truss example with failed members and a cantilever beam with a delamination have been used to study the problem experimentally and analytically.

REFERENCES

[1]Olsen, H.F., "Electronic Control of Mechanical Noise, Vibration Riverbarations," J. Acous. Soc. Am. pp. 966-972, 1956.

[2]Forward, R.L., "Electronic Damping of Vibration in Optical Structures," Applied Optics, pp. 690-697, 1979.

[3]Forward, R.L. and Liu, C.P., "Electronic Damping of Resonances in Gimbal Structures," AIAA paper no. 81-0556, Proceedings AIAA/ASME/ASCE/AHS 22nd Structures, Structural Dynamics and Materials Conference, Atlanta, GA 6-8, 1981.

[4]Forward, R.L. and Swigert, C.J., "Electronic Damping of Orthogonal Bending Modes in a Cylindrical Mast," AIAA 81-4017/40-18, J. Spacecraft and Rockets, 1981.

[5]Forward, R.L. and Swigert, C.j., "Electronic Damping of a Large Optical Bench," Shock and Vibration Bulletin, 53, pp. 51-61, 1983.

[6]Hanagud, S., Obal, M.W. and Calise, A., "Piezoceramic Devices and PVDF Films as Smart Sensors and Actuators for Intelligent Structures," Smart materials, structures and mathematical issues, Ed by C.A.Rogers, Technomic Publishing Co., pp. 69-80, 1988.

[7]Hanagud, S., Obal, M.W. and calise, A.J., "Optimal Vibration Control by the Use of Piezo Ceramic Sensors and Actuators," AIAA/ASME/ASCE/AHS 27th SDM Conference, 1986.

[8]Hanagud, S., Obal, M.W., and Mayyappa, M., "Electronic Damping Techniques and Active Vibration Control," AIAA/ASME/ASCE/AHS 27th SDM Conference, pp. 443-453, 1986.

[9]Crawley, E.F. and Luis, J. de,"Use of Piezoceramics as Distributed Actuators in Large Space Structures," AIAA no.85-0626, AIAA/ASME/ASCE/AHS

26th SDM Conference,1985.

[10]Fanson, J.L. and Caughey, T.K., "Positive Position Feedback Control for Large Space Structures," AIAA/ASME/ASCE/AHS 28th SDM Conference, pp. 588-598 1987.

[11]Hanagud, S., Glass, B.J. and Calise, A.J., "Piezoceramics and Artificial Intelligence Concepts in Time Varying Smart Structures," Proc. SPIE conference on fiber optic smart structures and skins, pp. 250-259, 1989.

[12]Hanagud, S. and Glass, B.J., "Artificial Intelligence Based Model- Adaptive Approach to Flexible Structures," AIAA J. Guidance Control and Dynamics. pp. 534-549, 1990.

[13] Nagesh Babu, G.L. and Hanagud, S., "Delaminations in Smart Composite Structures: A parametric Study on Vibrations," Proc. 31st AIAA S.D.M. Conference, pp. 2417-2426, 1990.

[14]Hanagud, S., NageshBabu, G.L., Stalford, H.L. and Won, C.C., "Robustness Issues in the Design of Smart Structures," Proc. 1st U.S.Japan Conference on Adaptive Structures, 1990.

[15]Fanson, J., Proc. 1st U.S. Japan Conference on Smart Structures, 1990.

ROBUSTNESS ISSUES IN MODEL ADAPTIVE CONTROLLERS

S.Hanagud G.L.NageshBabu and S.G. Savanur
School of Aerospace Engineering
Georgia Institute of Technology
Atlanta,Ga-30332

ABSTRACT

Smart, adaptive or intelligent structures can be used to actively control vibrations. For smart structures, to perform selected functions autonomously like an intelligent person, flaws or failures are to be accomodated. One way of accomodating flaws or failures is detect the flaw, incorporate the needed corrections and change the controller if necessary . The purpose of the present paper is to develop such a model adaptive controller and compare with the robust controllers designed using μ synthesis.

Preliminary Design of Optimal $\mathcal{H}_2$ and $\mathcal{H}_\infty$ Controlled Structures

Robert N. Jacques and David W. Miller
Space Engineering Research Center
Massachusetts Institute of Technology
Cambridge, Massachusetts

Abstract

Recent trends in spacecraft design place many flexible modes of the structure within the bandwidth of active controllers required to meet pointing and alignment requirements. A good structural design must take the presence of these controllers into account and hence the structural engineer must choose between designing the structure to improve performance directly, or designing the structure to improve the performance of the controller. One solution to this problem has been to formulate a cost function in terms of structural parameters and control gains, and then use a computer program to numerically search for an optimal design. While this approach produces mathematically optimal designs, it does not give the engineer any insight into the design problem. This insight is extremely important. It would be cumbersome to perform a numeric simulation of the system any time it is desired to understand the impact of a proposed change on performance. Furthermore, even if one chooses to use optimization to obtain a final design, this insight is necessary in correctly selecting the structural/control model and design parameters.

This work uses low order models to identify and study four basic mechanisms through which a structural change might influence the controlled performance. These mechanisms reflect how changes in the structure can affect:

1. how the disturbance enters the system.
2. how the response of the system enters the performance metric
3. how the actuators influence the system
4. the rate at which the open loop response decays

These mechanisms form the framework for understanding the interactions of the structure and the control.

Two different performance metrics are considered. The first is the $\mathcal{H}_2$ performance metric used in the standard Linear Quadratic Regulator (LQR). It is shown that the gradient of an LQR controlled structure with respect to its structural parameters can be divided up into four subgradients. These subgradients correspond directly to changes via the four mechanisms identified above. A cantilevered Bernoulli-Euler beam is analyzed, and its subgradients computed for different levels of state and control penalty. It is shown that the conclusions reached for the low order models apply most strongly to this higher order system for low levels of control. At high levels of control, it is found that the conclusions based on the low order models must be amended to include the effects of such things as the number of actuators used, the number of modes in the design model, and the rank of the state weighting matrix in the cost function.

The second performance metric considers the $\mathcal{H}_\infty$ norm of the system response and control effort. Usually, this type of performance metric is used to guarantee certain robustness properties of the system, however it is shown that this type of performance metric can better reflect the actual mission requirements than the $\mathcal{H}_2$ performance metric in some instances. It is discovered in a low order model that the relative sensitivity of the closed loop performance of the system to structural changes is identical to the relative sensitivity of the open loop performance of the system. The implies that a sequential design process, where the performance of the structure is optimized open loop and then the control is optimized for the resulting structure, will be just as effective as simultaneous control/structure design. This result is expanded to more general systems, and conditions under which this condition is valid are determined.

Control of Grumman Large Space Structure Using H_∞ Optimization

Chien Y. Huang and Gareth J. Knowles
Corporate Research Center
Grumman Corporation, MS A08-35
Bethpage, NY 11714, USA

Abstract

Studies on the dynamics and control of the large space structure (LSS) have been actively conducted since the Apollo program. With the upcoming launch of Space Station Freedom (SSF), there is a renewed interest in this area. This is partially due to anticipated experiments that require SSF modules to perform maneuvers that, because of the flexibility of their frames, can cause significant control-structure interaction. Furthermore, requirement to align and point the structure with high precision can only be met using active control.

Two major problems are associated with the control of an LSS. One of them is the modeling. High-fidelity mathematical description of an LSS inevitably leads to a high-order system from which, due to numerical considerations, a reduced-order model must be extracted for control design. The question becomes that of modeling just enough dynamics to account for all motions of interest without incurring control and observation spillovers (the former is excitation of uncontrolled modes and the latter is excitation of unobservable modes). A typical approach is to use modal decomposition to incorporate explicitly the desired modes, carry out the control design, and then perform tests to validate the result.

The other problem is control design itself. Several control techniques have been applied to LSS. They include pole-placement, modal control, LQG, LQG/LTR, output feedback, etc. The main difficulty in this phase is generation of a stabilizing feedback control to a large-order plant. Although most of these approaches have shown to be successful in computer simulations, none of them are conclusive as it is impossible to test an LSS in 1-g conditions (the strength of the frame is not sufficient to support the mass without deformations). Therefore, in an effort to validate the techniques, several experiments are being carried out using small-scale space structure.

Past LSS control design methods do not directly address modeling uncertainties. H_∞ optimization is a new control synthesis method that allows the designer to shape frequency response characteristics of the plant to achieve desired performance and, more important, robustness to account for unmodeled dynamics. The objective of this paper is to apply the H_∞ control design methodology to the control of the Grumman Large Space Structure (GLASS). The results show that bandwidth in excess of 300 Hz and an overall disturbance rejection of over 10000:1 can be obtained. Caveats regarding sensor and actuator dynamics are also included along with other issues relevant to the control design of a large space structure.

SYSTEM PARAMETERS OF OUTPUT FEEDBACK CONTROLLED FLEXIBLE STRUCTURES

James A. Fabunmi

AEDAR Corporation
Landover, Maryland

ABSTRACT

The combined system consisting of the baseline flexible structure modified by the system of active controllers is considered as a unified dynamical system. Techniques based on computer algebra are used to derive expressions for the transfer functions of the modified system, using the known transfer functions of the baseline flexible structure and the feedback gains of the active controller. The roots of the characteristic polynomial of this transfer function give the system resonant frequencies and damping parameters. Using the computer algebraic system MACSYMA, expressions for these parameters which are explicitly dependent on the output feedback gains of the active controller, are presented. These results permit the parametric study of the placement of the resonant frequencies and damping parameters of the combined system, as functions of the feedback gains.

SUBMITTED FOR ADPA/AIAA/ASME/SPIE CONFERENCE ON ACTIVE MATERIALS AND ADAPTIVE STRUCTURES,
November 5-7, 1991, Alexandria, VA

"Acoustic Waveguide Embedded Sensors for Submarine Structures"

R. T. Harrold and Z. N. Sanjana
Westinghouse Science & Technology Center
1310 Beulah Road, Pittsburgh, PA 15235

EXTENDED ABSTRACT

Acoustic waveguide embedded sensors are being investigated as part of the ongoing DARPA/MCAIR program called "Embedded Sensors for Submarine Structures", (ES^3)*. Other sensors, such as optical fiber lightguides are also being examined by other members of the ES^3 team. The major focus of the ES^3 program is on developing embedded sensors for composite submarine structures in order to monitor and control manufacturing processes, and also for assessing the relative structural integrity and load response of the structure. Acoustic waveguide embedded sensors are rugged and for large area inspection purposes, acoustic waves may be transmitted between two embedded waveguides.

At this stage of the waveguide program Nichrome waveguides 0.5 mm (20 mil) in diameter have been embedded within graphite-epoxy composite panels and used for cure monitoring prior to the panels being cut into specimens for mechanical tests. The mechanical tests, which consisted of interlaminar shear, flexural strength and strain tests, were carried out on specimens both with and without embedded waveguides and also with waveguides at different depths and in different orientations. SEM pictures of specimen cross-sections with embedded

*This DARPA submarine technology program which is managed by James Kelley is being integrated by McDonnell Douglas (MCAIR), St. Louis, Dan King, Program Manager. Besides Westinghouse, other subcontractors are McDonnell Douglas Electronic Systems, Martin Marietta A&N, Martin Marietta Labs, Stanford University and Virginia Polytechnic.

waveguides showed excellent bonding to the sensor surface and it was found that the Nichrome waveguides did not reduce the shear strengths of panels. With the flexural strength tests, a small (10-15%) reduction in strength was observed, but it is anticipated that the use of a smaller [125 μm (5 mil)] diameter Nichrome waveguide will eliminate this effect.

Another important area being investigated is the use of embedded acoustic waveguides for sensing strain, and some preliminary four point strain tests were carried out on epoxy-graphite composite specimens with embedded Nichrome waveguides. The waveguides were located between the 5th and 6th plies from the surface (close to maximum strain region) and also in the specimen center (minimum strain region). In the strain tests, as the specimens were subjected to loads up to 175 pounds, measurements of acoustic signals transmitted through the embedded waveguides were recorded for comparison with simultaneous measurements from conventional strain gauges bonded to the specimen upper and lower surfaces. Preliminary data analysis indicates that when the waveguide is located close to the specimen surface, the acoustic signal is proportional to the inverse cube root of the load, whereas the strain gauge readings are directly proportional to load. In addition, it was found that the waveguide response to a given strain was clearly different when the waveguide was located in the center (minimum strain region) compared with the cases of waveguides located near the specimen surfaces (close to maximum strain region).

Other areas to be investigated or pursued within this program, include pinpointing impact sites, identifying damage to composite panels; techniques for rapid re-connection at the waveguide terminations; and cure monitoring of thick composites. At the conclusion of this program it is planned to have a demonstration model consisting of a composite panel with an array of embedded acoustic waveguides coupled to a display, instrumentation and data acquisition system. This will allow demonstration of local and remote sensing of the location of events, such as impact and strain.

Submarine Mission Enhancement Using Active Materials and Adaptive Structures

K.J. Moore, J.V. Dugan, Jr., and J. Wilson

Cortana Corporation, 520 North Washington Street, Suite 200,
Falls Church, VA 22046

Enhanced survivability, reduced detectability, and improved vehicle performance have traditionally been the principle targets for the designer of advanced submarines. Current financial constraints add the seemingly inconsistent requirement that the design be superior to the anticipated threat, yet be produced at lowest possible cost. The use of advanced materials and structures engineered to adapt in a controlled fashion to a wide range of conditions has potential advantages in simultaneously meeting multiple performance criteria. Candidate areas for submarine mission enhancement utilizing active materials/adaptive structures technology are listed below. The importance of identifying synergisms among various submarine systems is highlighted with specific examples in order to demonstrate that significant technical advantages and cost reductions can accrue from an integrated approach to vehicle design. The ensuing discussion will be limited to mission enhancement potential. Manufacturing applications, such as the use of embedded sensors to monitor the processing of thick composite structures, will not be addressed except insofar as these sensors might subsequently be used for in-service vehicle health monitoring.

Given the "silent" nature of submarine operations, it is probable that many scientists and engineers may be unaware of the potential submarine applications of R&D in active materials/adaptive structures. Such a situation has been encountered by Cortana Corporation in its role as prime contractor for its DARPA-sponsored Advanced Submarine Technology and Integration (ASTI) Program. A significant number of ASTI subcontractors had no prior knowledge of submarine operations, beyond the fictional exploits of the *Red October*. However, with guidance from experienced submarine technologists on various aspects of submarine design, performance, and operations, the same subcontractors have been able to build on their existing expertise and to focus technology development for application to new generation submarines. It is hoped that a similar marriage of submarine experience and technical expertise will permit the full benefits of active materials/adaptive structures technology to be exploited on future submarines.

The principal pay-off areas for submarine mission enhancement using active materials and adaptive structures have been identified as follows:

- reduced detectability;
- improved hydrodynamic performance;
- in-service vessel health monitoring; and
- reduced cost through integration of both technologies and systems.

Some proposed submarine applications of active materials and adaptive structures are listed below. Although the proposals are grouped according to the main pay-off areas, it is important to note the interrelationships between submarine systems. For example, variable geometry appendages offer potential benefits in terms of reduced signature as well as reduced drag.

Reduced Detectability

- *Reduction in flow noise by counteracting turbulence.*

  An adaptive hull cladding would incorporate pressure sensors to detect the onset of turbulence, and actuators to control the compliance/damping of the coating.

- *Active cancellation of incoming sonar signals.*

  An adaptive hull cladding would incorporate pressure sensors to detect incoming acoustic waves, and actuators to generate destructive counter-vibrations.

- *Reduction in propellor signature.*

  A non-cavitating propellor would use pressure sensors to detect the onset of cavitation. Adaptive propellor blades would then undergo slight geometrical deformations to modify the flow path. Propellor blades might be fabricated from damping alloys which undergo small dimensional changes as a result of pressure-induced microstructural modifications.

  Active cancellation of propellor "singing" (tonal component rather than continuous noise spectrum) using an adaptive structure such as a propellor shroud is also envisaged.

- *Variable signature mast/antenna.*

  The physical properties of mast/antenna structures would be varied to reduce detectability as a function of background conditions such as sea state and luminance.

- *Reduction of machinery noise.*

  The noise emanating from rotating machinery inside the hull could be reduced using equipment foundations with controllable damping properties, e.g. electrorheological fluids embedded in composites, or by using active materials such as damping alloys for machinery components.

Hydrodynamic Performance.

- *Adaptable appendage geometry.*

  Modification of appendage geometry, e.g. sail/hull intersection, as a function of angles of attack and yaw offers potential reductions in drag and signature.

- *Adaptive hull cladding.*

  The injection of ordered, periodic vorticity into a turbulent boundary layer using a variable geometry active wall system may significantly reduce drag, (traveling wave concept).

In-service Health Monitoring

- *Structural test.*

  In-service structural test using embedded sensors will permit monitoring of any degradation in structural performance, particularly in high load-carrying regions. The effects of going to, or beyond, the limits of the operational envelope may be assessed.

- *In-service NDE.*

  Continuous monitoring of materials (composites) could reduce periodic inspection requirements, and hence operating costs.

- *Damage identification.*

  Embedded sensors may permit damage identification and assessment in a "combat" situation, thereby enhancing the capability to reconfigure systems and mitigate performance limitations.

Integration

- *Polyfunctional hull cladding.*

  An adaptive hull cladding may combine a number of functions such as suppression of flow noise, cancellation of incoming sonar, and reduction of radiated noise from machinery inside the hull.

- *"Smart" propellor.*

  An advanced propellor designed to optimize propulsion efficiency would incorporate adaptive structures for suppression of cavitation (and associated corrosion), and for active cancellation of "singing".

- *Survivability.*

  Data require to monitor various equipments and hull, mechanical and electrical (HM & E) systems is the same as that necessary for active vibration control (i.e. signature reduction). Identification of synergisms between sensor systems will prevent duplication and minimize costs.

The present paper attempts to identify certain advanced submarine technologies which may benefit from developments in active materials and adaptive structures, and to indicate a general approach to implementing these developments in the conceptual design phase. Lessons learned from the DARPA-sponsored Advanced Submarine Technology and Integration (ASTI) Program have demonstrated the importance of early technology integration, and of bringing together scientists and engineers from different disciplines in order to achieve a "value added" product. Many of the proposals listed above may only be fully exploited in submarine structures fabricated from advanced composite materials. Since composites are only slowly finding acceptance in the submarine community, the use of active materials and adaptive structures must be seen as a long-term development challenge. Realization of the design and manufacture of a "smart submarine" will require not just significant development of active materials and adaptive structures, but also more widespread use of composites in future submarines.

Acknowledgment.

The present study was funded by the Defense Advanced Research Projects Agency (DARPA) as part of Cortana's Advanced Submarine Technology and Integration (ASTI) Program under Contract No. MDA972-88-C-0064.

Active control of sound reflection/transmission coefficients using piezoelectric composite materials

R.C. Twiney, A.J. Salloway
GEC-Marconi Materials Technology Ltd.,
Caswell, Towcester, Northants.NN12 8EQ, UK.

This paper will outline the important operating principles of the new classes of formable piezoelectric materials: vinylidene fluoride (VDF:TrFE) copolymers and the so-called 0-3 composites. VDF:TrFE is closely related to the conventional polyvinylidene fluoride (PVDF) pure polymer and offers slightly improved figures of merit in piezoelectric applications. The major advantage of this copolymer mix is that it does not need the stretching process required to make PVDF active and so may be formed into structures using conventional polymer processing techniques such as vacuum forming or injection moulding prior to poling. The 0-3 composites are a mix of piezoelectric ceramic dispersed into an inert polymer carrier. This produces a highly robust piezoelectric material with large area capability and good piezoelectric figures of merit. The material also shares with the copolymer the ability to be formed by conventional processing techniques for filled polymers and so can be easily used to produce three dimensional structures. The paper will discuss both classes of material and their relevant performance parameters for use as structurally integrated transducer materials for both sensing and actuation applications.

The paper will describe the application of these materials in a simple underwater active sound control system in which the materials are used to provide both sensing and actuator functions. Combined with suitable control electronics a system capable of modifying the acoustic reflection or transmission coefficients of a metal plate barrier will be described. Theoretical modelling of the interaction between the control system, transducers and sound field will be described and compared with measured results made using a water filled pulse tube to determine the appropriate performance parameters. The system performance at low frequency (of the order of kHz's) is necessarily limited by the inefficient low frequency output of the available actuator materials/mechanisms. The output of available piezoelectric materials will be discussed in the context of the performance parameters needed for the working system. Ways in which the system could be optimised by the use of improved materials will be discussed.

This work has been supported by the Procurement Executive of the UK Ministry of Defence through the Admiralty Research Establishment (ARE) at Holton Heath. The assistance and support of Dr. R. Lane and Mr. D. Townend is gratefully acknowledged

Surface impedance modification of plates in a water-filled waveguide. Pieter S. Dubbelday. (Naval Research Laboratory, Underwater Sound Reference Detachment, P.O.Box 568337, Orlando Fl. 32856-8337)

The interaction of a plane harmonic wave propagating in a medium with density ρ_o and wave speed c_o with a plate of density ρ and dilatational sound speed c is mainly governed by the ratio $\rho c / \rho_o c_o$, for perpendicular incidence. For relatively narrow frequency ranges the reflection and transmission may reach high or low values, when the plate thickness is equal to a number of half wavelengths. [L.E.Kinsler and A.R.Frey, Fundamentals of Acoustics, second edition, Wiley, New York, 1962, section 6.5] Acoustical transparency is established when this ratio is equal to one. Total reflection is realized when this product is zero, (pressure release), or infinitely large, (velocity release). Echo reduction of waves in water by a plate under conditions far from transparency is possible by adding an elastomer layer; its necessary thickness, though, increases with reduced frequency.

Piezoelectric materials offer a different approach to the problem of modifying the impedance of a given surface, [Pieter S.Dubbelday and R.Homer, "Algorithm-based method for suppressing the transmission of sound in a water-filled waveguide", to be published in the J.of Intelligent Material Systems and Structures]. The wave suppression is accomplished by an actuator-sensor combination, under computer control. Its material is piezorubber (Trademark of NTK Technical Ceramic Division, NGK Spark Plug Co., Ltd, Nagoya, Japan), a composite of piezoelectric particles dispersed in a rubber matrix, The sensor measures the pressure of the transmitted wave, and on this basis the computer regulates the voltage on the actuator. The feedback algorithm follows from a technique for complex-root finding.

In the present study this algorithm is applied to the problem of reducing the reflection of waves by a given plate in a waveguide. One needs two sensors to distinguish the reflected wave from the impinging one. One may use a combination of two pressure sensors, a given distance apart, from which pressure and velocity may be deduced. The phase delay becomes smaller for lower frequency; to counteract this, one needs increasing distance between the two sensors. Another solution is to use a velocity transducer, (hot-film or laser-Doppler anemometer), in addition to the pressure transducer.

In the present study, to preserve the appearance of a "smart skin", a double layer of piezorubber was constructed. Into the computer, one enters a desired surface impedance z_i; to obtain the no-transmission condition one would choose $z_i = \rho_o c_o$, but other values may be selected. While a wave impinges on the plate, one measures the two voltages at the electrodes of actuator and sensor. From the six equations for the two layers (from the thin-disk approximation for wave-propagation in piezoelectric materials), one infers the acoustic impedance of the backing to the double layer. This impedance is used in the next step, where a voltage with prescribed amplitude and phase is given to the actuator, and the voltage from the sensor is measured. From the pertinent algebra one may compute the stress f and the velocity v at the front surface of the double layer. One realizes the requested impedance by finding the zero of the complex function $w = f - z_i v$, by varying the voltage on the actuator. The feedback algorithm is analogous to the iteration scheme known as the secant method in real-function root-finding. It takes one or two steps to reach close convergence to the desired z_i.

In this case, one should have knowledge of the material parameters of the sensor-actuator pair, which, for an elastomer material, are functions of temperature and frequency. Alternatively, one may calibrate the "smart skin" by using an external pressure and velocity measurement, prior to its deployment. Of course, it is important to know the sensitivity of the computed input impedance to deviations from the correct parameters.

(PACS) Subject Classification numbers: 43.20.Mv,43.85.Bh,43.88.Ew
Telephone number: (407) 857-5197

SHAPE MEMORY ALLOY ARTICULATED (SMAART) CONTROL SURFACES[1]

Charles H. Beauchamp
Richard H. Nadolink
and
Laurence Dean
Naval Underwater Systems Center
Newport, Rhode Island

Background

Numerous naval vehicles employ straight foils which are rotated at an angle of attack to control flight path. It is known that uncambered foils separate at relative small angles of attack. This limits the lift coefficient that can be obtained and disturbs the flow entering a downstream propeller. This paper presents a demonstrations of the SMAART control surfaces concept. Analysis and mesurements which show the improved fluid dynamic performance which can be obtained with articulated (variable cambered) foilsis discussed. Also a concept for fabricating an articulated fin using a spring steel backbone and molded elastomer shape with a shape memory alloy (SMA) actuator is demonstrated.

Approach

A typical control fin (for which a model was available) was selected as the baseline for this study. The selected fin is shown in figure 1A. The fin shape was a low aspect ratio swept wing with an end cap. A NACA 0012 foil section was used over its entire span. The performance of uncambered NACA 0012 foil was compared to an articulated NACA 0012 foil using two dimensional potential flow and boundary layer codes in the Transition Analysis Program System (TAPS) (ref. 1). The codes were used to predict lift, drag and separations on each foil as a function of angle of attack.

---

[1] Limited Data Rights This information is subject to secrecy order placed on U.S. Pattent Application No. 07591532 which has been filed for the shape memory alloy articulate control surface concept.

An operational articulated model with shape equivalent to the baseline model was fabricated as shown in figure 1B. The operational concept for articulation was achieved by a novel technique of combining a shape memory alloy with molded elastomers. The shape memory alloy provides the macro movement to a semi-flexible fin with essentially no moving parts or control actuator which normally penetrate the fuselage (aircraft) or pressure hull (submersible). The model shape was formed by molding elastomer over stiff leading edge and trailing edge plates which were joined by a spring mental backbone. The model was articulated by fixing the leading edge and applying a force to the trailing edge (via shape memory alloys) to bent the spring-back. This simultaneously applied both camber and angle of attack to the fin. The resulting foil shape has smooth continuous curvature surfaces. This concept has been patented by the authors in a filed patent (Navy Case No. 72282) and a patent disclosure (Navy Case No. 73152).

Both the baseline and the articulated fin models were tested in a water tunnel. Flow visualization techniques were used to investigate the separation on the models as a function of angle of attack. Both models were also mounted on a force balance in the water tunnel. Lift, drag, and torque on the mounting strut were measured as a function of angle of attack and speed.

Results

Figure 2 shows the predicted lift vs. angle of attack for both the uncambered and articulated foils. The straight foil reaches a maximum lift coefficient of approximately 1.5 at approximately 15° angle of attack (AOA). The articulated foils obtained a lift coefficient of 2.0 at an AOA of 7.5°.

Figure 3 shows the predicted percent of chord length which is separated vs. angle of attack for both foils. The articulated foil has a larger separation for a given AOA but at 7.5° it has only reached 25% separation. At 15° the straight foil has 95% separation. This analyses indicates the lift coefficient on the straight foil drops off at angle of attack larger the 15° due to the large degree of separation on the foil. Figure 4 shows the percent of separation vs. lift coefficient for each foil. This figure shows that at an equivalent lift the articulated foils produce less separation than the conventional.

Discussion and Conclusions

The concept for development of continuous curvature articulated foils for application to flight control fins on numerous hydrodynamic or aerodynamic vehicles has been presented. Analysis has indicated that the articulated foil produces higher lift (turning) force and less flow separations than the conventional uncambered foils currently used on numerous naval vehicles. An operational demonstration model of the articulated control fin concept has been fabricated. The model has been tested in a water tunnel to compare its performance to a conventional uncambered control fin. The results of flow visualization test and lift and drag balance measurements confirm the analysis presented here and will be presented in the full paper.

Sponsor
This work has been sponsored by Mr. Gary Jones at the Defense Advanced Research Projects Agency, Submarine Technology Program.

Reference

1. A.E.Gentry and A. R. Wazzan, "The Transition Analysis Program System, Vol. I - Theory", Report MDC J7255-01, McDonnell Douglas, Hunting Beach CA, 1976.

DISTRIBUTED CONTROL CONCEPTS USING NEURAL NETWORKS

J. J. Helferty
Temple University
Philadelphia, PA 19122

D. Boussalis and S. J. Wang
Jet Propulsion Laboratory
Pasadena, CA 91109

Abstract: The performance required of future precision large space structures such as orbiting interferometers and segmented reflector telescopes place very stringent requirements on the alignment and stability of optical components which are attached to large lightweight structural frameworks. The control system for the supporting structure, along with the rest of the system, must exhibit properties such as quietness (i.e., no vibration), high precision figure and position alignment, and system stability in the nanometer region. These performance requirements have motivated a new approach to spacecraft design, where decentralized feedback control principles coupled with advances in embedded sensing and actuation are applied to the design of high performance structural systems. Recent structural designs have employed the use of active-members which are structural elements with actuators and sensors built into them. These active members are then placed at various locations on the structure. Due to the distributed nature of the sensors and actuators, a distributed control system must then be designed to meet system requirements.

The achievable closed loop performance of large space structures is determined in a large part by the ability to synthesize controllers at various locations on the structure with guaranteed stability and robustness properties. Presently, control system designers rely heavily on an "accurate" model of the structure where the space structure is generally modelled as a linear time-invariant system. Although such a model may describe the physical system adequately, any model is only an approximation to the physical system. There is always some uncertainty present in the structure due to physical parameters not being known exactly, neglecting high frequency dynamics, or invalid assumptions made in the model formulation. In this paper we suggest a distributed control system architecture for large space structures based on multi-processor systems known as artificial neural networks (ANN) which do not require accurate knowledge of the structure to be controlled but rather learn to formulate control strategies from their operational experience. We shall exploit both the distributed nature and the learning capabilities of ANN for the design of a robust, distributed control system for precision space structures. A distributed architecture is developed in which an ANN controller is employed at each actuator for local control of a section of the structure. The objective is to determine the feasability of using ANN control strategies to provide accurate vibration suppression along with figure and position alignment over a wide range of frequencies for the entire structure.

Vibration Control of Cylinders Using Piezoelectric Sensors and Actuators

Hartono A. Sumali
Harley H. Cudney

Department of Mechnical Engineering
Virginia Polytechnic Institute and State University
Blacksburg, Virginia 24061-0238
Phone: 703-231-7088
FAX: 703-231-9100

Over the years, several researchers have examined vibration and acoustic control of cylinders for underwater applications. Recently, active control of sound using adaptive control methods has been an active area of research. This paper reports on experiments to perform active sound control on cylinders using piezoelectric sensors and actuators. Two control laws were used. The first control method was active wave cancellation, using a fully analog control system. The second method used an adaptive control technique, with the error measurements provided by piezoelectric film sensors mounted on the cylinder. Previous experiments in active control have, for the most part, used microphones located in the far acoustic field to provide error measurements--this paper demonstrates that structure-borne sensors can be used to provide the same function. The results further demonstrate that adaptive control is a practical and viable method for active noise control.

ON THE SEMI-ACTIVE CONTROL OF STRUCTURAL VIBRATIONS VIA VARIABLE DAMPING ELEMENTS

K. W. Wang and Y. S. Kim
Mechanical Engineering Department
The Pennsylvania State University
University Park, PA 16802
USA

Abstract Submitted to The
ADPA/AIAA/ASME/SPIE Conference On
ACTIVE MATERIAL AND ADAPTIVE STRUCTURES

Vibration in today's increasingly high speed, light weight, complex and flexible mechanical systems severely limits their performance. Excessive vibration, at worst, could be responsible for catastrophic failures or diminished life. To attain international competitiveness, the needs for effective vibration controls are rapidly increasing in many areas of engineering.

The most common and classical vibration control technique is to apply damping to the structure. In such passive control systems, dissipation mechanisms are being introduced, such as viscous dampers, frictional dampers, and composite damping materials. The damping parameters are synthesized through off-line design techniques [1] and no feedback measurements are required. The advantages of this approach are that the devices are usually relatively simple, and the system will always be stable. However, since it is a fixed design, there is no way that the damping can be optimal under all operating conditions.

Active vibration control of flexible structures has been of popular interest in recent years due to its increasing importance [2-17]. Force and torque inputs from actuators are usually used to suppress vibration amplitudes based on on-line measurements from sensors. The controllers are normally designed on the basis of a truncated model, which is an approximation to the distributed parameter system that has infinite degrees of freedom. The advantage of an active system is that the control can adapt for system changes, and thus more effective than a passive system. The major difficulty in this kind of approach is that the structure could be excited due to the interaction between the controller and the uncontrolled modes (residual modes), the so call spillover problem [5,6]. Various methods have been proposed to compensate for this undesirable effect, such as sensing devices relocation [7] and signal filtering [8]. Due to the physical nature of the problem, none of these methods can eliminate the problem completely. The vibration control is further complicated by the uncertainties in structural parameters and external disturbances. The controller is often designed using the Linear-Quadratic-Gaussian or pole-assignment techniques [9] where parameters are identified on-line [10]. The robustness of such controllers can not be guaranteed and the computation time for parameter identification may be too large to implement.

Another type of vibration control is the semi-active approach. The main idea is to combine adjustable dampers with feedback control laws. The dampers will be activated only if energy is being dissipated. They have been developed for buildings [19], vehicle suspensions [20-24] and engine mounts [25], where all the applications are on lumped-parameter systems. Some of these controls use a simple on-off strategy, others modulate the damper continuously.

It has been recognized that some features of the semi-active system could be attractive to the control of flexible structures. That is, combining the feedback characteristic of the active control approach with passive energy dissipation devices whose damping characteristics can be varied according to the controller commands. Since energy is always being dissipated, it is insensitive to the spillover problem, while at the same time reserves the benefit of feedback control. This approach will have the advantages of both the passive and active controls. With the recent development of smart damping materials [18,25,26] and intelligent variable dampers [19], on-line damping variation can be physically achievable. However, a direct extension from the lumped-system approach to flexible structure applications is not feasibie, and a novel feedback law is needed to utiiize these smart materials and actuators for structural control purposes.

The major consideration in applying semi-active action on structures, other than the distributed nature of the problem, is the nonlinear characteristic of the control system due to the state dependent damping parameters, the system uncertainties from unmodelled modes and external disturbances, and the constraints imposed upon the actuators (positive damping constants). This paper presents a novel strategy, based on the theory of sliding mode, for semi-active vibration control of flexible structures by on-line varying the damping characteristics of the actuators. The main advantage of employing the sliding mode control technique is that the system can be designed to be robust with respect to the unmodelled dynamics, system variations and external disturbances. These uncertainties are critical in vibrating flexible structures.

Although the feasibility of using an ad-hoc sliding mode control for semi-active structures has been shown [27], it is recognized that the important issue of actuator constraints needs to be addressed to fully utilize the theory. In this research, the ad-hoc semi-active sliding mode control law [27] is augmented and modified to compensate for the constraint problem. Both the proportional damping case and the general damping case are discussed and analyzed in the paper.

REFERENCES

1. K. W. Wang, "Indirect Damping Analysis and Synthesis of Band/Wheel Mechanical Systems", Journal of Sound and Vibration, Vol. 143(1), 1990.

2. R. G. Klein and C. L. Nachtigal, "A Theoretical Basis for the Active Control of a Boring Bar Operation", ASME Journal of Dynamic Systems, Measurements, and Control, pp.172-178, June 1975.

3. R. Ellis, C. D. Mote, Jr., "A Feedback Vibration Controller for Circular Saws", Journal of Dynamic Systems , Measurement, and Control, Vol. 101, 1979.

4. J. L. Nikolajsen, R. Holmes, and V. Gardhalekar, "Investigation of an Electromagnetic Damper for Vibration Control of a Transmission Shaft", Proc. Instr. Mech. Engrs., Vol. 193.

5. R. Cannon and E. Schmitz, "Initial Experiments on the End-Point Control of a Flexible One-Link Robot", International Journal of Robotic Research, Vol.3, No.3, 1984.

6. M. J. Balas, "Trends in Large Space Structure Control Theory: Fondest Hopes, Wildest Dreams" IEEE Trans of Automatic Control, Vol.AC-27, No.3, 1982.

7. S. Kumar and J. Seinfeld, "Optimal Location of Measurements for Distributed Parameter Estimations", IEEE Trans. Automatic Control, Vol. AC-23, pp. 691-698, 1978.

8. R. Skelton and P. Likins, "Orthogonal Filters for Model Error Compensation in the Control of Nonrigid Spacecraft", Journal of Guidance and Control, Vol.1, pp. 41-49, 1978.

9. L. Meirovitch, H. Baruh, and H. Oz, "A Comparison of Control Techniques for Large Flexible Systems", AIAA Journal of Guidance and Control, Vol.6, No.4, July-August, 1983.

10. D. B. Schaechter, "Adaptive Control of Large Space Structures", Proceedings of AIAA Guidance and Control Conference.

11. A. G. Ulsoy, "Vibration Control in Rotating and Translating Elastic Systems", ASME Journal of Dynamic Systems, Measurement, and Control, 106(1):6-14.

12. L. Meirovitch, D. Ghosh, "Control of Flutter in Bridges", Second International Symposium on Structural Control, Waterloo, Canada, July 1985.

13. L. Meirovitch, H. Oz, "Modal Control of Distributed Gyroscopic Systems", AIAA/AAS Astrodynamics Conf., Palo Alto, CA, 1978.

14. K. Nagaya, S. Takeda, "Active Control Method for Passing Through Critical Speeds of Rotating Shafts by Changing Stiffness of the Supports with Use of Memory Metals", Journal of Sound and Vibration, 113(2), pp.307-315.

15. Y. Chait, C. J. Radcliffe, C. R. UacCluer, "Frequency Domain Stability Criterion for Vibration Control of Euler-Bernoulli Beam", ASME 87-WA/DSC-22.

16. U. Ozguner, et al., "Decentralized Control Experiments of NASA's Flexible Grid", Proceeding of American Control Conference, Vol. 2, June 1986.

17. C. J. Radcliffe, C. D. Mote, Jr., "Identification and Control of Rotating Disk Vibration", ASME Journal of Dynamic Systems, Measurements and Control, 105(1):39-45.

18. T. G. Duclos, "Applications of Smart Materials in the Fields of Vibration Control", Proceedings of Smart Material and Smart Structure Workshop, VPI/SU, Blacksburg VA, Sept. 1988.

19. D. Hrovat, P. Barak, M. Rabins, 1983, "Semi-Active versus Passive or Active Tuned Mass Dampers For Structural Control", ASCE Journal of Engineering Mechanics, vol. 109(3), pp. 691-705.

20. D. C. Karnopp, "Active Damping in Road Vehicle Suspension Systems", Vehicle System Dynamics, Vol.11.

21. D. Hrovat, D. L. Margolis, M. Hubbard, "An Approach Toward the Optimal Semi-Active Suspension", ASME Journal of Dynamic Systems, Measurements, and Control, pp.288-296, September 1988.

22. S. Kimbrough, "Bilinear Modelling and Regulator of Variable Component Suspensions", ASME-WAM-1986, AMD Vol.80.

23. D. Margolis, "The Response of Active and Semi-Active Suspensions to Realistic Feedback Signals", Vehicle Systems Dynamics, Vol.11, No.6.

24. T. Butsuen, J. K. Hedrick, "Optimal Semi-Active Suspension for Automotive Vehicles: The 1/4 Car Model", ASME-WAM-1989, DSC-Vol.13.

25. K. P. Nicholas, R. J. Goudie, and F. P. Boyle, "Actively Controlled Damping in Electrorheological Fluid-Filled Engine Mounts", SAE Paper 881785.

26. B. Stanway, J. L. Sproston, N. G. Stevens, "Non-Linear Modelling of An Electro-Rheological Vibration Damper", Journal of Electrostatics, 1987.

27. K. W. Wang, Y. S. Kim, " Semi-Active Vibration Control of Flexible Structures", Proceeding of ASME Computers In Engineering Conference, vol.1, pp.449-454, August 1990.

Beam Vibration Control through Strain-Actuation and Bending-Twist Coupling

Gregory S. Agnes* and Sung W. Lee**

Department of Aerospace Engineering
University of Maryland
College park, MD 20742

Abstract

Adaptive control of structures through strain actuation shows promise in pushing the performance envelope of aerospace systems. Within a range of materials that can be used for strain actuation, piezoelectric materials have been studied extensively for application to adaptive structures. Piezoelectric materials respond quickly to applied voltage and are suitable for active vibration control.

On the other hand, composite materials offer an opportunity to realize benificial stuctural tailoring in aircraft and spacecraft design. This can be done through the stiffness couplings introduced by using lamination with different ply angles. For example, in rotary wing aircraft applications, bending-torsional stiffness coupling can be used for pitch-flap stability of rotor blades while extension-torsional stiffness may be used to change the twist distribution in two-speed tilt rotors. These stiffness coupling of laminated composites can also be exploited to improve aeroelastic or vibration characteristics of aerospace structures in conjuction with strain actuation using adaptive materials.

Accordingly, the objective of this paper is to study vibration control of composite beams through the combination of piezoelectric strain actuation and bending-twist stiffness coupling.

In order to model composite beams, a finite element model that allows transverse shear deformation and cross-sectional warping is developed.This is necessary because both the transverse shear effect and out of plane warping are very important for composite beams with stiffness couplings. The cross-sectional warping is taken into account by superimposing small displacements normal to the cross sections. These displacements are assumed to be continous functions of the cross-section weighted by a parameter which is a function of axial coordinate only. This allows integration through the cross-section to be carried out *a priori*.

* Lt. USAF, Formerly grauaduate student
** Professor

As an illustrative example, a cantilevered angle ply graphite epoxy beam as shown in Figure 1 is used. Piezoelectric crystals are assumed bonded to the beam. The example beam was modeled using eight three-node beam elements. Using the mass, stiffness and actuation matrices obtained from the finite element model, a three mode modal appproximation was formed with a model reduction technique, resulting in a four state design model. The controller was designed using the H-infinity control algorithm. The controller performance was then verified by performing time simulations to step and impuse responses.The results show that vibration control is possible through the use of strain actuation in combination with stiffness couplings. A typical time histories of the uncontrolled and controlled cases are shown in Figures 2 and 3 for a beam with 30° ply angle. The increase in damping is apparent for the controlled case.

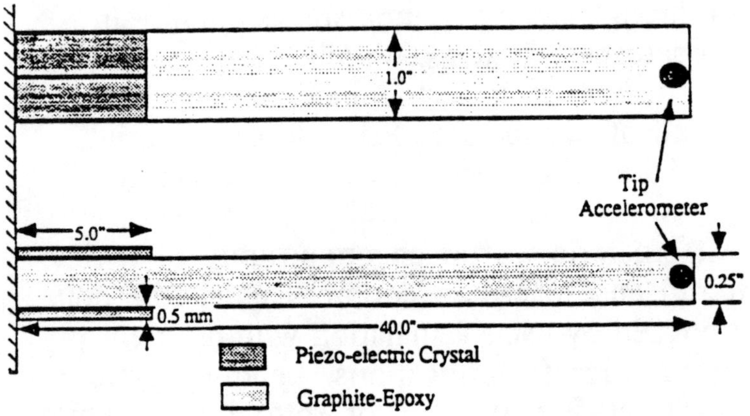

Fig. 1. Cantilevered angle ply composite beam

Fig. 2. Uncontrolled response

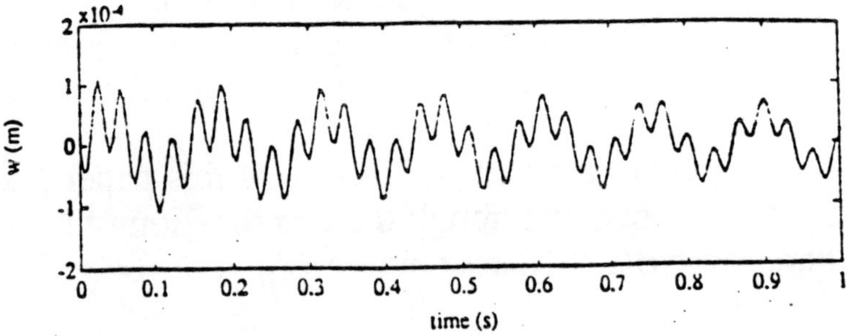

Fig. 3. Controlled response

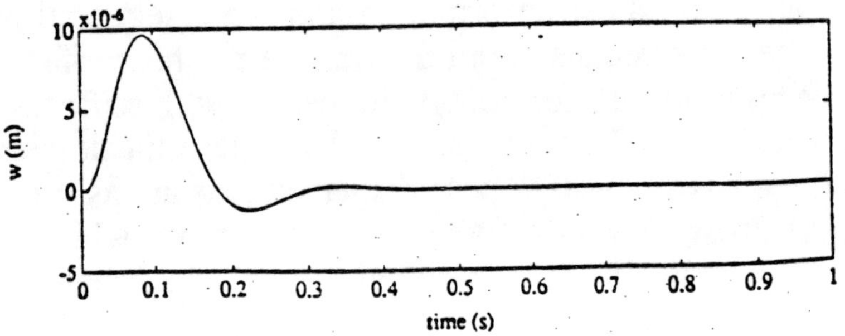

Analytical and Experimental Studies on Adaptive Control of Flexible Structures

Anthony P. Tzes
Polytechnic University
Department of Mechanical Engineering
333 Jay Street
Brooklyn, NY 11201

Farshad Khorrami
Polytechnic University
Department of Electrical Engineering
333 Jay Street
Brooklyn, NY 11201

Extended Summary

Introduction

The application and effectiveness of an indirect adaptive control scheme to the vibration damping control problem of a flexible structure is addressed and investigated in this study.

The growing interest in the incorporation of indirect adaptive controllers in the control of large flexible structures is motivated by several factors. For example, modeling uncertainties will be increased with current and envisaged (composite) materials due to aging and exposure, with potential structure deformation and parameter variation due to human interaction, varying payloads and so on. From the control viewpoint, the error caused by such modeling uncertainties may cause control designs to destabilize the system.

Indirect adaptive control schemes attempt to decrease the effects of plant uncertainty by identifying the system on-line, thereby yielding a closed loop system with reduced sensitivity and improved performance over non-adaptive algorithms. Such adaptive control systems therefore consist of two components–the controller and the estimator. The estimator, or adaptive filter must provide a reliable system estimate in the presence of unmodeled dynamics and unmeasurable disturbances. Several algorithms have been developed for the implementation of adaptive filters and can be classified into either Frequency Domain Methods, or Time Domain Methods. The main characteristic of frequency domain identification methods is that the input signal is transformed to the frequency domain before adaptive filtering is applied. The common objective of all methods, for control purposes, is a time domain parameterization of the system transfer function. The choice of one algorithm over another is determined by factors such as rate of convergence, misadjustment, computational requirements, numerical properties, and minimum information requirements. Self–tuning controllers utilize the information provided by the identification scheme to update controller parameters. The issues associated with implementation of self–tuning schemes based on time domain identification methods are particularly critical in flexible structure control problems. This fact is due primarily to characteristics such as lightly damped, closely packed modes, the presence of unmodeled high frequency components, impending requirements of fast sampling, and several others.

The proposed indirect decentralized adaptive controller will be applied in theoretical studies on an example flexible structure system. Moreover, experimental results will be presented to compliment the analytical analyses. A series of undergoing experiments performed in our testbed facility will be utilized to illustrate and validate the feasibility of the advocated algorithm.

NON-OBSTRUCTIVE PARTICLE DAMPING
NONLINEAR CHARACTERISTICS

By

H. V. Panossian, Ph.D.
Rockwell Int'l/Rocketdyne
6633 Canoga Avenue
Canoga Park, Ca. 91303
(818) 773-5633

ABSTRACT

Presented in this paper are the non-linear characteristics of Non-Obstructive Particle Damping (NOPD). NOPD is a new passive vibration damping technique that consists of making small diameter holes (or cavities) at appropriate locations inside the main load-paths of a vibrating structure and filling these holes to appropriate levels with particles that yield the maximum damping effect. Metallic or non-metallic particles in powder, spherical or liquid form (or even mixtures) with different densities, viscosities and adhesive and cohesive characterisitics can be utilized.

A cubic of aluminum was machined to form a contilevered beam on one side. The block is 5" x 5" x 4.5" in dimensions and the beam has 0.5" x 2.5" x 1" dimensions and is sticking out from the middle of one side of the block.

A hole of 0.16" (4mm) diameter was drilled from the center of the side opposite the beam all the way to 1/4" from the tip of the beam through the center. This hole was filled up to 2" full (80%) with steel shots, Tungsten, Nickel and Zirconium Oxide powder and tested for damping under no compaction force, and under, 20 lb, 40 lb, and 70 lb of compaction force. The results indicate that, although compaction reduces the damping effectiveness, significant damping is achieved with such a small amount of particles. Moreover, the damping value increases with increasing vibration amplitudes. The compaction forces were applied by inserting a piston over the particles and pressurizing over the piston with gaseous nitrogen at 1000 psig, 2000 psig, and 3500 psig (see figure). Thus, the damping ratio of bare aluminum is about 0.04% and damping with tungsten powder was 4.73% without compaction force, and 2.17% with 3500 psig compaction. Similar results were obtained with the other particles. These studies were carried out to assess the effectiveness of NOPD under centrifugal force effects at high speed rotations. There were characteristic behaviors typical of non-linear systems. These will briefly be discussed and heuristic explanations will be presented to account for some of these non-linearities.

STRUCTURAL MOTION CONTROL BY ANALYTICAL DETERMINATION OF OPTIMUM VISCOELASTIC MATERIAL PROPERTIES

Harry H. Hilton[1] and Sung Yi[2]

University of Illinois at Urbana-Champaign

ABSTRACT

Metals at elevated temperatures, polymers, rubbers, concrete, solid propellants and composites all exhibit viscoelastic behavior, i.e. they dissipate energy during creep and/or stress relaxation [1]. These materials are used in primary and secondary flight vehicle structures and as sound and other energy absorbers, such as shock mounts, and motion and flutter control dampers. Viscoelastic damping reduces stress and displacement amplitudes, structural fatigue and attenuates structure borne sound due to for example aerodynamic and engine noise. Improperly chosen materials may lead to early failures of the flight structure or of some of its components and additionally may unnecessarily increase flight weight, performance and fabrication costs. It is, therefore, important to select optimum materials which have properties to provide the necessary and desirable structural service for given environments and constraints.

Thermo-viscoelastic isotropic and anisotropic stress and strain analysis including the effects of moisture is well formulated in terms of elastic viscoelastic analogies and numerical analyses [1-8]. Creep and relaxation behavior as well as its analytical characterization are fully understood from a general over all point of view. However, the equally important theoretical understanding how actual shapes of viscoelastic moduli or creep and/or relaxation functions in either time or frequency domains influence the

[1] Professor Emeritus of Aeronautical and Astronautical Engineering, Associate Fellow AIAA, 104 S. Mathews Ave., Urbana, IL 61801-2997, Phone 217-333-2653, FAX 217-244-7705, E-Mail: HILTON AT UIUCVMC.BITNET.

[2] Doctoral Candidate in Aeronautical and Astronautical Engineering

potential and dissipative energies has not been investigated. In the frequency domain, the complex representations of these material functions consists of two parts, namely the storage (real part, multicolor curve in Fig. 1) and the loss (imaginary part, red bell shaped curve in Fig. 1) components. The parameters affecting viscoelastic behavior are the ratio of the maximum to minimum values of storage modulus, the shapes of the two knees, the slope, the areas under the storage and loss modulus curves and the maximum value of the loss modulus as compared to the maximum value of the storage modulus. In Fig. 1, for the storage modulus, the blue portions are represented by straight horizontal lines, the knees (red and green) are third order polynominals and the central part (black) is a straight line. For the loss modulus (red lower curve) a beta function representation is used with A and B as exponents. This allows for parametric skewing of this curve. For both curves the various junction points along the horizontal axis are also considered as parametric material properties and their influence is studied.

For each set of parameters, the analytically formulated complex modulus (or creep/relaxation function) curves are converted into a Prony exponential series by a least square fit. Such series are derived from corresponding generalized Kelvin bodies and lend themselves readily to the determination of potential and dissipative energies for combinations of particular material properties, structural geometries and loading conditions [8].

In the present pilot study the inverse problem is investigated. Various specific problems, such as viscoelastic bending of beams and plates and bending torsion flutter [9] are systematically considered and the shape parameters are varied one by one to obtain fundamental understanding of their influence on the response of viscoelastic structures and on their damping capabilities. The integrated results identify candidate material properties for various specific tasks and provide analysts and designers with detailed catalog information on the selection of optimum materials. This also leads to a knowledge base of what real materials need to be fabricated to obtain properties for best performance, i.e. manufacturing real materials with suitable modulus specifications to conform to specific parameters and performances. Reference [10] treats the relation of viscoelastic material chemical composition to fabrication.

Selection of these materials has direct applications to the design and analysis of sound proofing, shock absorbers, helicopter blades, lifting surfaces flutter, etc.

REFERENCES

1. Hilton, H. H., "An Introduction to Viscoelastic Analysis," *Engineering Design for Plastics,* Reinhold, New York, 1964, pp. 199-276.
2. Christensen, R. M., *Theory of Viscoelasticity,* Academic Press, New York, 1982.
3. Hilton, H. H., and Dong, S. B. "An Analogy for Anisotropic, Nonhomogeneous, Linear Viscoelasticity Including Thermal Stresses," *Proceedings 8th Midwestern Mechanics Conference,* 1964, pp. 58-73.
4. Hilton, H. H., and Yi, S., "Finite Element Formulation for Thermo-Viscoelastic Analyses of Composite Structures Subjected to Mechanical and Hygrothermal Loadings," *Proceedings First NCSA Conference on Finite Element Applications in Computational Mechanics,* 1990, pp. 1-13.
5. Hilton, H. H., and Yi, S.,"Bending and Stretching Finite Element Analysis of Anisotropic Viscoelastic Composite Plates," *Proceedings Third Air Force/NASA Symposium on Recent Advances in Multidisciplinary Analysis and Optimization,* 1990, pp. 488-494. Submitted for publication to *Journal of Composite Materials,* 1991.
6. Hilton, H. H., and Yi, S., "Dynamic Finite Element Analysis of Viscoelastically Damped Composite Structures," to be published in *Proceedings of Applications of Supercomputers in Engineering,* 1991.
7. Yi, S., "Thermoviscoelastic Analysis of Delamination Onset and Free Edge Response in Epoxy Matrix Composite Laminates," *Proceedings of AIAA/ASME/ASCE/AHS/ASC 32nd Conference on Structures, Structural Dynamics and Materials, Part 2,* 1991, pp. 1016-1026.
8. Hilton, H., H., "Viscoelastic and Structural Damping," to be published in *Proceedings of Air Force Conference on Damping '91.* Submitted for publication to *AIAA J.,* 1991.
9. Hilton, H. H., and Vail, C. F., "Bending-Torsion Flutter of Linear Viscoelastic Wings Including Structural Damping," submitted for publication to *AIAA J.,* 1991.

10. Kaelbe, D. H., *Computer-Aided Design of Polymers and Composites,* Marcel Deckor, New York, 1985.

BUILDING VIBRATION DAMPING INTO TUBULAR COMPOSITE STRUCTURES USING EMBEDDED CONSTRAINING LAYERS

S. S. Sattinger
Mechanics & Tribology Department
Z. N. Sanjana
Advanced Materials Technology Department
Westinghouse Science and Technology Center
Pittsburgh, PA 15235

ABSTRACT

The effectiveness of active control systems for suppressing structural vibration is heightened when they are used in conjunction with passive damping. Providing passive damping can augment active vibration and noise control performance by reducing algorithm convergence time, reducing the controlled responses at system resonances, ameliorating control spillover, and providing attenuations at high frequencies. This paper deals with a novel approach to the task of adding passive damping to tubular structural components whose vibrations may be of concern.

Add-on treatments using viscoelastic materials have been successfully used to introduce passive damping in many types of structures, but there are a number of factors that have tended to limit their use. It may be difficult to gain access to interior regions of a structure that would otherwise be plausible locations for the installation of damping treatment. The treatment, once installed, may also create obstructions or mechanical hazards to personnel. The treatment may also cause toxicity, flammability, or other environmental problems due to outgassing and combustibility of the damping material. Conversely, exposed areas of this material may be susceptible to the effects of moisture, lubricants, oxygen, or vacuum conditions. The limited static and creep strength properties of many of these materials may also pose obstacles to the use of these treatments, as in the case of rotating components having high centrifugal acceleration levels.

Embedding damping materials inside the walls of organic-matrix composite beam or shell structures during fabrication shows promise for overcoming these limitations. Described is a construction that provides high damping of all vibration modes, including those which tend to be difficult to damp. The latter modes include the long-wavelength, global beam-bending and torsional modes that occur at low frequencies and the in-plane shear and column modes that are observed at higher frequencies. This paper discusses attained and projected damping and structural performance, various advantages of the construction, and some candidate product applications.

We have designed and fabricated a set of filament-wound tubular fiberglass/ epoxy specimens, each 36 inches in length by 5 inches in diameter, for proof-of-principle bench testing of this construction. A graphite/ epoxy embedded constraining layer and adjoining viscoelastic material layers are built into two of the specimens; this constraining layer is segmented in one case and continuous in the other. The segmentation of the first specimen is both axial (cuts running perpendicular to the tube axis) and circumferential (cuts running parallel to the tube axis). A third specimen is a control specimen featuring conventional fiberglass/ epoxy construction.

All vibration modes are strongly damped in the specimen with the segmented, embedded constraining layer. Peak measurements of damping loss factors in the vicinity of 0.08 for the extensionally stressed vibration modes of this specimen (the global bending modes and column modes which produce vibratory extensional stresses that are nearly uniform through the wall thickness) represent high damping of these hard-to-damp modes. The good overall agreement obtained between predictions and measurements confirms that the damping of these modes is predictable in segmented-layer constructions. A torsional-mode loss factor of about 0.05 has also been measured on this specimen.

In the specimen having the continuous, embedded constraining layer, neither the global beam-bending modes nor the lower-order column modes receive significant damping additions. However, the local shell-bending or lobar modes are strongly damped in both the segmented-layer and the continuous-layer specimens, with loss factors in the range from 0.1 to 0.4.

We expect that combinations of high damping and high specific modulus will be obtainable in similar constructions with stiffer reinforcing fiber combinations such as segmented constraining layers of high-modulus, pitch-based graphite fibers and load-carrying layers of PAN-based graphite fibers. In addition to surmounting application obstacles associated with access, strength, and environmental factors, composite structures damped by embedded constraining layers offer much higher local-mode damping performance than is attainable with add-on treatments of comparable thickness. Likely applications using the passive damping of this construction in conjunction with active vibration control include marine and aerospace structures for both military and commercial uses. Passive damping applications for this construction may include shafting, robotic devices, turbomachinery structures, sporting goods, and damped structural shapes.

MULTILAYER COMPOSITES - MATERIALS FOR VIBRATION AND NOISE REDUCTION

Edward J. Vvdra, Pre Finish Metals, Elk Grove Village, IL

ABSTRACT

Multilayer composite materials are defined as laminated structures consisting of two skins of metal separated by a layer or polymeric material, applied as a coating, film or prepreg, and adhesively bonded together. Such structures can be successfully manufactured in a continuous manner and this article describes such composite materials utilized for sound and vibration damping over a wide range of frequencies and temperatures. With high damping properties (loss factor η (nu)$\geq$0.1, these materials can be formed and fastened by using existing equipment and technology (including welding).

BUSINESS OPPORTUNITIES IN SMART MATERIAL SYSTEMS

by

Mohammad Usman, Ph.D.
Director, Innovative Technologies
Quantum Consultants, Inc.
Hannah Technology and Research Center
4660 S. Hagadorn
E. Lansing, MI 48823

ABSTRACT

The impending revolution in smart materials and structures technologies will very significantly influence numerous industries including the aerospace, defense, advanced manufacturing, and biomedical sectors of the economy. The various smart materials technologies that are currently being developed and exploited for commercial applications in these industries are reviewed in this paper. These technologies include shape-memory alloys and plastics, piezoelectric ceramics and plastics, electro-rheological fluids, magnetostrictive materials, and fiberoptics. A succinct summary of significant technical contributions to the field of smart material systems will be presented prior to providing a state-of-the-art technological assessment of this rapidly evolving area.

The technological and economic impact of the diverse smart materials technologies on the commercial sector in the U.S., Europe, and the Pacific Rim will be enunciated. A discussion of the evolving market topology will be presented, in addition to providing projections for market growth in this innovative discipline. Impediments to the commercialization of products featuring smart material technologies will be highlighted, and recommendations for addressing these impediments will also be discussed. Several industrial case studies will be presented in order to distill guidelines for industrial practices and to further motivate the directions for future research and development.

IMPLEMENTING SMART COMPOSITES:

ORGANIZATIONAL/ENVIRONMENTAL ISSUES

by

Michael J. Martin
Director, Industrial Development Institute and Technology Transfer Center
Michigan State University
East Lansing, MI 48823

"Smart" Composites Technology will not automatically be implemented simply because it provides mechanical and economic advantages. There are significant organizational and environmental issues which need to be understood and dealt with. This paper will focus on implementing smart composites in a mature organization and cost modeling/economics.

"Traditional" cost accounting cannot provide sufficient operating information for the "new" world, "new" customer, "new products", and "new" manufacturing organizations. Even though the U.S. government may have positioned advanced materials/composites as a potential "leap frog" technology, the existing metal-based manufacturing/design firm needs "new" techniques of cost analysis to justify the shift to a team-based "designed for manufacturing" organization utilizing smart composite materials.

The Michigan State University Composite Cost Comparison (MSU C^3) Model attempts to provide an evaluation tool for these "new" cost justification techniques. Composites are unique from traditional materials of construction because they require an integrated approach to "design for manufacturing." Utilization of these new materials can be the basis for the initiation of a cultural change in a firm.

Title: **Activities of the Smart Structures Research Institute**

Authors: P.T. Gardiner, B. Culshaw, A. McDonach, W.Craig Michie, R. Pethrick.

Address: Smart Structures Research Institute
The University of Strathclyde
204 George Street
GLASGOW G1 1XW
Scotland, U.K.

Tel: 041 552 4400 ext.2886
Fax: 041 552 2487

ABSTRACT

The 'Smart Structure' concept is emerging as a critical contributor to 21st century technology. The idea is simple; sensors and actuators are incorporated within the structure to enable it to respond to changes in both its surrounding environment and its loading conditions. Responding to the environment implies the ability to recognise, discriminate, notify and adapt.

Smart Structures technology will undoubtedly yield a wide range of new materials plus new material sensing and actuation technologies and these will have a radical effect on current approaches to structural design. This is a truly multi-disciplinary research challenge, best met by an organisation capable of integrating the appropriate technologies.

On 14th February 1991 the Smart Structures Research Institute, the first such institute in Europe, was officially inaugurated on the campus of the University of Strathclyde. There are a number of reasons why the Smart Structures Research Institute was established and these were uncovered during a comprehensive feasibility study. This identified a strong level of interest from industrial companies and government laboratories. This study also showed that research into Smart Structures applications needs to be supported by the fundamental science base which a University is ideally suited to provide. In addition a detailed internal review established that the key technological and science-based building blocks were already resident within the various departments of the University. Ongoing research at the University over the last few years has established the basic proof-of-concept relating to embedded sensors and considerable expertise in signal processing, materials and controls.

Our ref: SSRI/PTG/035

7 May, 1991

The technologies that are being integrated through the Institute include:

- Materials Science
- Pure and Applied Chemistry
- Sensors and Instrumentation
- Structural Design and Testing
- Control and Signal Processing
- Manufacturing and Process Engineering
- Systems Engineering

The mission of this Institute is to become a world centre of excellence and leader in Smart Structures research. Applications for Smart Structures technology fall into short, medium and long term timescales, all of which need to be supported by a sound science base. The Fundamental Science Programme is viewed as the 'backbone' of the Institute, and a number of projects have already been initiated:

- Embedded Sensor Systems
- Adaptive Structures
- Molecularly Smart Materials

Another major activity of the Institute is that of collaborative research with industry. Examples of such programmes include:

- Cure Monitoring/Smart Processing Programme - aimed at significantly improving the manufacturing process of advanced composites.

- Damage Detection/Structural Integrity Programme - aimed at detecting the onset of in-service damage of a structure, using the continuous monitoring capability of an in-built sensor network.

- Distributed Sensing Programme - key to many Smart Structure programmes and applications. The concept in values monitoring of parameters such as strain and temperature as a function of position through the structure.

- OSTIC (Optical Sensor Techniques in Composites) Programme - this has set out to prove the feasibility of imbedding optical fibres in composite structures and to explore what measurement capability then exists.

Managing multi-disciplinary research is a major challenge, and the University has taken great care to establish the appropriate framework. Effective programme management is the key, and this is the responsibility of the Director and his management team.

Our ref: SSRI/PTG/035 7 May, 1991

This paper will describe the organisation and activities of the Smart Structures Research Institute and will review the relevant research activities that are currently underway.

Smart Materials – The Center's Agenda for Implementation

Vasundara V. Varadan and Vijay K. Varadan
Research Center for the Engineering of Electronic and Acoustic Materials
Department of Engineering Science and Mechanics
The Pennsylvania State University
University Park, PA 16802

"Smart materials" can sense changes in the environment either physical or chemical and can respond in an optimal, desirable manner. This talk specifically addresses the current and future research of the Center with respect to the problems of structural vibration control and structural acoustics control. The former refers to damping of unwanted structural vibrations and the latter to the control of radiated sound from structures due to external excitation or due to fluid-structure interactions or structural vibrations. Every smart material must incorporate both sensor and actuator functions with appropriate communication links for feedback. The sensors and actuators that are currently in use for such problems are piezoelectric and our center has taken a four prong approach for the development of smart materials which can then be integrated into smart structures.

(1) Theoretical modeling of the integrated sensor – actuator functions of several different types of piezoelectric transducers mounted on structures immersed in water. A hybrid finite-element technique has been developed to evaluate the performance of a composite transducer mounted on a structure immersed in water. This technique developed is capable of solving complex geometry, anisotropic material and the effect of fluid loading. Future plans include simulation of smart material performance by inputting the sensor response to a given input field into a control algorithm that can then provide the appropriate voltage to excite the actuator. Numerical simulation of the actuator function can then be used to find the response of the structure to the original input.

(2) For each type of application customized sensors and actuators have to be developed in order to get optimal results. Recent work on the development of composite transducers, shear transducers and air coupling transducers will be discussed. Specially designed piezoceramic shear transducers have been used for sensing and actuating shear strains for active control of pure torsional vibrations.

(3) The Center has experimentally implemented discrete piezoelectric sensors and actuators for the vibration control of several types of structures – truss structures, plates, torsional members. Both analog and digital control has been implemented. Multimode control and control of radiated sound from a vibrating structure such as a clamped plate is an ongoing study. Active vibration control has also been implemented in the precise positioning of a robotic arm controlled by a piezoelectric motor. Finite element modeling of the structures with and without mounted transducers ensures that the transducers do not significantly affect the natural structural dynamics of the system. This is of significant advantage relative to conventional damping methods like added mass or constrained layer damping.

(4) Active acoustic coatings have been developed for 'acoustic stealth' applications which incorporate large area sensors and 1–3 piezoelectric composite actuators with a digital delay electronic feedback system. Test results from the acoustic pulse tube facility of our Center will be reported. Over a frequency range of 3.8 kHz to 11.0 kHz at intervals of 200 Hz, a reflection reduction of 40 dB was achieved over the whole frequency band. The control of radiated sound from a vibrating structure is also being tested in this facility and the results will be reported. The use of acoustically chiral coatings for suppressing the acoustic response of a submerged structure is also under study. A bilaminate 1–3 actuator has been constructed and implemented along with a dual sensor in the front and a single sensor at the rear to control both sound reflection and transmission. The reflected loss measured was 12 dB while the reduction of the transmitted signal was 22 dB.

ABSTRACT

Paper for the "Active Materials and Adaptive Structures" Conference

Arlington VA

November 4-5, 1991

TITLE: Technology Integration Requirements for Adaptive Structures in Space

AUTHORS: J.P.Henderson, Materials and Vibration Engineering (MVE), Fairborn, Ohio
P.E.Stover, Nichols Research Corporation, Dayton, Ohio

Successful application of adaptive structures technology to future space systems will require the integration of several technologies, in addition to those associated with adaptive materials and active control methodology. Specifically, the authors discuss characteristics required of support electronics systems in space systems and the relationships between these requirements and the structural dynamics issues associated with the control of vibration and jitter. On-going developments in sensors, force generators, signal conditioning and processing, memories, and signal and power distribution, are reviewed. Parameters must be defined to take advantage of these developments in the design and optimization of electronic systems for adaptive structures. Vibration suppression parameters for these electronic systems will depend on the degree of successful integration of passive damping and active control strategies. Approaches for achieving higher levels of passive damping are described, and the necessity of doing trade studies involving temperatures, frequencies of excitation, modal densities and other environmental factors is discussed.

EXPLORATORY STUDY OF THE ACOUSTIC PERFORMANCE OF PIEZOELECTRIC ACTUATORS

O. L. Santa Maria
NASA Langley Research Center
Hampton, Virginia

and

E. M. Thurlow, M. G. Jones
Lockheed Engineering & Sciences Co.
Hampton, Virginia

The unique noise generating properties of the proposed ducted fan engine are forcing the aeroacoustician to consider new and novel techniques to reduce the sound levels of these engines to acceptable levels. Among the techniques being considered are those involving active noise control. One of the primary difficulties of providing an active noise control system for this application is the need for extended spatial control inside a duct. Therefore, a distribution of highly efficient, lightweight sound sources is required.

A current program at NASA Langley Research Center is involved in evaluation of alternative sound sources and their application to the noise control problem. As a first step in the development process, exploratory tests were conducted to evaluate the acoustic transduction properties of three piezoelectric samples. Piezoelectric materials vibrate and produce sound when driven by an input voltage. If large acoustic outputs can be obtained directly from these piezoelectric samples, an extended in-duct source distribution may be installed using simple geometries. However, more complex geometries may be required to amplify the acoustic outputs if the piezoelectric direct outputs cannot be significantly improved. One sample evaluated was a 28-µm-thick Polyvinylidene Flouride (PVDF) piezo film sample embedded in a block of foam. The other two samples were wafer-type composites of Lead Zirconate Titanate (PZT) rods embedded in fiberglass. Two tests were performed on the piezoelectric samples.

The first test was to ascertain the acoustic efficiency of the sample, i.e. the acoustic output as functions of frequency and excitation voltage. This was done by mounting the sample normal to the axis of a 2-in by 2-in duct with a nonreflecting termination and measuring sound levels approximately one meter from the sample. Figure 1a shows a schematic of the setup described above.

Figure 2 shows a frequency response plot of the acoustic output for one of the piezoelectric samples. The peak in the response curve represents a resonant response of the sample and is a function of the diameter of the embedded PZT rods. The test also showed that the acoustic output is directly related to the excitation voltage or:

$$\Delta dB = 20 \log \frac{V2}{V1}$$

for this range of excitation voltage.

The second test was conducted to investigate the ability of a piezoelectric sample to alter a simple standing wave field by controlling the termination boundary condition or impedance using existing facilities. A standing wave tube was used to determine if the active sample could act as a nonreflecting termination for various phase settings relative to a primary sound source.

The sample was mounted at one end of a 2-ft long standing wave tube, as shown in Figure 1b, with acoustic drivers on the other end to produce an incident sound field. The acoustic drivers and the sample were adjusted to produce the same output levels at each frequency tested. The change in the standing wave ratio (*SWR*) was then measured as the secondary source was driven at various phase settings relative to the primary source. A reduction in the SWR indicated that the reflected wave from the sample was being minimized.

Reduction of the SWR was achieved at all frequencies tested. Figure 3 illustrates the most significant reduction. The maximum standing wave ratio for this frequency was 18 dB when the sample was at a phase setting of 180 degrees. When the phase was set to 90 degrees, the *SWR* was reduced to 1.2 dB.

A survey of *SWR* for phase settings at 1900 Hz, in increments of one degree, determined that the optimal phase setting was 82 degrees, which provided a *SWR* of less than 1 dB. This result indicates that the reflected wave from the sample is nearly eliminated.

These tests demonstrate that the reflected wave from an active piezoelectric sample can be eliminated, fulfilling one of the initial goals of the experiment. Since the termination impedance can be controlled using a piezoelectric sound source, a set of small PZT patches may be placed around the circumference of the nozzle to obtain the desired acoustic environment. However, the acoustic output of the secondary source must be equal to that of the primary source. The response of these samples needs to be increased by at least one order of magnitude in order to match sound levels predicted for the ducted fan engine.

The paper will provide a more detailed description of the piezoelectric sample characteristics and the test setup. Results will be given for three piezoelectric samples.

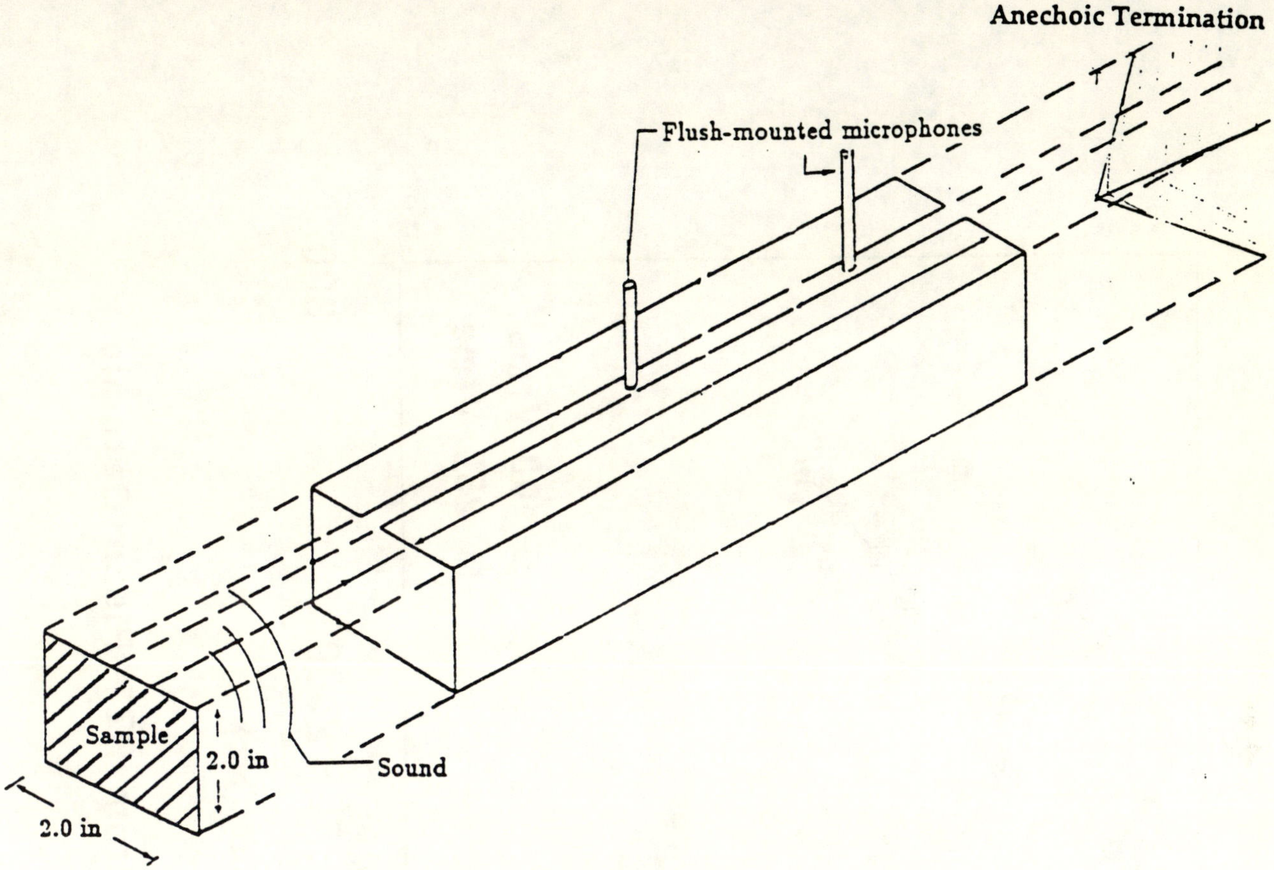

(a) Acoustic Output Measurement Setup

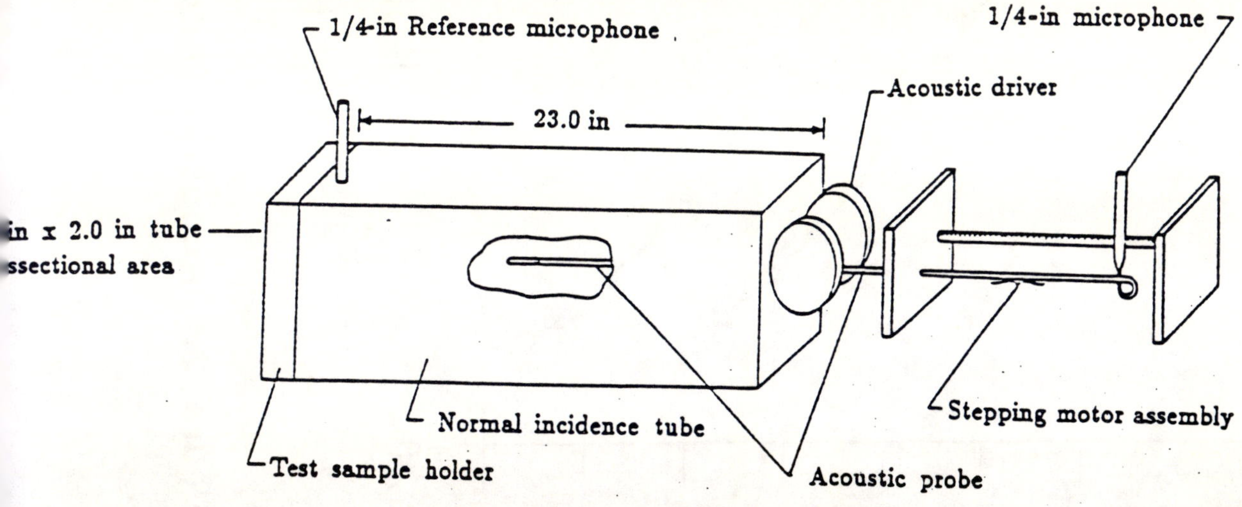

(b) Standing Wave Tube

Figure 1. Schematic diagram of test setup (not to scale).

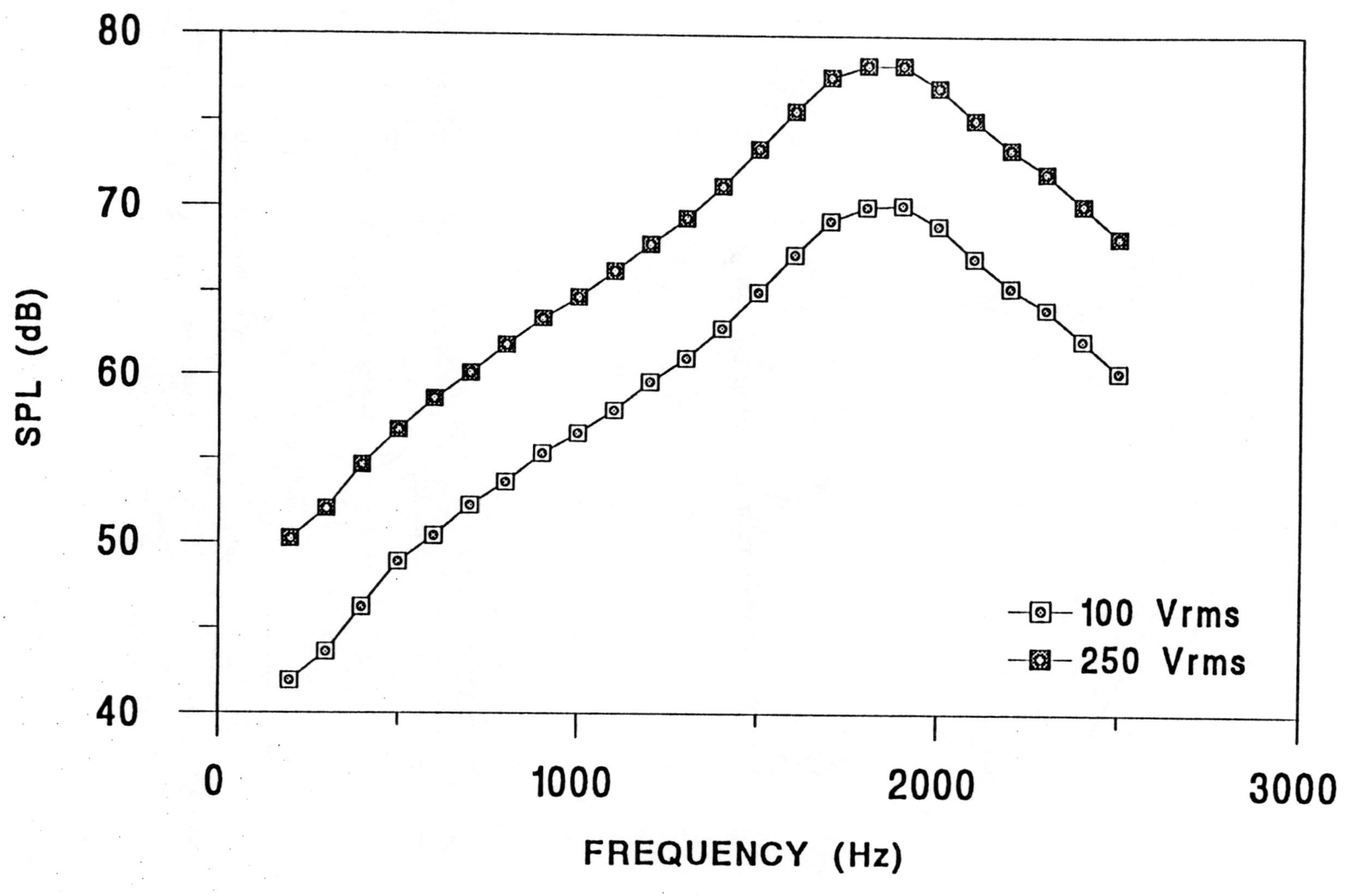

Figure 2. Acoustic Output Data from PZT Piezo Ceramic.

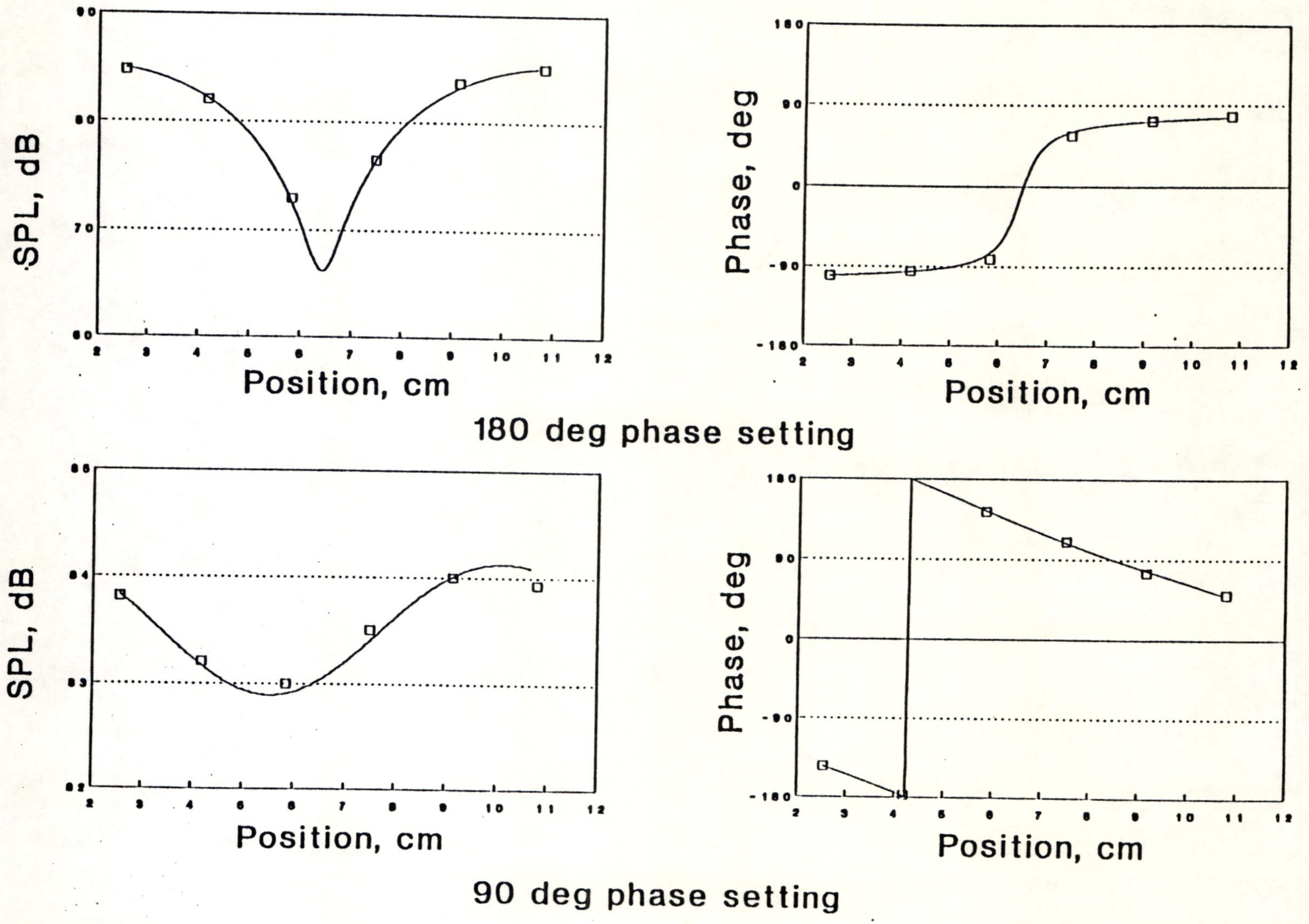

Figure 3. Standing Wave Amplitude and Phase Angle at 1900 Hz Frequency vs. Distance from face of Sample in the Standing Wave Tube

COMPARISON OF FEEDFORWARD VS FEEDBACK DESIGN IN SOUND RADIATION SUPRESSION

James Thi and Erdal Unver

AT&T Bell Laboratories, Arlington, VA 22202

ABSTRACT

Much of the research in active control of sound radiation employs adaptive filtering techniques or the so-called feedforward system. Adaptive filtering methods for acoustic suppression has been applied successfully in a number of practical problems, such as noise contol in a fan duct, cabins of aircraft and the silencing of an engine exhaust system. If the active control system is to suppress the acoustic response at locations where acoustic measurements are available, then adaptive filters can be used to cancel the measured sound from an input microphone. This type of active control system is refered to as a feedforward design with system identif·cation. An alternative approach which does not require acoustic measnrements is to control the vibrational modes that contribute the most to the radiated acoustic energy.

It has been shown that modal based active control can minimize the number of actuators, sensors, and control energy. This is especially true if the vibration problem is known to involve only a few significant radiating modes. The advantage of modal based active control strategy is that the vibration motion from individual modes can be estimated without introducing time delay into the cancellation path. A time delay will be introduced by the acoustic propagation where acoustic measurements are used. Therefore, it is possible to develop modal based active control system using feedback design since acoustic propagation delay can present stability problems in the feedback design.

There are many factors that will influence the success of an active control system in any given application, one of the main factors is the selection of either feedforward or feedback control approach. The objective of this paper is to compare the feedforward and feedback approaches in the design of a modal-based active control system. There are major differences between feedforward and feedback approaches. Reference signals without the effects from the control signals are required in the feedforward approach, whereas feedback approach requires the availability of residual error signals. Furthermore. It can be shown that if the effects on the reference signals from the control signals can not be removed perfectly, then a feedback loop exists in the cancellation path which can degrade cancellation performance and can present stability problems.

The effects of cancellation path delay on the bandwidth performance will be compared between the feedforward and feedback approaches. The analysis and simulation results use the dynamics of a simply supported plate in air. The study described here illustrates the fundamental performance differences between the modal-based feedforward and feedback active control systems that are constrained to act causally. Comparisons of the controlled and uncontrolled vibration energy and the radiated acoustic energy will be presented.

Post-It™ brand fax transmittal memo 7671	# of pages ▸ 1
To Dr. Gareth J. Knowles	From Dr. James Thi
Co. Grumman Aerospace	Co. AT&T Bell Labs
Dept.	Phone # (703) 271-7336
Fax # (516) 575-7716	Fax # (703) 271-7676

Conference: ADPA/AIAA/ASME/SPIE Active Materials and Adaptive Structures, November 5-7, 1991 Radisson Mark Plaza Hotel, Alexandria Virginia

Presenter: Gary H. Koopmann
Director, Center for Acoustics and Vibration
Penn State, University Park, PA 16802

Abstract for paper entitled : Active Control Of Acoustic Radiation From Structures

The control strategy to actively force a structure to respond as a weak radiator to an arbitrary force field proceeds as the follows.

1. The frequency bandwidth over which the weak radiator condition is required is specified.

2. A modal frequency response is performed on the structure to determine its modal characteristics (frequency, damping, and mode shape). Note that the structure could be a passively-designed weak radiator.

3. A series of weak radiator structural modes that fall within the prescribed frequency bandwidth is computed via numerical models.

4. Motion transducers (accelerometers, piezoelectric films, embedded optical fibers, etc) and actuators (electromechanical devices) are distributed on or within the surface of the structure to measure the response of each of its dominant modes in the given frequency bandwidth to an external, arbitrary force.

4. The forces at each actuator that drive the structure as a weak radiator are determined a priori via the modal frequency response of the structure at each motion transducer with inverse matrix methods. This force vector represents the optimum force distribution for a given series of weak radiator modes.

5. When the structure is subjected to an arbitrary, external force field, its frequency response at each motion transducer is used to calculate an equivalent external force at each actuator location using inverse matrix methods.

6. To achieve the prescribed weak radiator modal response, the controller computes the difference between the measured, equivalent, external force field and the prescribed optimum force distribution, and drives the actuators such that the optimum force distribution and hence, a weak radiator condition is realized.

The technologies under development for this research effort include
1) combining computational methods of structural finite elements, acoustic superposition methods, and design optimization to address coupled, structural/acoustics problems and 2) writing adaptive control algorithms. Examples of the applications of the above design method will include active control of radiation from planar and curved structural radiators.

Structural Acoustic Control using Optical Fiber Sensors and Piezo-Electric Actuators

R. Clark and C. Fuller
Vibration and Acoustics Laboratory
Department of Mechanical Engineering

B. R. Fogg, W. V. Miller, A. M. Vengsarkar, and R. O. Claus
Fiber & Electro-Optics Research Center, Bradley Department of Electrical Engineering

Virginia Tech, Blacksburg, VA 24061 - 0111

The thrust toward quieter submarine structures and the development of smart structures for aerospace applications has placed a demand on embedded sensors and actuators. Sound radiation can be effectively sensed using conventional techniques that use microphones. The search of embeddable sensors has led us to the investigation of optical fiber sensors that will yield a sensor response similar to microphones placed in the far field of a vibrating structure.

In this paper, we will present preliminary results obtained from the use of elliptical-core, two-mode fiber sensors attached to a thin, simply-supported baffle plate. Piezoelectric patches are used as actuators. Two-mode, e-core fiber sensors operate on the principle of differential phase modulation between the LP_{01} and LP_{11}^{even} (optical) modes. As the two modes propagate through the length of the fiber, an interaction between the symmetrical LP_{01} mode and the asymmetrical LP_{11} mode leads to a spatially alternating two-lobe pattern that evolves along the longitudinal direction. Monitoring the intensity from one of the lobes in the far-field pattern results in a sinusoidal output signal that accurately reflects the vibration of the host structure. We place such sensors on the thin plate such that only some of the acoustic modes are detected by the fiber sensor. The output from the sensor is used as an error signal and fed back for active control.

We will review the operation of the two-mode fiber sensors for active structural control and evaluate the efficiency of different types of fiber sensors for similar applications.

ACTIVE ACOUSTIC ECHO REDUCTION USING PIEZOELECTRIC COATING

Xiao–Qi Bao, Vasundara V. Varadan and Vijay K. Varadan (Department of Engineering Science and Mechanics, Center for the Engineering of Electronic and Acoustic Materials, The Pennsylvania State University, University Park, PA 16802), Thomas R. Howarth (HVS Technologies, 820 N. University Dr. State College, PA 16803)

The traditional solution to the problem of reduction of sound echoes from objects has been through the use of highly damped materials which have acoustic impedance well matching the impedance of surrounding medium. In recent years, active control technique such as active noise control and active vibration control received growing attention. This technique also offers a new solution to cancel the echo from the objects. In this paper, an experimental research of active echo control is presented. A multilayer piezoelectric coating, which is able to sense the incident wave and to work as an actuator to send a wave was used in the experiment.

The active echo control testing system containing piezoelectric coating, phase shifters and a feedback circuit was developed and tested in a water filled acoustic pulse tube. The coating includes two layers of piezoelectric polymer and a layer of piezoceramic composite which are encapsulated in a host polymer. The phase shifters are applied to compensate the signals received by the polymer layers. Through the feedback circuit, the outputs of the phase shifters are fed to the composite layer to cancel the echo. The system is able to reduce the echo of normally incident acoustic wave from air backed plate. The primary results, which proved the feasibility of this technique, were reported[1]. Improvement of the system including replacement of the phase shifters by digital delay lines and new design of the coating has been done recently. In this paper, testing results of the value of the echo reduction and bandwidth of the improved system are presented in frequency range of 4 to 11 KHz. The performances of the system with various parameters of the coating and the compensator are compared and discussed.

In addition, theoretical analysis pointed that the reflected echo from a plate and the signal transmitted through the plate into backing medium are able to be canceled simultaneously by a bilaminate actuator when it is properly exited[2]. This concept is proved by an experiment done in the pulse acoustic tube. The backing medium in the experiment is water. Two layers of 1-3 type piezoceramic composite were used as the actuator. The sensors are the same type as mentioned in last paragraph. The compensator is similar but has a additional channel. 12 dB reduction of reflection and 22 dB reduction of transmitted signal were achieved at the same time by the active control technique.

[1] Howarth, T.R., Bao, X.-Q., Varadan, V.K. and Varadan, V.V., Colloque de Physique 51 (1990) C2-801

[2] Bao, X.-Q., Varadan, V.K., Varadan, V.V. and Howarth, T.R., J. Acoust. Soc. Am. 87(1990) 1350

Damage Detection in Smart Structures Using Neural Networks and Finite Element Analyses

J. N. Kudva, N. Munir and C. Marantidis
Northrop Corporation, Hawthorne, CA 90250

An important aspect of the smart structures concept is structural health monitoring. This requires detection of various types of damage to a structure and assessing the effect of the damage on structural performance. Damage detection and definition involves processing signals from sensors; assessing the effects of the damage requires appropriate structural analysis. This paper presents a new approach to detecting, defining and assessing the impact of large area damage (simulating battle damage) on a structure. The approach is based on using a neural network to deduce the damage size, location etc., from measured strain values at discrete locations. The neural network is trained using results from finite element analyses.

For a typical structural component under a given loading condition, the first step involves determining the effects of canonical damage (say a circular hole) of various sizes and at several locations using finite element analyses. The results of these analyses are represented by sets of strain values at 'L' locations as follows:

$$\text{Strain-pattern}(i,j) = (S1,S2,S3,...SL)_{ij}$$

where i and j represent damage size and location respectively and S1, S2, etc., are the corresponding strain values. The 'L' locations represent arbitrary but convenient and judiciously chosen strain sensor locations on the structure. For each loading condition, these sets of strain values are determined from M*N finite element analyses (where M is the number of locations considered and N is the number of different damage sizes considered at each of the locations). The finite element analyses are also used to determine the effect of the damage on the structure in terms of the buckling loads .

The next step involves training the neural network using the finite element results. The strain patterns are used as inputs and the damage location, size and effects of the damage as outputs to train the neural network to a desired level of accuracy. The trained network can then be used to determine the location, size and effects of any unknown damage using measured strain values (at the same locations as before) as inputs.

To validate the approach, several examples of flat, stiffened panel structures under in-plane and shear loading conditions were considered. For a typical panel with five stiffeners in both directions forming sixteen bays, canonical damage in the form of circular holes of various diameters located at each of the bays were modeled. The finite element analyses were performed using NASTRAN and a fine mesh was used to accurately model the holes. Strain values at a total of forty locations, at equal distances along the stiffeners, were evaluated in each run. A commercially available neural network program on a 386PC was used for developing the neural network. A total of 48 patterns for each load case was considered and several options on the neural network including hidden layers, functional links and normalization were investigated.

For the cases considered, all with forty input nodes, it was found that two hidden layers(with forty nodes each) and without functional linking provided the best results. The times taken to train the network were minimal; on the 386PC, in most cases, the network was trainable in less than 10 minutes to within 1% of the desired outputs. After training, the network was able to identify damage locations correctly for several examples of simulated unknown damage. In all cases damage sizes were determined within 15%. While the buckling loads could also be deduced from the network, it was found that they could be determined from the damage sizes more accurately by interpolation.

The examples studied demonstrate the basic feasibility and usefulness of the approach. The paper presents the details of these studies as well as the extension of the approach to general damage states in complex flight structures. Also discussed is a modification of the approach in which variable subsets of the input set are used to train the network.

EXPERIMENTAL DETERMINATION OF MICRO-DAMAGE AND INTERACTION MICRO-MECHANICAL STRAIN FIELDS NEAR ACTIVE AND PASSIVE INCLUSIONS EMBEDDED IN LAMINATED COMPOSITE MATERIALS

by

J. S. Sirkis, H. Singh, A. Dasgupta, and C.C. Chen
University of Maryland
Department of Mechanical Engineering
College Park, Maryland 20742
301-405-5265

Some micro-mechanical interaction issues associated with developing intelligent structures with embedded passive or active inclusions are addressed in this paper. The emphasis is on experimentally identifying possible damage and failure mechanisms caused by embedded optical fibers acting as passive inclusions within an intelligent composite structure, with some effort devoted to identifying damage mechanisms caused by embedded piezo-electric actuators acting as active inclusions. Once identified, methods to combat the failure mechanisms can be devised. Previous investigations of optical fiber sensors embedded in uniaxial tension specimens under quasi-static tensile loading have presented seemingly contradictory conclusions. High strain concentrations near the embedded optical fiber have been detected through experiment and finite element modeling, but ultimate strength tests have not indicated any structural degradation attributable to the presence of the optical fiber. The low velocity impact investigations which have been reported to date are generally in agreement that the presence of the embedded optical fiber does not induce additional failure mechanisms. Fatigue investigations have been lacking in the literature.

What has also been lacking in the mechanics related smart structures literature is a unified approach to identifying failure mechanisms, and high spatial resolution experimental methods required to provide a detailed description of the interaction mechanics leading to those failure mechanisms. The literature generally has dealt with natively coated optical fibers embedded in graphite/epoxy or Kevlar/epoxy laminates. Often such details as optical fiber diameter, coating material, or even the presence of coatings are omitted from the reported results. This paper addresses quasi-static, low velocity impact, and fatigue characteristization of graphite/epoxy composite specimens with embedded optical fibers. The specimens which are tested have embedded within them 80μm, 125μm, and 350μm diameter uncoated optical fibers, and 200μm, 250μ, and 650μm diameter coated optical fibers. The specimen lay-ups which are investigated include $[0_2/90_4/OF/90_4/0_2]$, $[0_6/OF/0_6]$, and $[\pm45_6/OF/\mp45_6]$, where OF refers to the optical fiber. In the quasi-static tests, a compressive load is applied parallel to the major axis of the lenticular resin rich zone created by the presence of the embedded optical fiber. This loading induces a delamination type of failure in the specimen and is deemed the most critical of the fiber induced failure mechanisms under quasi-static loading conditions. The low velocity impact tests reproduce earlier work by striking clamped-clamped square composite plates with embedded optical fibers, but the test matrix is expanded to include the various fiber diameters, coating materials, and specimen lay-ups. The fatigue tests in this study are motivated both by the large quasi-static strain concentrations induced by the optical fiber, and the history of the in-service structural fatigue failures common to the aerospace industry. Finally, embedded piezo-electric actuators are subjected to a cyclic voltage history to identify damage mechanisms induced by internal fatigue loading.

A high-resolution moire' interferometry method which combines standard moire interferometry with image processing based spatial homodyning techniques has been developed to measure strain fields in and around passive and active inclusions. The spatial resolution of this technique is on the order of .25 microns, and has thus lead to a wealth of information regarding the actual interaction mechanics on the micro-scale. This high resolution method has been used to quantify the interaction strain fields in 100 micron square regions surrounding structurally embedded optical fibers, optical fiber/host interfaces, and structurally embedded piezo-electric/host interfaces. Qualitative data is correlated with the qualitative damage information in an attempt to develop some damage laws which can be incorporated into both structural health monitoring and/or structural control algorithms. The discussion of the impact damage is qualitative in nature and draws into context the results already existing in the literature. The fatigue results are limited, and also qualitative in nature; however, this very new data provides insightful information into the reliability of the smart structures with structurally embedded active and passive inclusions.

Active Materials and Adaptive Structures - Meeting , Virginia, Nov 1991

Damage Assessment within Composite Material Structures with Embedded - Tailored Optical Fibers

R. M. Measures, M. LeBlanc,K. McEwen, K. Shankar and R. C. Tennyson

University of Toronto Institute for Aerospace Studies
4925 Dufferin St., Downsview, Ontario, M3H-5T6, Canada
and
The Ontario Laser and Lightwave Research Centre

Abstract

Damage assessment based on the fracture of specially damage sensitized optical fibers has been undertaken within a number of different composite material structures. These include: a full scale Kevlar/epoxy *aircraft leading edge*; fiberglass shells, and carbon/epoxy coupons. In each case we have been able to demonstrate that the same general rules apply and that arrays of tailored optical fibers can be used to assess the location and extent of *threshold impact damage*. We have also shown that such systems could also be used to map load induced growth of regions of delaminations.

The *aircraft* composite *leading edge* was instrumented with a " fiber optic damage assessment system" comprising a trilayered, embedded grid of 250 - tailored optical fibers. We shall report on the extensive series of impact tests undertaken with this system and discuss the technological fall out of this program including some of the critcal issues raised.

MICRO-DAMAGE ANALYSIS WITH EMBEDDED SENSORS IN MACRO-COMPOSITES

May 10,1991

Gregory P. Carman, John J. Lesko, Kenneth L. Reifsnider [1], Ashish Vengsarkar, Bill Miller, Brian Fogg, & Richard Claus [2]

[1] Materials Response Group, Engineering Science & Mechanics
[2] Fiber & Electro-Optics Research Center, Electrical Engineering
Virginia Polytechnic Institute & State University
Blacksburg Va., 24061

Abstract

Micro level damage events which occur in composites are extremely important issues when addressing the remaining strength and life of the material. These phenomenon include fiber fracture, matrix cracks, and fiber end effects. The ability to understand and quantify these events during the life of the composite require the use of internal sensors in conjunction with accurate micromechanical analysis. Internal sensors are a necessity to monitor local damage events which occur in composites under in service use. The data generated by the sensors can subsequently be analyzed with the appropriate micromechanical representations of the damage in an attempt to accurately derive a quantitative measure of the remaining structures life. We have developed a unique system which is capable of providing base line information on measurements made with embedded sensors in the vicinity of local damage events. This methodology is shown to provide accurate quantitative measurements of local point-wise strains in a composite, and is used to validate current micro-mechanical representations of this damage. This provides a unique capability to understand and incorporate the correct physical mechanisms in the analytical developments of micro-models. These two capabilities give the scientific community a method to understand embedded sensor technology and a technique to directly verify analytical micromechanical models. The quantitative results presented will demonstrate the relationship between internal measurements, micromechanical analysis, and laminate level strength predictions.

We have developed in our laboratories a unique test methodology to study the local stress redistribution which occurs in composites containing local anomalies (e.g. fiber fracture). This techniques involves an experimental macro-model (i.e. scaled up version) of the composite with the appropriate constituents; fiber, matrix, and interphase. With the use of Fabry-Perot fiber optic sensors and resistance strain sensors embedded in the macro-model, direct measurements of internal strains at the fiber diameter level in the composite are achieved. These measurements have been verified with an appropriate comparison to classical external measurement techniques (see Figure 1). By introducing into the composite a highly controlled internal damage (e.g.fiber fracture) of known magnitude and location accurate representation and measurements of the local strain redistribution is achieved. This provides base line data on the response exhibited by embedded sensors in the vicinity of localized damage. The tests performed demonstrate the sensor's ability, not only to measure the initiation of damage, but to also provide accurate data in the presence of the internal anomaly. We show that the fiber-optic sensor actually detects the the compressive wave generated during the development of a local damage event (i.e.fiber fracture) at locations outside the affected region. In fact, we suggest that embedded sensors should provide the capability to locate the damage event with the use of triangulation methods and to measure the extent of damage with appropriate energy methods. In addition to providing baseline data on measurements with the embedded sensors in the presence of internal anomalies, the test generates data on local strains in the damaged region which are utilized to verify and improve current analytical micromechanical representations (see Figure 2). This latter capability is a necessity to predict real time composite strength and suggest the resulting failure modes.

The methodology presented permits the ability to vary independent parameters which may influence the stress redistribution in the neighborhood of the anomaly. Some of these parameters discussed in this paper are local fiber volume fraction, fiber/matrix interphase, and variable fiber spacing. The ability to systematically change these quantities enables the researcher to analyze the effect each parameter has on stress redistribution based on direct measurements with the embedded sensors. This work could be extended to include a study of interphase coatings on the fiber optic sensors to provide an optimal response in the presence of the localized damage event.

We will present in this paper direct experimental strain measurements with embedded sensors in the vicinity of a fiber fracture. This technique will be validated with classically accepted external measurements. The results generated during the parametric studies will be utilized to validate a micromechanical model presently under development to describe the point wise stress state in the vicinity of an internal fiber fracture for a general class of materials. These predictions will than be utilized to formulate a laminate level strength prediction based on knowledge of the localized damage in a composite. This will provide a methodology to utilize internal sensors in actual composites to predict the remaining life of the system.

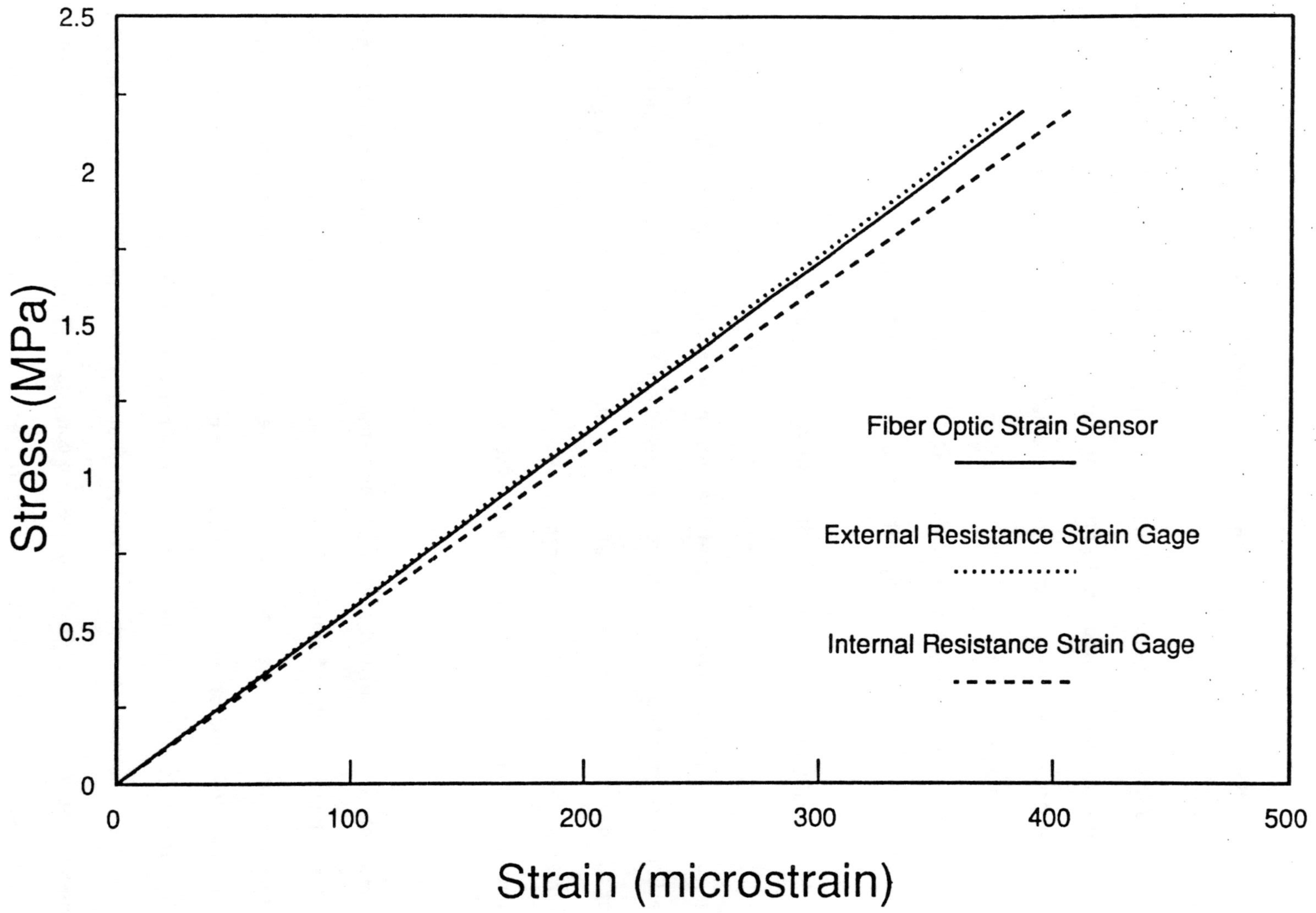

Figure 1: Illustration of the comparison between internal sensors and a classically accepted external method.

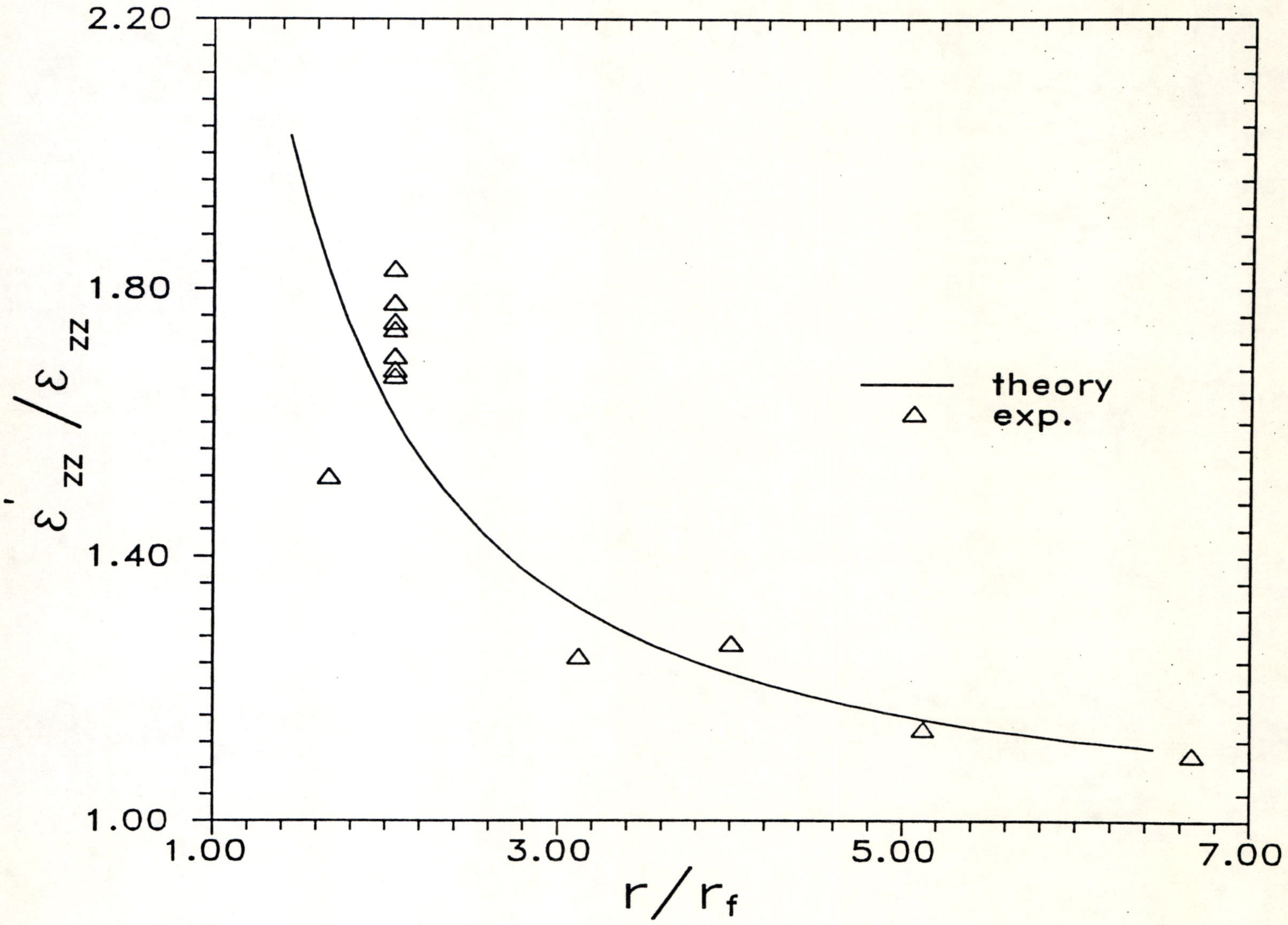

Figure 2: Figure depicting the comparison of theoretical predictions to experimental results of strain concentration versus normalized radial position for an Glass/E-poxy system.

Intelligent Sensor Systems for Smart Aerospace Structures

by Jeff Schoess
Systems & Research Center

Future advanced aerospace vehicles such as next generation aircraft, advanced launch system vehicles and spaceborne platforms will require smart embedded sensor systems and data links to monitor structural integrity and flight environment characteristics to observe and initiate corrective actions. This health management system, referred to as a *smart structure* will be capable of assessing vehicle structural damage in real-time and then ultimately reconfiguring flight controls to ensure mission performance and flight safety.

Several potential applications for smart structures include advanced aircraft such as the National Aerospace Plane (NASP), ATF, ATA and National Launch System (NLS) vehicles. These advanced aerospace vehicles will assess structural integrity prior to takeoff/launch as well as during flight. Intelligent embedded sensor systems will be used to measure in-flight environmental performance parameters and actuator performance. The major benefits are expected to be reduction in ground time and maintenance and improved performance capability.

This paper describes an innovative health management approach for smart structures which achieves the goals of high reliability, low life cycle cost, and automated vehicle checkout. This approach is to integrate smart sensor and conventional sensor technology with dedicated sensor supervisory management processors to effectively interpret and manage information provided by multiple sensors. Multisensor integration is the key to providing integrated health monitoring for high integrity systems such as smart structures. The approach supports the idea of multisensor integration both at the sensor level and condition assessment level through the implementation of key, modular building blocks such as smart sensors and sensor supervisory processors. The description of this sensor architecture approach, implementation of sensor hardware including optically powered sensors and optical sensor avionic interfaces, and potential issues for embedding avionics in aerospace structures are discussed.

RESEARCH ACTIVITIES ON ACTIVE CONTROL TECHNOLOGY OF AIRCRAFT IN JAPAN

Yuji MATSUZAKI* and Hiroshi MATSUSHITA**
Nagoya University
Chikusa, Nagoya, Japan

National Aerospace Laboratory
Mitaka, Tokyo, Japan

The research activities carried out and now going on in Japan on the aircraft active control technology are described. The theoretical and fundamental works, the experimental verification studies and the practical application efforts which have been done in every research sectors; the universities, the research institutes and the aircraft manufacturers, are summarized.

Researches in Japan on aircraft active control technology were initiated around mid 70's. Stimulated by the pioneering research works undertaken in the United States, some theoretical researches related to this topics were performed by some individual researchers in National Aerospace Laboratory; their titles being the synthesis method of active aeroelastic systems, the finite state modeling of the unsteady aerodynamics, the subcritical response identification.

Based on these theoretical studies NAL conducted the experimental verification studies on gust load alleviation and active flutter suppression using the aeroelastically simulated wind tunnel model. They could succeeded in alleviating the gust load response by 45% and increasing the flutter speed by 13%. Their studies have continued through 80's.

The fundamental research which have been done in the universities are briefly reviewed, which includes the series of basic experimental works done by the University of Tokyo on the two dimensional wing flutter suppression and the application investigation carried out by Nagoya University of the advanced control theory such as robust control to this topics.

In mid 80's Japanese aircraft manufacturers were involved in the advanced aircraft technology studies sponsored by the Society of Japanese Aerospace Companies Inc. Within that program major manufacturers conducted the active control related topics; such as the wind tunnel test study on active load alleviation by Kawasaki Heavy Industries Ltd., the conceptual design study on the high reliability active control system by Mitsubishi Heavy Industries Ltd. or the feasibility study on the hydrodynamics load alleviation of the airship by Shin Meiwa Industry Co. Ltd.

Based on these previous studies NAL and Mitsubishi are planning the extended researches. NAL plans to make flight test of their load alleviation system using their experimental aircraft, while Mitsubishi are to treat the low aspect ratio wing flutter suppression wind tunnel test.

---

* Professor, Department of Aerospace Engineering.
** Senior Researcher, Group Leader, Advanced Aircraft Research Group.

Active Stabilization of a Beam under Nonconservative Force

Junji Tani and Yuzhou Liu
Institute of Fluid Science
Tohoku University
Katahira 2-1-1, Aoba-ku
Sendai, Japan

This paper presents a study on the active stabilization of a cantilevered beam under nonconservative forces. A pair of piezoelectric devices are used as an actuator. A digital optimal regulator technique is introduced to stabilize the motion of the beam. The nonconservative forces are induced by fluid flow. The beam with rectangular cross section exposed to fluid flow gives rise to the galloping which may lead to the rupture of the structure.

The analytical model is shown schematically in Fig.1. A pair of piezoelectric devices is fixed below the clamped end of the cantilever beam. The governing equation of the beam in the uniform smooth wind stream can be obtained in mode coordinate system as follows.

$$\ddot{q}_k(t)+c\omega_k\dot{q}_k(t)+\omega_k^2 q_k(t)=b_k\theta(t) \qquad (1)$$

$$(k=1,2,....,n)$$

where $\theta(t)$ is an incline angle produced by the piezoelectric actuators as shown in Fig.1, c is a constant which depends on internal damping of the beam and velocity of the wind stream, ω_k is the natural angle frequency of the beam, $q_k(t)$ is an unknown function of time and b_k can be determined by using the natural vibration modes of the beam and the distance of two piezoelectric actuators.

In order to establish the state equations of the control system, a state variable vector is introduced and then a continuous time state equation is obtain. Further, discretizing the continuous time system with a sampling interval ΔT, the difference equations of the discrete time system are given by

$$X[(i+1)\Delta T]=AX(i\Delta T)+Bu(i\Delta T) \qquad (2)$$
$$Y(i\Delta T)=CX(i\Delta T) \qquad (3)$$

in which Y represents the displacement of the beam at the free end, which

U is called control input, that is, the displacement of the piezoelectric actuator and determined by means of the digital optional regulator theory.

The equipment used in the experiment is depicted schematically in Fig.2. The test model was cantilevered vertically in a low turbulent wind tunnel and made of aluminium (Young's modules is 63. 18Gpa and densi t_y is 7.7 X $10^3/m^3$). The length, width and thickness of the beam are 200mm, 2.4mm and 2.0mm, respectively. The experimental and theoretical values of the natural frequency of the cantilever beam used in the experiment are shown in Table 1.

When velocity of the wind was set up V_0=8.0m/s, the response of the beam displacement measured with a laser sensor are shown in Fig.3. The first vibration mode was predominant. In this figure, displacement 1mm is equivalent to the output voltage, 0.309V, and an arrow indicates the starting time of the control. It can be found that the self-excited oscillation of the beam is sufficiently suppressed and the beam is stabilized when the control is applied.

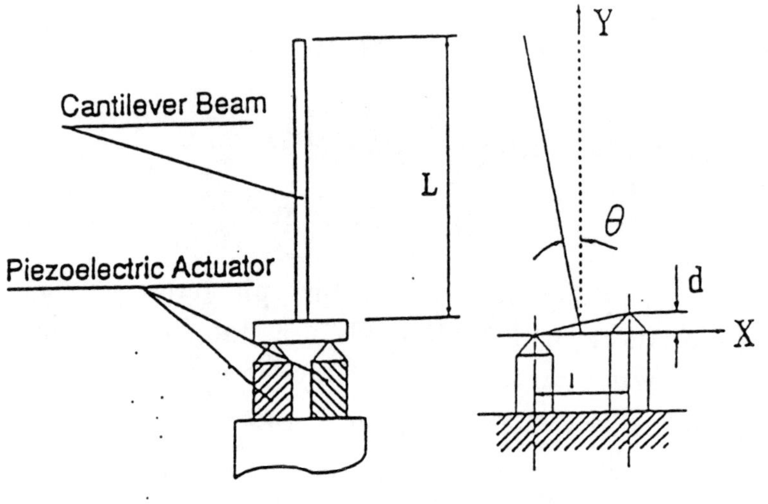

Fig.1 Analytical model and coordinate system

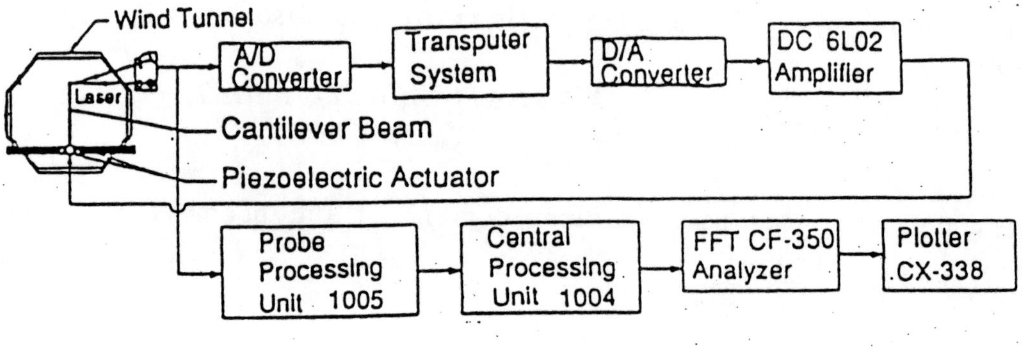

Fig.2 Schematic diagram of the experimental equipment

Table 1. Frequencies of the beam

	Frequency (unit)	
Mode	Theory	Experiment
1st	40.7 (Hz)	41.25 (Hz)
2nd	225.6 (Hz)	226.25 (Hz)

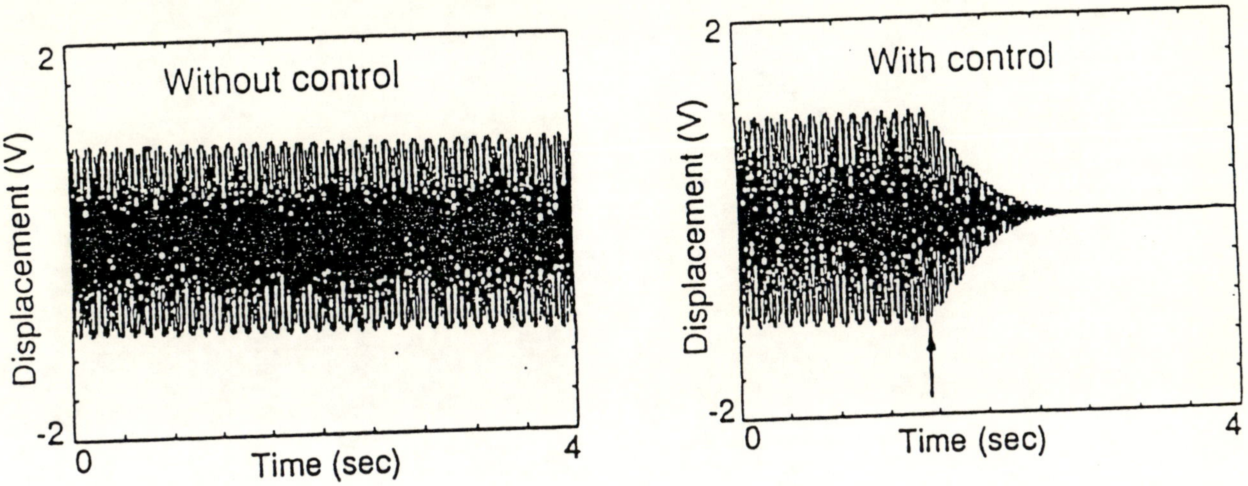

Fig.3 Effect of vibration suppressed with the optimal control

RECENT RESEARCH STATUS ON ADAPTIVE STRUCTURES IN JAPAN

Michihiro Natori
Institute of Space and Astronautical Science
Yoshinodai, Sagamihara 229, Japan

Adaptive structures are related with three different research fields such as structures, mechanism, and controls. These concepts are expected to adapt to more precise mission requirements even in an uncertain space environment. They cover necessary shape and vibration control functions for space structures. Some adjustable functions in orbit to compensate the uncertainty of ground testing for relatively large space structure systems are available from these concepts, and other active functions for in orbit identification are followed. These new concepts of structures also directly leads to a new construction scheme for large space structure systems in future.

In this paper, various aspects of adaptive structure concepts are summarized, and especially overall view of research status on this new technology in Japan is presented. Research topics and establishments in Japan on intelligent materials, actuators, shape control, vibration control, space robotics, space construction, optimization, etc. are introduced.

CONFIDENCE INTERVALS IN MODAL IDENTIFICATION USING THE ERA/DC ALGORITHM

Dong-Huei Tseng
Columbia University, New York, NY 10027

Jer-Nan Juang
NASA Langley Research Center

Richard W. Longman
Columbia University, New York, NY 10027

Abstract

Recent publications by the authors have developed methods of computing the variance and bias present in modal parameters identified by the Eigensystem Realization Algorithm (ERA) and the modified version of this algorithm based on data correlations (ERA/DC). The results allow the algorithm user to establish confidence intervals for the frequencies and damping factors of identified vibration modes in structures, based on an assumed noise level in the measurement data. The first objective of the present paper is to extend these results to consider models that include both measurement and plant noise, again establishing ways to calculate confidence intervals of modal results. In other work by the authors, direct methods were developed to identify not only the system differential equations but also simultaneously the Kalman filter gain, directly from an input-output measurement history. A second objective of this paper is to extend these results in order to establish confidence intervals of the identified modal parameters, based totally on the modal test data used in the identification, i.e. based on the experimentally determined measurement noise covariance and steady state Kalman filter gain.

STATISTICAL ESTIMATES OF IDENTIFIED MODAL PARAMETERS FOR A SCALE MODEL PRECISION TRUSS[1]

Lee D. Peterson,[2] Steven J. Bullock,[3] Scott W. Doebling[3]

Purdue University
School of Aeronautics and Astronautics
West Lafayette, Indiana 47907

To achieve high performance from an active structure, it will be necessary to obtain more than simply an accurate, nominal model of the structural dynamics. At high performance levels, the controller becomes very sensitive to small, almost imperceptable errors in the dynamic model used to synthesize the control law. To design the controller to be insensitve to these errors, experimental modal analysis must estimate the variance and bias in the identified mode shapes, frequencies, and damping ratios.

One approach to obtaining the variance and bias in the modal parameters is to propagate measurement statistics through the (nonlinear) modal identification algorithm. This requires the analysis of the sensitivity of each modal parameter to perturbations in each individual time or frequency domain data sample. Such an analysis has been done for the Eigensystem Realization Algorithm (ERA) [1,2], but little experimental experience is available with this methodology.

The purpose of our paper is to present the results of an experimental investigation which is studying this approach as a possible method for estimating the statistics of a modal model derived from experimental data. Our experimental test structure is a 1/7 scale model of an interferometer orbiting truss. This structure has a three-sided imaging plane, with an extended vertical truss which extends to the focal plane. Each leg of the structure is approximately 3.6 meters long, and is made from a repeating pyramidial truss arrangement. The high modal density of this structure makes it an appropriate test article for modal identification studies. Approximately 30 piezoelectric accelerometers are distributed throughout the structure to measure the dynamic response. Force inputs are applied through a modal shaker at several locations in the structure. In the modal analysis results reported in this paper, we have not installed any active struts, as our intention is to compare with and revise a finite element model.

Our experimental method measures impulse responses (Markov parameters) using frequency response function estimates from cross-spectra measurements. The noise in the impulse response can be estimated from the coherence of the transfer function method, but requires a modification of the IFFT algorithm used to obtain the estimates. We then obtain an identified model using the ERA and the corresponding variance and bias estimates.

Our experimental study is concentrating on several issues. One goal is to investigate how the variance and bias estimates depend on the form of the applied input force (random,

---

1 Submitted to the ADPA/AIAA/ASME/SPIE Conference on Active Materials and Adaptive Structures

2 Assistant Professor, Member AIAA
(317) 494-4782

3 National Science Foundation Fellow

impulse, periodic random, etc.), the number of averages used to compute the cross spectra, and the number of reference inputs used in the modal identification. Another objective is to study how the selection of ERA parameters affects the variance and bias in the modal parameters. Eventually, we are interested in observing how nonlinearities (intentionally introduced to the structure) affect the modal accuracy. This would prove useful in monitoring the structural health. Another eventual goal is to observe how close the variance of the ERA results approaches the Cramer-Rao information bound for the measured data.

REFERENCES

[1] Longman, R.W., Bergmann, M., and Juang, J.N. "Variance and Bias Confidence Criteria for ERA Modal Parameter Identification" AIAA Paper No. 88-4312. AIAA Astrodynamics Conference, 1988.

[2] Bergmann, M., and Longman, R.W., "Variance and Bias Computation for Enhanced System Identification" 28th Conference on Decision and Control, Tampa, December, 1989.

Approximation of Parameter Uncertainty in Weighted Least Squares Parameter Estimation Schemes - Case Study of a Truss Structure

W. R. Witkowski† and J. J. Allen
Divisions 1425 and 1545
Sandia National Laboratories
Albuquerque, NM 87815-5800

This paper investigates the approximation of second order statistics of parameters estimated using weighted least squares schemes. A process model is only as good as its estimated parameters. The ability to accurately model physical processes mathematically is vital in all areas of scientific research. In the structural engineering field, the ability to predict a system's behavior leads not only to the efficient design of new structures, but also to the optimal operation and control of existing units. Unfortunately, an exact model is usually unknown and an approximate one must be used. Model parameters are typically evaluated by fitting the model to experimental data. The procedure of formulating a mathematical model, estimating its parameters, and validating its structure is referred to as *model identification*, which involves both state and parameter estimation.

Nonlinear optimization schemes have become very popular in parameter estimation schemes. This popularity stems from the fact that the analysis is independent of both the model form and the type of experimental data available. The minimization of the summed squared error between model predictions and experimental data is commonly used as the criterion of optimality. Least squares theory states that at the optimum, linearized statistics can be

† Address all correspondence to this author

approximated for the estimated parameters - assuming that the measurement errors have zero mean, are independent, and have constant variances.

Unfortunately, these linearization schemes provide reliable confidence intervals only when the objective function is linear at the optimum. When this is not the case the approximation schemes can provide confidence intervals that are severely underestimated providing a false measure of parameter reliability. Several investigators have devised techniques to evaluate the objective function curvature and how these linearization schemes degrade.

This paper further analyzes the model identification results achieved by Allen and Martinez [1] for a truss structure. In that analysis, parameter estimates were found using integrated commercial software packages for finite element analyses (MSC/NASTRAN), mathematical programming techniques (ADS), and linear system analysis (PRO-MATLAB). The parameter estimation techniques and the software for controlling the overall system were programmed in PRO-MATLAB.

System parameters that were estimated, using eigenvalue data, included Young's modulus of the members and the stiffness value of the truss support structure. Initial investigations found an improvement from 15% to 1% in the first 8 modal frequency predictions using the improved model. Mode shapes were used for qualitative comparisons between experimental and theoretical predictions. Confidence intervals for the parameter estimates are calculated using different approximation schemes. Reliability of these approximations with respect to objective function curvature is also discussed.

References

[1] J. J. Allen, D. R. Martinez, "Techniques for Implementing Structural Model Identification Using Test Data," To appear AIAA Journal, Oct. 1991.

Abstract

EFFECT OF MODEL VERIFICATION ON THE PREDICTIVE ACCURACY OF STRUCTURAL DYNAMIC MODELS

by

Timothy K. Hasselman and Jon D. Chrostowski
Engineering Mechanics Associates, Inc.

The predictive accuracy of structural dynamic models has been evaluated on the basis of prior analysis and test experience. Modeling uncertainty derived from the difference between predicted and measured natural frequencies and mode shapes for generically similar structures is the basis for this evaluation. Generic modeling uncertainty is applied to specific structures to determine covariance matrices of both parameter uncertainty and response uncertainty. Bayesian parameter estimation uses the parameter covariance matrix to refine initial parameter estimates, thereby minimizing bias errors and reducing parameter uncertainty. The predictive accuracy of the model is improved accordingly.

Once a model has been tuned to match test data from a particular test article, it should be representative of that test article under the conditions it was tested. To a lesser degree, one would expect it to be representative of similar hardware built to the same design specifications, under similar environmental conditions. The degree of representativeness, or predictive accuracy, will depend on the degree of similarity in the hardware and its environment.

The true accuracy of a model cannot be known until after the fact, i.e. until model predictions can be compared with actual performance. These comparisons will eventually provide the basis for evaluating predictive accuracy. Until such data are available, however, other means for evaluating predictive accuracy are required. Two are presently available. One is to propagate the revised parameter covariance matrix (from Bayesian estimation) through the model to obtain a response covariance matrix. The revised parameter covariance matrix and corresponding response covariance matrix represent uncertainties in mean estimates inasmuch as they tend to zero as more and more data are used. They do not necessarily represent the true predictive accuracy of the model in the sense that comparison of model predictions and actual in-service performance would provide.

Another way to evaluate the predictive accuracy of a model is to take the statistics of comparisons between posttest models and ground test data for generically similar structures and propagate these statistics through a "verified model" to estimate a response covariance matrix. This covariance matrix may be interpreted as a sample covariance as opposed to the covariance of the mean.

This paper will present the theory and application of these concepts to practical examples.

ABSTRACT

Control Structure Optimization of Active/Passive Damping
in Large Flexible Structures

G.L. Slater
Department of Aerospace engineering
& Engineering Mechanics
University of cincinnati
Cincinnati, Ohio, 45221

(513)556-3223

for presentation at
ADPA/AIAA/ASME/SPIE Conference on
Active Materials and Adaptive Structures
Nov. 5-7, 1991

Control-structure optimization involves determination of optimal structural elements and the control system so as to simultaneously minimize the total weight of the structure while implementing constraints on structural displacements, velocities, etc, and also constraints on the control forces allowed. Many control-structure optimization approaches utilize full state feedback solutions with quadratic penalties or constraints on the controls and displacements. Such a structure leads to a parameter optimization of the structural elements coupled with a linear quadratic regulator (Riccati equation) solution. For realistic structures, this approach is too computationally intensive and impossible to implement. Recently, McLaren and Slater[1] utilized a stochastic optimization approach with quadratic penalties to solve a variety of control structure optimization problems, including direct output feedback, dynamic compensation, and the full state feedback solutions. Using this framework, it is possible to consider directly, as a control optimization problem, the issue of active versus passive damping, and to determine the optimal blend of active and passive components. See [2] for our initial approach to this area.

To determine the optimal combination of active and passive control the control optimization seeks to optimize simultaneously two classes of controllers, one a centralized controller, with or without controller dynamics, and the second, is a set of decentralized output feedback controllers. These latter controllers are in fact the passive damping component. For just a controller optimization with a fixed structure, the effect of passive damping may be to increase or decrease the active control energy requirement. For given specifications or requirements on mass, energy requirements, etc., an optimal combination of components can be defined. For example in [2], precise pointing requirements often lead to the conclusion that only small amounts of passive damping should be included, and that additional damping causes increased energy requirements from the active controller to meet mean square displacement constraints. If a total structure-control optimization is performed, then by appropriately weighting the controllers in the cost function, a minimal mass including the effect of control mass can be implemented. In addition to optimizing for a fixed controller structure, other related issues can be included. For example, for a fixed amount of amount capacity, what is the optimal distribution on the structure. similarly the effects of sensor and actuator placement on the active control problem can similarly be studied, while simultaneously determining the structural components of the members to meet performance constraints in terms of pointing and displacement constraints.

The proposed paper will present some of our numerical results applied to a tetrahedral type of truss structure. Optimal controller gains and structure parameters are obtained numerically through an optimization code. Some representative results from the references are included with the current Abstract.

References

[1] McLaren, M., Slater, G.L., "A Covariance approach to Integrated Control/Structure Optimization", Paper 90-1211, AIAA Dynamics Specialist's Conf., April, 1990.

[2] Slater, G.L., McLaren, M., "Active versus Passive Damping in Large Flexible Structures", 4th NASA Workshop on Computational Control of Flexible Aerospace Systems, Williamsburg, VA, July, 1990.

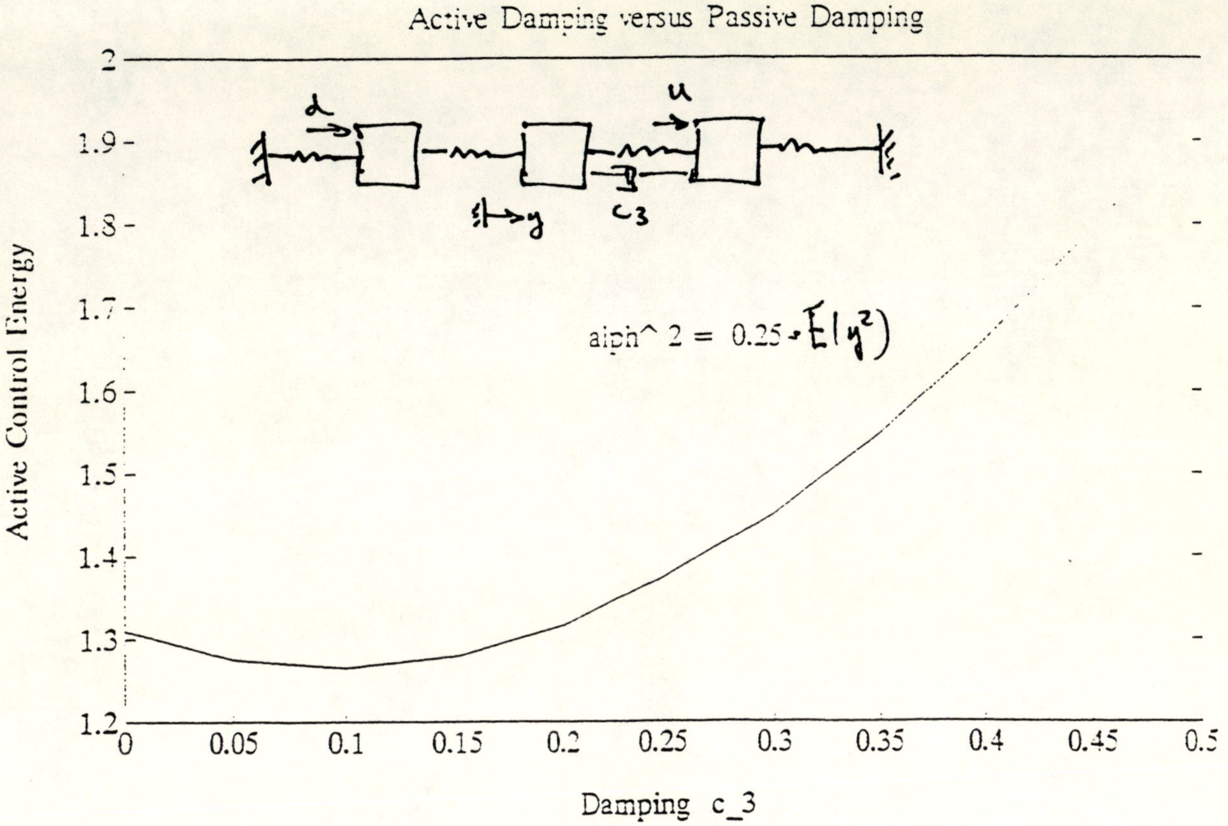

Figure 2: Active control energy required to achieve mean square response constraint

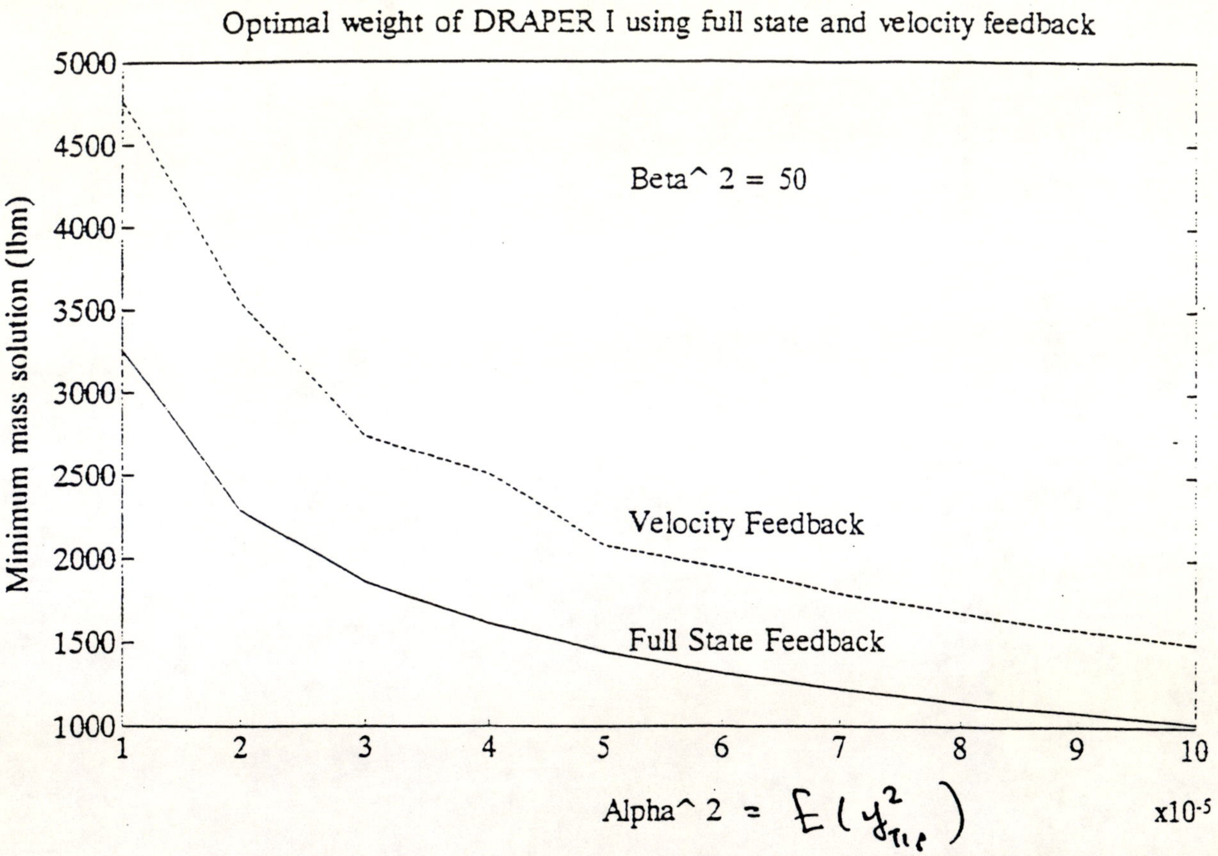

Figure 4: Comparison of optimal DRAPER I mass for full state vs. velocity feedback

Synergism of Passive Viscoelastic Damping and Active Control in High Modal Density Structures

L. Rogers, USAF/WL/FIBG - Structural Dynamics Branch
Area B - Bldg 24C - Rm 220
Wright-Patterson AFB, OH 45433-0653

ABSTRACT

Generalizations of control-structure-interaction are drawn for the case of passive viscoelastic damping and active control for challenging vibration suppression in high modal density structure. The results of several analytic studies will be summarized together with selected experimental results. Passive damping has been shown to remove large amounts of vibratory energy, lower the response of structure to excitation, stabilize an active control system, and make an active system more robust. For extremely challenging vibration suppression requirements, an optimum blend of active and passive is probably the only practical approach.

THE OPTIMAL MIX OF PASSIVE AND ACTIVE CONTROL AND ACTUATOR SELECTION

Jae H. Kim and Robert E. Skelton

School of Aeronautics and Astronautics

Purdue University

West Lafayette, IN 47907

Abstract

This paper shows how to redesign a structure to make it easier to control while satisfying multiple output RMS performance requirements. Given an initial structure which is to be redesigned, an optimal set of noisy actuators is selected to minimize the control effort subject to the output RMS inequalities. The selection procedure is an iterative algorithm composed of two parts : Output Variance Constrained(OVC) control and Input Variance Constrained(IVC) control. The initial structure and the OVC controller constitute a closed-loop system which yields the required performance. The OVC controller will not be synthesized as an active controller but regarded just as an initial controller. The controller will be synthesized into two parts, which we shall call the active and passive parts. The active part is synthesized as a feedback (static or dynamic) controller using electronic sensors and actuators mounted on the structure. On the other hand, the passive part is synthesized by changing the physical parameters of the structure (such as cross-sectional areas of members, masses, stiffnesses of elements, etc.). The structure is redesigned (this is the passive control) to minimize the active part of the control effort, while preserving all the charateristics of the original closed-loop system

Optimal Passive Damper Placement and Tuning Using Ritz Augmentation Model Reduction Method*

Cheng-Chih Chu   Mark H. Milman   Andy Kissil

MS 198-326
Jet Propulsion Laboratory
California Institute of Technology
Pasadena, CA 91109

Abstract

Structural vibration control is necessary to satisfy the stringent pointing and shape requirements for future large precision flexible structures where the vibrations could be introduced into the structure by both internal and external disturbances. It is also known that introducing passive damping is a reliable way to reduce peak responses in the vicinity of resonant frequencies. The resulting system is therefore more stable. However, the effectiveness of passive dampers is dictated not only by the locations where these dampers are placed, but also by the damper's physical parameters. This has motivated us to study the optimal placement and tuning problem.

For the passive viscous dampers, the tuning parameters are the damper stiffness (k_p) and damping rate (k_v). The objective of the tuning process is to identify the optimal values for k_p and k_v such that certain performance measures are optimized. Our approach to this problem is to treat these two tuning parameters as the gain factors for the colocated position and velocity feedback respectively. Therefore, the tuning process of passive dampers is simply to "optimally" adjust these two feedback gains given the "undamped" structural model. Based on this framework, the feedback gain for the relative velocity measurement is simply the damping rate k_v. However, the feedback gain for the relative displacement measurement is the difference between the stiffness of the undamped structural element at the location where the passive damper is placed and the damper's stiffness. This is due to the fact that the stiffness of passive dampers needs not to be the same as the structural element stiffness. As demonstrated in this study, the passive damper typically performs much better if the corresponding stiffness is low.

*Abstract for Submission to ADPA/AIAA/ASME/SPIE Conference on Active Materials and Adaptive Structures, November 5-7, 1991, Alexandria, VA

Unfortunately, the success of this approach depends heavily on the accuracy of the undamped structural model. It is unrealistic, if not impossible, to use the "full-order" model to perform such a tuning process. A reduced low order model must be used to make the computation involved much easier. In this study, the Ritz augmentation method is applied to derive a useful reduced low-order model. This reduced order model can then be used to reliably predict the performance of closed-loop (damped) system.

The Ritz augmentation method can be best described as a projection method where the projection matrix is formed by a subset of eigenvectors and certain specific Ritz vector(s). These eigenvectors typically correspond to a set of lower frequency modes and the Ritz vector(s) is generated to provide static corrections for the truncated higher frequency modes. It is noted that this method has been used to derive the reduced order model for computing the dynamic response of the open-loop (undamped) system and also has been a standard technique in component mode synthesis methods for many years (known as the static correction modes).

It is shown that the Ritz augmentation method can be used very effectively as a tool for approximating the poles of the "parameterized" family of models that naturally arises with the passive damper tuning problem, and for predicting the closed-loop (damped) system performance with very good accuracy. In addition, since the reduced order model can typically be made much smaller than the full order model, the computational aspects of the passive damper tuning and placement process become much more manageable. In particular, the Ritz augmentation method has been successfully applied to generate the specifications for the stiffness and damping rate values of Honeywell D-struts which will be used on the JPL's CSI Phase B Testbed. Both numerical and analytical results will be presented for this design specification problem.

FINITE ELEMENT METHODS FOR THE NUMERICAL SIMULATION OF THE ACTUATOR AND SENSOR PERFORMANCE OF COMPOSITE TRANSDUCERS IN THE FLUID

L. C. CHIN, V.V. VARADAN, X.Q. BAO AND V.K. VARADAN Research center for the Engineering of Electronic and Acoustic Materials and Department of Engineering Science and Mechanics, The Pennsylvania State University, University Park, PA 16802

Abstract

Piezoelectric composite transducers combine a piezoelectric ceramic and a polymer host material. This kind transducers offer several advantages in a variety of applications which have been well documented. Many composite transducers are made of very hard ceramics and most of them are shaped into very small dimensions, so that it becomes quite difficult to examine their performance during experiments. To avoid the unnecessary time-consuming and trial-and-error steps in the course of an experiments, a numerical simulation technique to predict transducer performance through a wide range of their characteristics is necessary. Also the actual performance of the transducer will vary depending on the arrangement of the piezoelectric phase,the materials used, and the frequency bandwidth desired. In addition, the structure on which the transducer is mounted and the environment in which it operates, gas or liquid, should also be considered. Since there are several variables that can affect performance, it is advantageous to have a numerical simulation scheme that can predict transducer performance for several combinations of the variables. A hybrid finite element method in conjunction with a boundary element technique to model the field radiated into water by a periodic, 2-D piezoceramic transducer array mounted on an immersed structure. The properties of the backing and matching layers as well as the host material in which the transducer elements are embedded are considered in the modeling. The normal modes of vibration of such a configuration are first studied for the actuator case, and then the transmission efficiency and the input electric admittance value are computed while a single element of the array is excited. The effect of cross talk and influence of neighboring elements is studied by taking into account the effect of neighboring elements in increasing order. The cross talk is also studied as a function of element spacing and material properties. For the sensor function of composite transducers, we give an incident acoustic wave in the fluid and the voltage excited on the electrodes of a sensor transducer can be obtained. Such performance simulation can be used in the design of transducers and transducer arrays for underwater applications, NDE and medical applications.

COMPO: CAD of Active Composite Materials

G.L. Blankenship L.G. Lebow
Techno-Sciences, Inc.
USA

A. Hassim A. Dutoya
SIMULOG, S.A.
FRANCE

Abstract

A software system, COMPO, based on finite element methods, has been developed which implements the homogenization methodology for evaluation of macroscopic effective thermal and elastic moduli of continuous and short fiber reinforced composites. Homogenization is a systematic procedure for analytical evaluation of the properties of heterogeneous materials. It permits one to estimate the precision of the approximations to the field equations of the composite. Moreover, it retains interaction effects in the macroscopic approximations due to the microstructural geometry of the composite. The homogenization method produces an effective parameter model of the macroscopic behavior of the material (conductivity, longitudinal and transverse Young's moduli, Poisson coefficients, etc.), and a description of the microscopic distributions of heat or stress within the cell – especially at the fiber-matrix interface.

COMPO computes not only the macroscopic characteristics of the composite, but also the microscopic thermal or stress and strain or force fields acting at the fiber–matrix interfaces.

COMPO can also compute the effective material properties of composite materials containing active elements – specifically imbedded piezoelectric components. We shall emphasize this feature in the presentation.

The COMPO software system is designed for easy use by engineers who need not be familiar with the underlying analytical techniques. Interaction takes place through a a graphics user interface which requires specification of the material properties of fiber and matrix and of the "geometry" of a typical cell of the composite. Both thermal and elasticity properties are treated for both continuous and short fiber reinforced materials. The system can treat composites with various fiber shapes and packing arrangements. It can also treat multi-ply laminates. COMPO supports two additional modules for the evaluation of the thermal performance of electronics support structures and for the production of input code to the ANSYS structural analysis program.

COMPO is fast and efficient. It has been ported to both workstations and 386 based PC's. It will be demonstrated in the presentation.

PIEZOELECTRIC FINITE ELEMENT FORMULATION APPLIED TO DESIGN OF SMART CONTINUA †

H. S. Tzou [1]

C. I. Tseng [2] and **H. Bahrami** [3]

[1] Department of Mechanical Engineering

[1] Center for Robotics and Manufacturing Systems
University of Kentucky
Lexington, Kentucky 40506–0046

[2] SDRC, Detroit, MI

[3] IBM, Lexington, KY 40511

ABSTRACT

Studies on smart continua with integrated sensor/actuator for structural identification and control have drawn much attention in recent years. This paper is devoted to a new piezoelectric finite element development with applications to smart continua. Hamilton's principle and variational equation are used to derive the thin piezoelectric solid element. In order to soften the transverse stiffness, three internal degrees of freedom are used, which are condensed from the system equations using Guyan's reduction scheme. Proportional feedback – constant gain – feedback control is also implemented in the finite element program. A square plate with four–pair colocated segmented distributed sensors and actuators is studied. Finite element results are closely compared with analytical solutions.

---

† Supported by NSF, Army Research Office, Kentucky EPSCoR, and CRMS.

(C/PzFEA–ADPA.PzShl3)

Reduction of Stress Concentration in a Plate with a Hole by Applied Induced Strains

Markku J. Palanterä, Pradeep K. Sensharma and Raphael T. Haftka
Department of Aerospace and Ocean Engineering
Virginia Polytechnic Institute and State University
Blacksburg, Virginia 24061

Introduction

Recently there has been much interest in adaptive structures that can respond to a varying environment by changing their properties. Piezoelectric materials and shape-memory alloys are often used to create such adaptivity. These materials can be viewed as a means to induce strains in the structure, and such induced strains can also be used to reduced stresses in regions of stress concentration. For example, Rogers (Ref. 1) has demonstrated substantial reduction in stress concentration near a crack by the use of shape-memory alloys.

The proposed paper employs optimization to effect the maximum possible reduction in stress concentration with a minimum of applied induced strains. In particular, the problem of reducing stress concetration in a plate with a hole subject to uniaxial loading is investigated. The induced strains are simulated by applying temperature change to a small region of the plate. By having a coefficient of thermal expansion which is zero in one direction it is possible to restrict the applied induced strain to a single direction. The present investigation is limited to axisymmetric induced strain distributions.

Analysis Methods

The plate shown in Figure 1 is treated analytically as an infinite plate. For this case the tangential stresses are given by

$$\sigma_\theta = \frac{S}{2}\left[1 + (\frac{A}{R})^2\right] - \frac{S}{2}\left[1 + 3(\frac{A}{R})^4\right] cos2\theta$$

with similar expressions for the other stress components. Our goal is to reduce the stress concentration as measured by the Von Mises or the maximum shear stress criteria by adding an axisymmetric induced strain field over a minimal region of the plate. Without the induced strain field, σ_θ varies from $3S$ to $-S$ around the edge of the hole. Since it is the only nonzero stress component there, the stress concentration factor is 3. An axisymmetric induced strain can lower that stress concentration no lower than 2 (when the variation will be between 2 and -2). Our goal was to effect that reduction with the smallest possible region of applied induced strains.

An analytical solution was found for the simpler problem where the stress concentration is posed in terms of the tangential stress instead of the Von Mises stress. Then that analytical solution was used as a starting point for a numerical optimization that employed a finite element model of the plate. The numerical optimization was based on sequential linear programming.

Results

The temperature distribution that reduces the tangential stress concetration factor with the minimum heated area was found to be

$$T(r) = \frac{S}{E\alpha}\left[2\frac{A}{r} + (\frac{A}{r})^4 - 2\right], \qquad r \leq 1.25A,$$

where α is the coefficient of thermal expansion. The corresponding tangential and Von Mises stresses are shown in Figure 2.

Next the numerical optimization was carried out with the finite element model. The design variables were the temperatures of five rings of increasing radii around the hole. The objective was to minimize the energy associated with the applied temperatures. The optimal applied temperatures for the three stress criteria are shown in Figure 3.

Finally, the same problem was solved with the additional requirement that only tangential induced strain can be applied. This was accomplished by setting to zero the radial coefficent of thermal expansion α_r. The optimal temperature distributions are shown in Figure 4, and are only slightly higher than the temperatures required for the isotropic expansion case.

The results obtained so far indicate that substantial reductions in stress concentration factors are possible with the application of induced strains only in the immediate vicinity of the hole. The proposed paper will present additional results for composite plates where the stress concentration factor is much higher

Reference

1. Rogers, C.A., Liang, C. and Li, S., "Active Damage Control of Hybrid Material Systems Using Induced Strain Actuator," Proceedings, AIAA/ASME/ASCE/AHS/ASC, Structures, Structural Dynamics and Materials Conference, Baltimore, MD, April 8–10, 1991.

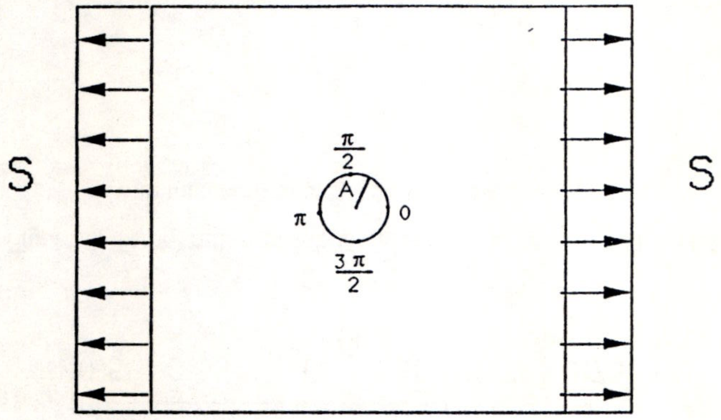

Figure 1: Plate with a hole of radius A subject to uniaxial tension

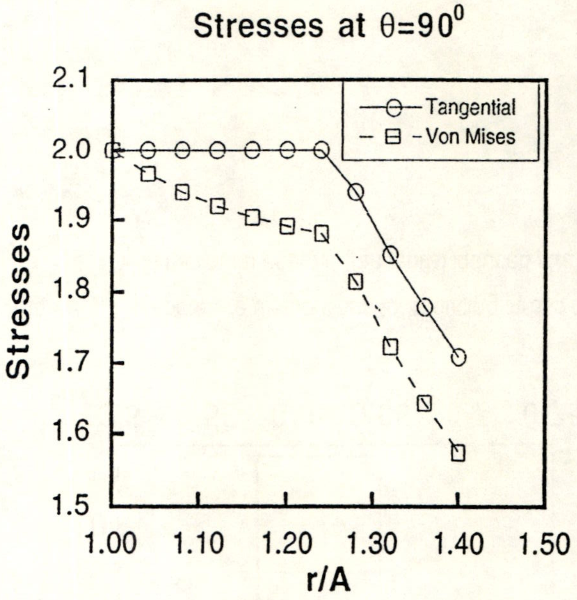

Figure 2: Stress distribution with temperatures optimized to reduce σ_θ (analytical solution).

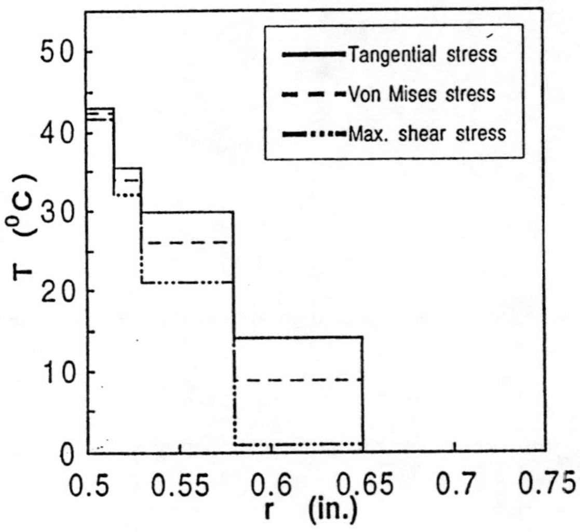

Figure 3: Temperature distribution for reducing stress concentration to 2 with minimum energy (isotropic expansion).

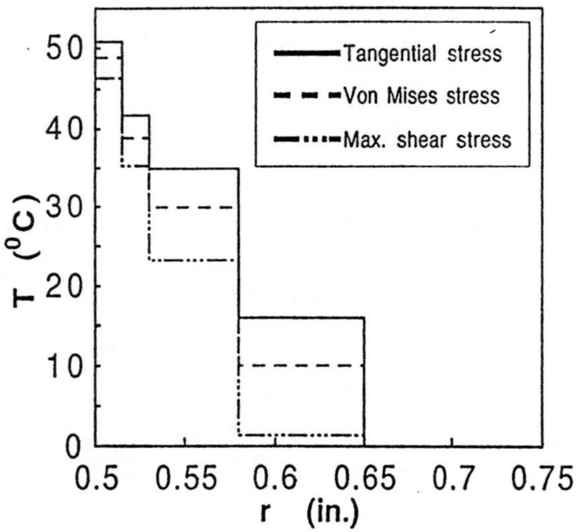

Figure 4: Temperature distributions for reducing stress concentration to 2 with minimum energy (tangential induced strain only).

U. V. Induced Length-limited Bragg Reflections Filters with Smart Structure Applications

by
Donald R. Lyons

ABSTRACT

The development of composites, fiber optics as sensor elements and communications media, and microprocessors has progressed sufficiently to permit integration of these technologies into aircraft and space structures. Thus, a smart structures system can be viewed as the result of integrating sensor technology into a workable and economic system for providing "real time" measurements of dynamic events.

Our corporate research and development areas are in the process of creating and embedding fiber optic sensors into composite structures in order to sense and possibly modify, through information feedback and processing, their mechanical and structural characteristics. These optical sensing devices have shown promise for measuring a multiplicity of local and distributed strain, stress, pressure, temperture, etc., depending how the sensor is configured during manufacture and embedment.

The photosensitivity of germanium-doped optical fiber has been known since 1980. Exposing the core of a single-mode optical fiber to an intense 488 nm laser beam results in the creation of an interference filter. This is the result of a nonlinear process which produces periodic modulation of the index of refraction as a consequence of the interaction of forward and reverse traveling waves in the Ge-doped or core region of the fiber. As an outgrowth of this idea experiments have been performed at 244 nm to simplify the fabrication of more efficient (localized) filters.

Bragg sensors were written in single and multi-mode optical fibers. They were purposefully distributed along given regions of certain optical fibers in known or referenced positions. The physics of these filters is such that they are extremely responsive to changes in their (individual) wavelengths. I.e., a change in their resonant wavelengths results in a corresponding change in their reflectivities. Therefore, if N gratings of wavelengths varying from l1 to lN are distributed along the length of a fiber and the fiber is illuminated with a multiple wavelength source then in response to strain at a given location along the fiber the reflection spectrum of the fiber is modified according to the degree of strain and the strain location and changes its intensity at that wavelength. This reflection spectrum then becomes information feedback data for active control and monitoring of the mechanical structure in question.

The experimental arrangement consists of an argon-ion laser operating at 488nm which has been intracavity frequency doubled to 244nm using an uncoated BBO crystal. The output of approximately 400mW, at 244nm, is then used to produce a U. V. interference pattern across an optical fiber. After exposure the fiber will have a sharp reflected wavelength response occuring at twice its corresponding Bragg wavelength. Any "localized" strain of sufficient magnitude will cause a frequency shift of the grating away from this resonance. This condition will be observed as a distinctive drop in the reflected light amplitude at the resonant wavelength.

The present report is a description of an imbedded strain sensor which can be used to sense distributed strain along a structure.

Optical Fiber Fabry-Perot Sensors for Smart Structures

by

C. E. Lee, Y. Yeh, W. N. Gibler, R. A. Atkins,
and H. F. Taylor
Department of Electrical Engineering
Texas A&M University
College Station, Texas 77843

Fiber optic sensors are being developed for smart structures primarily because they provide immunity from all forms of electromagnetic interference (EMI) and they are amenable to multiplexing. Numerous studies of the embedding of fiber sensors in composites and plastics have been conducted. However, the sensor designs reported in most of these studies have not provided both high sensitivity and the ability to measure the parameter of interest in a localized region.

Interferometric sensing schemes generally provide much higher sensitivity than other fiber sensors. Most of the interferometric sensing work has utilized Mach-Zehnder, Michelson, or Sagnac interferometers. These interferometers require the use of fiber couplers and (except for the Sagnac) reference fibers, which present major impediments to both miniaturization and to embedding. The Fabry-Perot configuration, in which the interferometer is formed by two mirrors arranged in series, has been shown to provide high sensitivity and "point" sensing capability.

Over the past several years, a novel fusion splicing technique for producing internal dielectric mirrors in optical fibers has been developed at Texas A&M University. Reflectances of the individual mirrors ranging from < 1% to > 85% have been demonstrated, with excess losses of only a few percent. Two of these mirrors separated by a length of single mode fiber form the cavity of a fiber Fabry-Perot interferometer (FFPI). Experiments with these interferometers have been carried out using pulsed 1.3 μm laser diode light sources, with the output signal monitored in reflection. A FFPI temperature sensor with a 1.5 mm-long cavity has been operated in air from -200°C to + 1050°C. We believe that this is still the greatest range of operating temperatures reported for any fiber optic sensor.

The internal mirrors have good mechanical properties (tensile strength ≈ 40 kpsi), which allows them to withstand the thermal and mechanical stresses experienced during the process of embedding them in a variety of materials. FFPIs produced at Texas A&M have been embedded in graphite-epoxy composites in materials laboratories at McDonnell Douglas and Stanford University. Recently, we have also succeeded in producing aluminum parts containing embedded FFPIs. In this case, the temperature sensitivity of phase shift is almost 3 times greater than for the same sensor in air because of aluminum's high thermal expansion coefficient.

The embedded FFPIs have been used for sensing ultrasonic pressure produced by a PZT transducer mounted on the sample surface. In one experiment, a 1.45 rad phase modulation index was obtained at an acoustic frequency of 1.85 MHz using a 1 cm-long interferometer embedded in aluminum. Nondestructive testing applications are forseen in which embedded FFPIs are used for the detection of ultrasonic waves in the interior of structural materials.

Signals from reflectively monitored FFPI sensors are well suited for processing by digital means under microprocessor control. A processing system which samples reflected waveforms from a sensor FFPI and a reference FFPI to produce a digital temperature readout has recently been demonstrated in our laboratory. This design can easily be extended to the processing of the signals from a number of time-division-multiplexed sensors using a single light source and photodetector.

In summary, FFPI sensors have been embedded in structural materials of interest for "smart skins" applications - graphite epoxy composites and aluminum. High sensitivity and point sensing capability for measurement of temperature, strain, and ultrasonic pressure have been demonstrated. A FFPI temperature sensing system which utilizes a digital processor with time-division multiplexing capability has been developed.

Hybrid Fiber Optic Strain Sensor Resolves Directional Ambiguity of Time Multiplexed Fabry-Perot

J. P. Andrews
Martin Marietta Aero & Naval Systems
Baltimore, Maryland 21220

Abstract

The sensor described in this paper is a hybrid design combining the air-gap Fabry-Perot interferometer with the longitudinal misalignment strain sensor. Using two techniques simultaneously resolves the directional ambiguity of the interferometer and the lead-in fiber sensitivity of the intensity sensor. Further, this design is easily time-multiplexed. The sensor uses the double reflection of an air gap splice to form a Fabry-Perot interferometer. The far end of the lead-out fiber is cleaved so that it creates a third reflection. The lead out fiber is of sufficient length so that the fiber's far end reflection is resolvable in time from the splice double-reflection using a high resolution OTDR system. Thus the optical output consists of two pulses. The amplitude of the first will oscillate through fringes as the strain changes. The amplitude of the far end reflection will decrease with increasing strain and vice versa. Thus relative strain will be measured by counting fringes and the strain direction will be determined by the direction of intensity change of the far end reflection of the lead-out fiber. The basic principle of operation is proven in a laboratory experiment and strain in a unclamped cantilever beam was measured.

ABSTRACT

Active Structural Control for Damping Augmentation and Compensation of Thermal Distortion

S. W. Sirlin
Jet Propulsion Laboratory
California Institute of Technology

At the Jet Propulsion Laboratory (JPL), we have been experimenting with the use of active structural members as part of the Control/Structure Interaction (CSI) program. The program goal is to develop and demonstrate the technology necessary for future large space structures,some of which require extremely precise dimensional stability.

One such candidate future mission is a space-based interferometer. This has been chosen as a focus of the CSI effort, and so is termed the Focus Mission Interferometer (FMI) [1]. It requires stabilization of the optical pathlength to the nanometer level. Such high performance requires a layered or hierarchical control system including augmentation of structural damping, isolation of disturbance sources, and high bandwidth compensation using articulated optical elements. This paper focuses on replacement of certain structural members with active piezoelectric members [2]. These active members have embedded force and displacement sensors, and may be used for a variety of purposes, including compensation for thermal distortion and reduction of structural vibration. This work addresses both of these issues.

Active Compensation for Thermal Distortion

As the FMI is heated by the Sun in different orientations, displacements of the structure occur on the order of 60 μm. Displacements along the optical pathlength can be compensated for at high bandwidth. Displacements transverse to the optical pathlength cannot be compensated for by the optical system however. In this case an on-board optical metrology system can be used to determine the current structure shape, and corrections at critical points using the piezoelectric elements. A control system for reduction of these transverse distortions will be presented. Practical issues such as the forces and displacements achievable by the active struts will be addressed.

Suppression of Structural Vibration

In addition to the thermal control, the active elements can be used in a variety of ways to augment the damping inherent in the structure. Two local strategies will be presented and discussed. First, simple proportional and derivative feedback for softening the element and adding damping will be considered. Next we consider the use of Dial-A-Strut control. This scheme uses sophisticated Bode design methodology for local feedback of both force and displacement in the active strut to obtain high performance and robustness to plant parameter variations. This scheme has had some success in a ground testbed [3].

Additional Vibration Control

In addition to the active struts, passive struts will be used to isolate disturbance sources (in this case the reaction wheels) from the rest of the spacecraft. The struts must allow low frequency signals to pass while stopping high frequency noise. Results using isolation will be discussed briefly.

To complete the hierarchical controls picture, high bandwidth control of the optical pathlength is done directly using a staged system. Large, voice coil actuated trolleys and small, pzt actuated mirrors are used to cover a large range of motion over a wide bandwidth.

In addition to the above controls results, the model reduction problem will be mentioned briefly as it pertains to this problem. Modal truncation and correction of the truncated model to match the true transfer function behavior at low and high frequencies will be mentioned. Simple modal truncation is shown to be inappropriate for active structural elements.

ACKNOWLEDGEMENT

This work was performed at the Jet Propulsion Laboratory, California Institute of Technology under a contract with the National Aeronautics and Space Administration.

REFERENCES

1. R. A. Laskin, A. M. San Martin, "Control/Structure System Design of a Spaceborne Optical Interferometer," AAS/AIAA Astrodynamics Specialist Conference, August 1989.

2. S. W. Sirlin, R. A. Laskin, "Sizing of Active Piezoelectric Struts for Vibration Suppression on a Space-Based Interferometer," 1st Joint U.S./Japan Conference on Adaptive Structures, November 1990.

3. J.L. Fanson, B.J. Lurie, J.F. O'Brien, C-C Chu, R.S. Smith, "System Identification and Control of the JPL Active Structure," AIAA/ASME/ASCI/AHS/ASC Structures, Structural Dynamics and Materials Conference, April 1991.

Conference on Active Materials and Adaptive Structures
Alexandria, Virginia

Inertial Decoupling in the Application of Actuators to Flexible Structures

by

Ephrahim Garcia
Assistant Professor

Smart Structures Laboratory
Department of Mechanical Engineering
Vanderbilt University
Nashville, Tennessee

Abstract

[illegible] implementation of structural control systems in future space structures. This work addresses the effects of actuator inertia on the dynamics of a flexible structure and the overall effects on the control system performance.

Often in the control of flexible structures, the open loop plant of the system is considered to be the structure, without consideration for the effects the actuator has on these dynamics. The act of adding actuators to the flexible system changes the boundary conditions for the plant and, hence, the open loop system dynamics of the structure. Two examples will be given that show that the performance of the control law can be degraded when actuators are placed on structures without regard for the issues governing this aspect of control structure interaction. The first of these is the slewing control of a flexible structure. Here the degree of interaction between the actuator and the flexible structure is determined by the effective servo inertia of the driving servo actuator with respect to the inertia of the flexible structure. The second system under consideration is the application of proof mass actuators to flexible systems. In this case it is the parasitic mass of the actuator that determines the degree to which the actuator and structure interact.

MODIFICATION OF DAMPING IN A STRUCTURE WITH COINCIDENT MODES

Steven G. Webb

Daniel J. Stech

Jeffrey S. Turcotte

M. Scott Trimboli

ABSTRACT

This paper presents the results of analytical and experimental studies of the interactions between a low-order structure and a reaction mass actuator acting as a passive vibration absorber. The structure has tunable modes; the first bending and torsion structural modes can be adjusted mechanically to vary between closely spaced and coincident configurations. For this study, the actuator was passively tuned to the structure's first resonant frequency and both uncontrolled and passively damped structural responses were compared. For the structure with separated modes the actuator significantly reduced the peak amplitude ratios of the first two structural modes. This effect was observed at all points on the structure even if the actuator and disturbance were not collocated. On the other hand, when the structure's first two modes were coincident, the actuator did not damp all points on the structure. When the actuator and disturbance were collocated, the structural vibrations were damped at all points on the structure. However, with the disturbance occurring at a point away from the actuator, the magnitude of one of the structure's first two resonant modes wer enhanced at certain points on the structure.

Shear mode piezoceramic sensors and actuators for active torsional vibration control

Chia–Chi Sung, Xiao–Qi Bao, Vasundara V. Varadan and Vijay K. Varadan (227 Hammond Bldg., Department of Engineering Science and Mechanics, Center for the Engineering of Electronic and Acoustic Materials, The Pennsylvania State University, University Park, PA 16802)

There is growing interest in application of piezoceramic sensors and actuators in active vibration control. Most published works in this area are focused in beams, trusses and plates. This paper presents a investigation of pure torsional vibration control of tubes using piezoelectric sensor/actuators.

Circular shear mode transducers were designed to be sensors/actuators of shear stress/strain exhibited on pure torsional vibration. The transducers were made of piezoelectric ceramics (PZT), which has higher electromechanical coupling coefficients than other piezoelectric materials. The 'd_{15}' constant of the piezoceramic was utilized. The coupling factors of the sensors/actuators to the resonant modes when they are mounted on the surfaces of the controlled tubes were formulized. The sensors/actuators are only sensitive and effective to the torsional vibration. The output current of the sensors is proportional to the modal velocity of the vibration. The output equivalent force of the actuators is proportional to the applied voltage. By the help of finite element approach, the effect of the sensors/actuators to the dynamic performance of the tubes such as resonant frequencies, mode shapes were also analyzed. A theoretical model to predict the performance of the sensors/actuators with a velocity feedback control system was developed.

To demonstrate the ability of the shear mode sensors/actuators, experiments were performed on circular tubes which were clamped at one end. A velocity feedback control algorithm was applied. One sensor and one actuator, which are closely located and both are near the clamped end, were utilized in the control system. The output current of the sensor is amplified and converted to voltage through a current preamplifier. Then, the signal go to a compensator which consists of a phase shifter and a filter. It is fed to the actuator through a power amplifier. Two different methods were applied to judge the control performance: one is to measure the damping factor from the waveform of free vibration, another is to compare the frequency responses of transfer function from an exciter to the sensor in the cases of the tube with and without active control. Several different cases were tested: 1) structure - with and without end mass; 2) material of the tubes - phenolic and pyrex; 3) sensor/actuator - ring type and strip type; 4) control range - fundamental vibration mode and multiple modes. The damping factor increase of 2 to 3.4 times or the response curve down from 6 to 12 dB at resonant frequencies were obtained by the active control technique. The developed theoretical model is confirmed by the experimental data.

This investigation shows the feasibility of using the piezoceramic shear mode sensors/actuators to actively control the torsional vibration of tubes.

SOME CONSIDERATIONS IN THE FABRICATION TECHNOLOGY FOR SMART STRUCTURES

C. S. CHEN
YALE UNIVERSITY

The constructional details of a smart structure, or smart composite material, can affect the detection of external stimuli. Such details can also affect the excution of the required responses. Proper design and manufacturing of the physical structure and all attached active components including sensors, actuators, and avionics are thus very important. Similarly, integration of those active components into the structure is also critical in determining the overall performance. Therefore, the associated processes for faborating the required smart structures become important issues and deserve close attention. As the smart structures are made to be more intelligent, probably more and more active components will be incorporated into the system. How to fabricate such complex structures and how to provide reliable designed functions will become very challenge tasks.

Conventional fabrication methods including integration of the prefabricated active components and physical structive can be inefficient for the more sophisticated smart structures. Some fabrication processes may even be incapable of providing the designed smart structures. Innovations in manufacturing process are needed to meet the constant challenge resulting from continuous development in smart structures.

We propose to explore a family of fabrication process for the formation of the physical structures, and their integration with active components. We propose to fabricate the smart structures in modules. Selected active components can be made by applying some decal processes. Within each module, formation of the structure and its integration with the active components are accomplished in a single molding process. Such process can produce the complex structure, and can provide the needed reliability. Furthermore, many intermediate assembly operations are eliminated which can lead to processing cost reduction. We will review the status of the three-dimensional molded packaging which the proposed processing technology is based on. We will summarize the advantages and disadvantages of various approaches used in the technology. Some modifications will be suggested to make the technology more useful to smart structures.

Fabrication of Multilayer Ceramic Actuators

A. P. Ritter, A. Bailey, F. Poppe, N. Shankar and B. Rawal* (Martin Marietta Laboratories, Baltimore, MD; * AVX Corporation, Myrtle Beach, SC)

Piezoelectric and electrostrictive actuators have application in micropositioning, vibration damping and other types of electromechanical transduction where relatively high force loads over a broad frequency range are required. Depending on a given application, material selection and actuator design can accent different types of performance. Primary issues usually include displacement; hysteresis; operating voltage and power; stability over temperature, prestress and frequency; and size and weight. None-the-less, actuators can be made using high volume, low cost automated fabrication techniques, incorporating advanced electromechanical materials and computer-based designs. These methods produce reliable devices that can be integrated with sensors, drive electronics and processors with great flexibility for "smart", active structure control.

Ceramic actuators must be made using a multilayer approach in order to produce appreciable stroke (displacement) at practical operating voltages -- electrically parallel and mechanically series. State-of-the-art ceramic materials produce field-induced strains of ~ 0.1% at 1 MV/m, however, thinly layered devices, with 150 μm individual layers, operate at only 150 V to achieve such strains. Multilayer ceramic actuators are, to first approximation, capacitors, and thus can be modeled in preliminary circuit analyses. However, unlike conventional capacitors, ceramic actuators are <u>mechanically</u>, as well as <u>electrically</u>, active. Proper actuator design must take both properties into account. Approaches to provide insulation / interconnection of alternate layers must accommodate the differential motion and resultant stresses that occurs between layers and at interfaces between actuators and other structural components. Properly designed, internal stresses in actuators are low enough to prevent degradation during repeated cycling, and differential motion (distortion) can be minimized to obtain reliable, peak performance.

The presented talk will review actuator designs, fabrication techniques (laboratory and production scale), and testing approaches relating to uniformity and longevity. The primary focus will be fabrication techniques based upon those used for the high volume, low cost production of multilayer ceramic capacitors. The tape casting process that is the basis of that approach will be reviewed, highlighting flexible manufacturing benefits -- and limitations. Actuators have been made with cross sectional areas ranging from 10's to 1000's mm^2, and incorporate 100's of active layers, and 1000's of individual sheets of ceramic. Several actuator examples will be discussed that have application in "smart systems", including small components suitable for embedded structures, large discrete devices and new multi-element devices that integrate sensor and actuator within a single monolithic structure. An assessment of current performance and cost drivers that impact practical utilization of multilayer ceramic actuators will also be discussed.

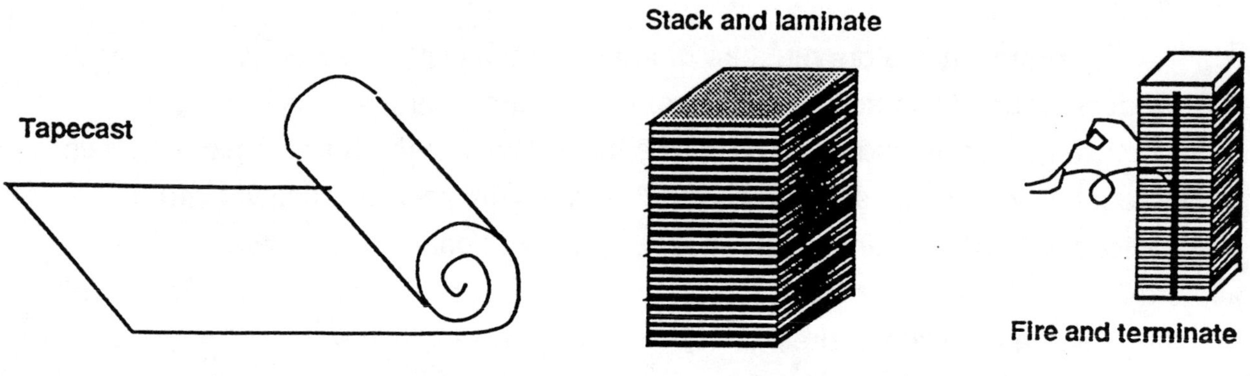

Actuator fabrication steps

Smart Structural Composites with Inherent Sensing Properties

Nisar Shaikh
Center for Material Research and Analysis
University of Nebraska-Lincoln

Carbon epoxy composites are the most important structural material for the present and future needs of the Department of Defense. The fundamental research in progress is leading to structural composites that have intelligent properties. This research is being pursued by synthesizing "sensitive" carbon filaments deposited with thin films of piezoelectric materials. The carbon fibers are intercalated to enhance their electrical and thermal conductivity. Through these synthesized fibers, features of sensing, actuation, and controllability can be imparted to composites. Smart materials with a capability to carry out desired functions as demanded by their particular applications can be developed. For example, in certain structural applications, smart materials can continuously monitor the structure's health (vibration), make decisions on the proper response, and act (actuate) accordingly. Thus, smart structural composites can themselves carry out functions that are now the aim of smart skin patches and fiber optics sensors.

The feasibility of inducing strain sensing properties by an integral thin piezoelectric film has been demonstrated [Shaikh] on a beam sample made from stainless steel strips deposited with a thin film of piezoelectric ZnO. In the past, thin film coating techniques were limited to flat surfaces. Fibers and whiskers offer new challenges. Research is in progress on sputtering carbon fibers with piezoelectric materials. The technique of sputtering fibers with ZnO was initially developed by coating Ni-Cr wires of 143 micron diameter. Some of the general problems of dealing with cylindrical piezoelectric constituents as well as electrical connections were successfully dealt with. We have been successful in making "intelligent" strands of carbon fibers coated with ZnO. These conductivity of these carbon fibers was enhanced by intercalation with bromine.

The synthesis of intelligent carbon strands progressed systematically, starting with simple stainless steel strips and eventually moving to carbon fibers through models of exceeding sophistication. A cantilever beam specimen was fabricated from each of the model constituents. Figure 1 shows cross-sections of these types of beam samples. Model 1 has ZnO coated Ni-Cr wires sandwiched between two brass strips. Model 2 has ZnO coated Ni-Cr wires embedded in Carbon-epoxy. The brass layer in this model is only for aid in handling and vibration testing. Model 3 has synthesized carbon strands dispersed in the Carbon-epoxy. Again, the brass is for handling and vibration testing. In all three of the above cases, the brass is also used for a ground electrode. Each of the beams was tested for its inherent ability to sense vibration. The test consisted of forced vibration through a shaker as well as natural vibration by impulse.

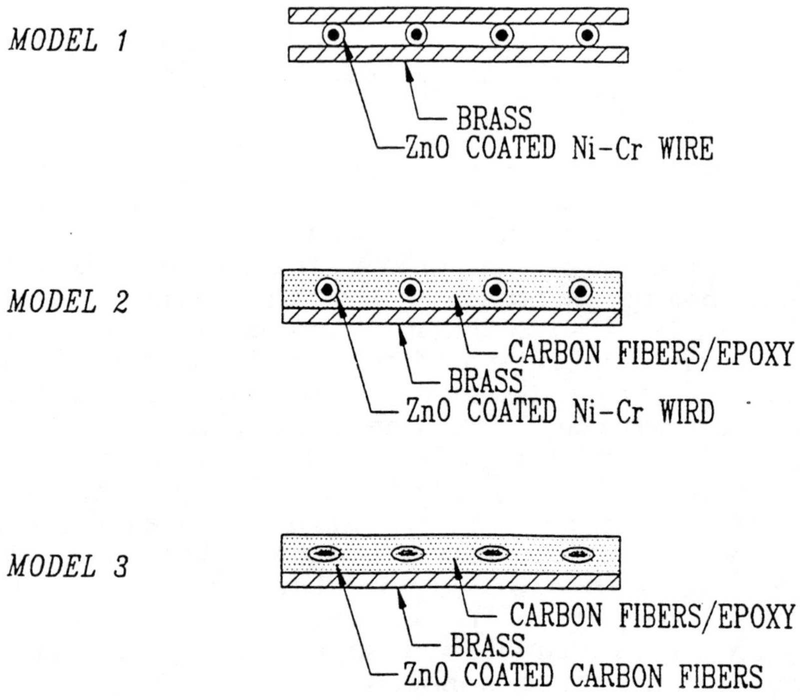

Figure 1

Figure 2a shows results from a test on a Model 1 type beam. A damped natural vibration containing both low and high frequency modes is seen. Figure 2b shows results from a forced vibration test on a Model 2 type beam. The top signal is from the strain gage and the bottom signal is the voltage signal induced in the material itself. Figure 2c shows results from a similar test on a Model 3 type beam. The top signal here is from an accelerometer mounted on the fixed end of the shaker.

Work has also started on applying the Pulsed Laser Deposition (PLD) technique for coating piezoelectric film on wires and carbon fibers. The technique was successfully demonstrated by deposition of ZnO on a flat substrate of stainless steel and silicon. The technique offers a great advantage for carbon fibers in that it permits low temperature deposition and allows for continuous fiber deposition for larger samples.

(This work is supported by the Material Science Division of the U.S. Army Research Office. Their encouragement and support is greatly appreciated.)

REFERENCE

1. Shaikh, Nisar and Dillon, Rod," Smart Structural Composites", U.S./Japan Workshop on Smart/Intelligent Materials and Systems, Hawaii, 1990, pp 287-293.

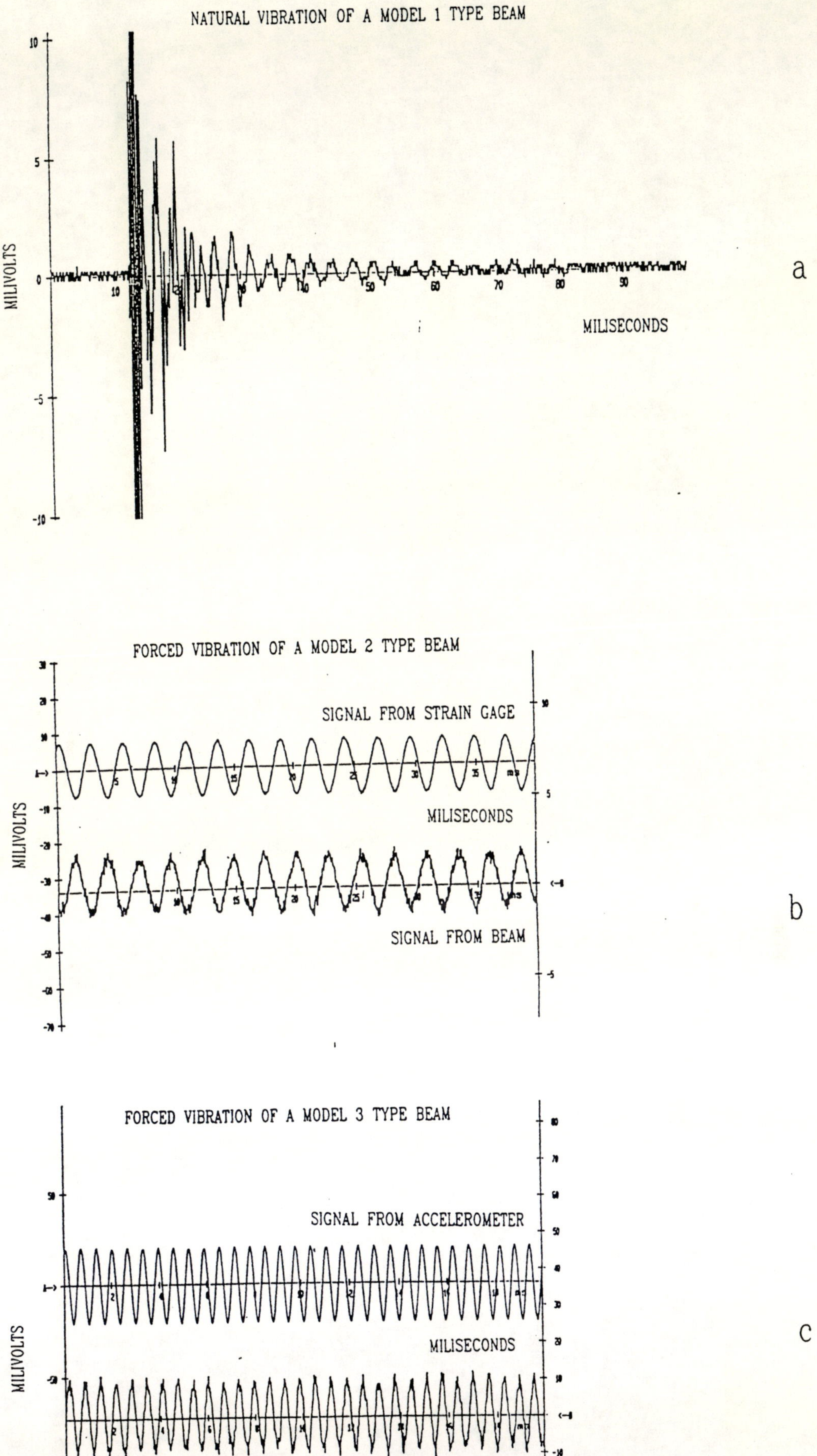

NATURAL VIBRATION OF A MODEL 1 TYPE BEAM
MILIVOLTS
MILISECONDS
a
FORCED VIBRATION OF A MODEL 2 TYPE BEAM
SIGNAL FROM STRAIN GAGE
MILIVOLTS
MILISECONDS
SIGNAL FROM BEAM
b
FORCED VIBRATION OF A MODEL 3 TYPE BEAM
SIGNAL FROM ACCELEROMETER
MILIVOLTS
MILISECONDS
SIGNAL FROM BEAM
c

Figure 2

FABRICATION AND CURING OF LAMINATES WITH MULTIPLE EMBEDDED PIEZOCERAMIC SENSORS AND ACTUATORS

Shiv P. Joshi and W.S. Chan

Center for Composite Materials
Aerospace and Mechanical Engineering Departments
University of Texas at Arlington, Arlington, Tx 76019

EXTENDED ABSTRACT

Fabrication and curing processes for advanced laminated composites have been optimized for the last two decades. Curing processes involve control of pressure and temperature over the curing time. Recent research in this area has increased our understanding of the curing processes and resulted in real time control of curing parameters (Kranbuehl et.al.,1988). Real time control of curing processes has reduced the costs and time associated with a trial and improvement procedure for thick laminated composites (Tam and Gutowski, 1988).

The fabrication and curing processes for smart composite laminates involve additional parameters arising from the placement of sensors and actuators. This paper describes various aspects of fabrication and curing processes for laminates with multiple embedded piezoceramic sensors and actuators. Graphite epoxy laminates with embedded piezoceramic wafers are discussed in detail.

Chapin and Joshi (1991)considered residual thermal stresses due to curing process in deciding the optimum placement of piezoceramics in a laminate. Shaw et.al. (1990) discussed edge stress distribution due to presence of piezoceramic layers in a laminate. Crawley and Luis (1989) briefly discussed manufacturing aspects of intelligent structures with embedded piezoceramics. The following is a comprehensive discussion of fabrication and curing methodology.

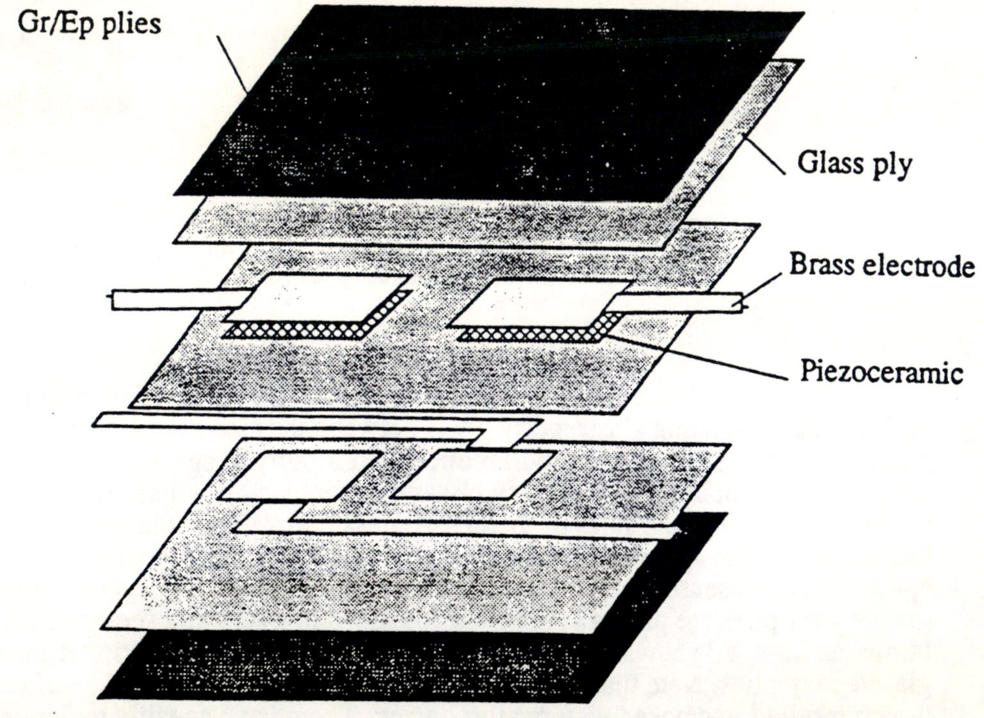

FIGURE 1. Schematic of a laminated plate lay-up with multiple piezoceramic wafers

Fabrication

We know the lay-up and geometry of the structural elements. In addition, the planer positions and through-the-thickness placement of actuators and sensors in the laminate is known. Figure 1 shows a schematic of a laminated plate lay-up with multiple piezoceramic wafers. The brass ribbon electrodes (thickness 0.001 inch) cut in a required pattern are bonded to piezoceramic wafers. Echo bond 57C conductive epoxy is used for bonding. The bonding epoxy should be applied uniformly. The brass electrodes should cover the piezoceramic area completely but should not extend over the edges. Electrodes extending over the edges may short-circuit the piezoceramic wafer. One should also be careful in applying the conductive epoxy to avoid short-circuiting. An uneven application of the conductive epoxy will result in cracked piezoceramic wafers after curing. An x-ray radiograph of cured laminates showing cracked piezoceramic wafers as a result of the uneven application of the conductive epoxy is shown in figure 2.

FIGURE 2. X-ray radiograph showing undamaged and damaged embedded piezoceramic wafers in the insulating glass/epoxy plies. The cracks initiated at the edges of thick patches of cold-solder epoxy

The brass electrode pattern is chosen to avoid overlap of reverse polarity electrodes. Although an insulating glass/epoxy ply is between the electrodes, even in the case of cracked glass/epoxy ply, such a pattern will reduce a possibility of sparking. Cut-outs of same size as the wafers are cut in the glass/epoxy ply and wafers are placed in the cut-outs with electrodes running on the opposite sides of the ply. A glass/epoxy ply is placed on each side to encapsulate piezoceramic wafers and electrodes to avoid a connection between conductive graphite/epoxy plies and sensors and actuators. These multiple sensors and actuators encapsulated in glass/epoxy plies are placed at required locations in a stacking sequence of a graphite/epoxy laminate. Figure 3 shows nine piezoceramic wafers with brass strip terminals in insulating glass/epoxy plies. Note that brass electrodes do not cover the whole surface of the wafers which resulted in cracked wafers after curing. There is no specific requirement on the width of the brass strips extending out from the wafers to the edges of the laminate.

The vacuum bagging procedure is the same as the commonly used bagging procedure. However, extra precautions should be used to avoid breaking electrode terminals extending out of the laminate.

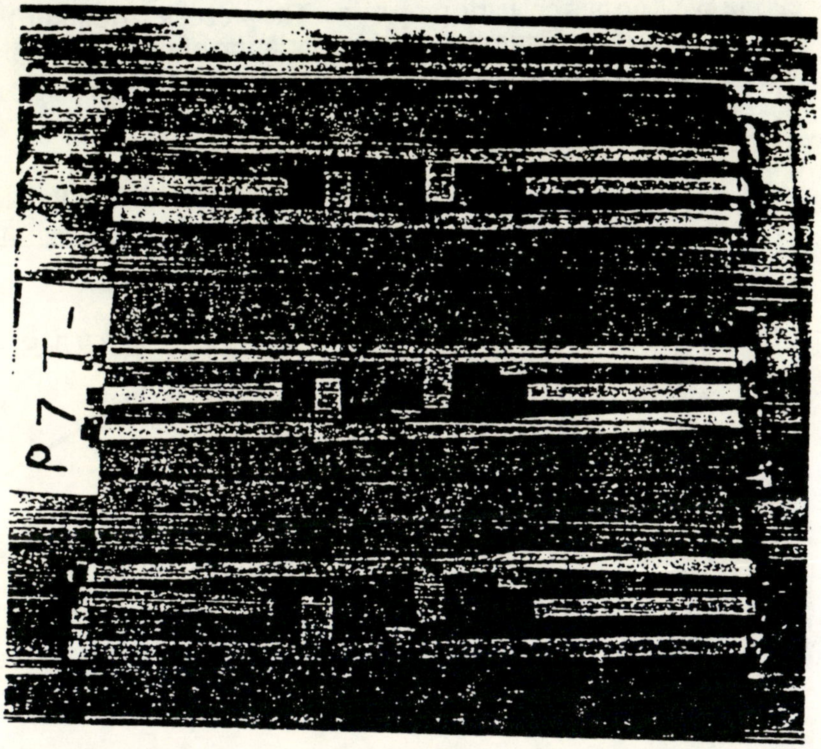

FIGURE 3. Nine piezoceramic wafers with brass terminal strips encapsulated in glass/epoxy insulating plies.

Curing Procedure

Two curing procedures are tried. The first one is a recommended curing procedure for graphite/epoxy (AS4/3501-6) laminates of moderate thickness. In our case the laminate also contains glass/epoxy and piezoceramic wafers (G-1195). The maximum temperature and pressure applied in this cycle are 350 degree F and 80 psi, respectively. The 350 degree F temperature is slightly higher than the one half of the curie temperature. Curie temperature is the temperature at which ceramic lose piezoelectric properties. Piezoelectric properties may deteriorate if the piezoceramic is subjected to temperatures higher than one half of the curie temperature. Effects of curing on piezoelectric behavior of the G-1195 wafer is discussed later in this paper.

The second curing procedure uses the recommended curing cycle for glass/epoxy laminates to separately cure glass plies containing piezoceramics. These cured plies are then placed in graphite/epoxy pre-preg lay-up and cured according to the first cycle. Second curing procedure doubles the curing time. However, it has certain advantages over the first curing procedure. The second curing procedure minimizes the piezoceramic wafer cracking and shorting between brass terminals and graphite fibers. The shorting between graphite fibers and brass terminals occurs frequently in the single cure cycle procedure because graphite fiber penetrated the epoxy rich areas of the glass/epoxy pre-preg during co-curing. Cured glass/epoxy plies with embedded piezoceramics can be screened for cracks in piezoceramics and electric connections before curing them with graphite/epoxy lay-up. This results in a higher percentage of undefective specimens.

The two cycle curing procedure is now routinely used in preparing test specimens at the University of Texas at Arlington laboratory facilities.

Effect of curing on piezoceramic wafers

Piezoceramic wafers are subjected to the curing cycle's pressure and temperature without embedding them in a laminate. These wafers are tested to evaluate the effect of the curing cycle on piezoelectric properties. The piezoelectric strain coefficient (d_{31}) is compared with virgin wafers. Figure 4 shows the variation of the in-plane normal strain induced by the electric field applied in the thickness direction of the wafer.

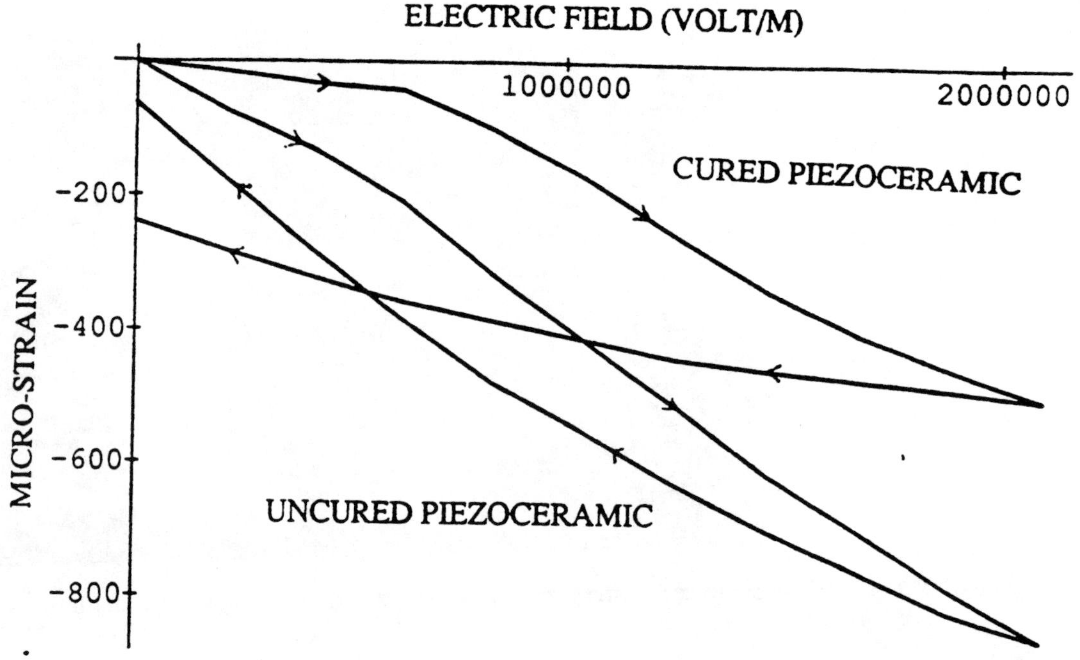

FIGURE 4. Variation of the in-plane normal strain due to application of electric field in the thickness direction

The strain data is recorded after they stabilize over time. Strain readings are stable for virgin wafers. However, the wafers subjected to the cure cycle become stable after a minute or so. It is clear that the curing cycle deteriorates piezoelectric properties. Detailed evaluation of the piezoelectric properties are performed at present and will be included in a full length paper.

Acknowledgments

The authors acknowledge the help from students D. Shaw, S. Subramanian, A. Howard, V. Jacklin, C. Bauer, J. Stephens and R. Lawrence. The research is sponsored by the Army Research Office. Dr. G. Anderson is the project monitor.

REFERENCES

1. Kranbuehl, D.E., Hoff, M., Eichinger, D.A., Loos, A.C., Freedman, W.T., Jr., "Monitoring cure of Composite Resins Using Frequency Dependent Electromagnetic Sensing Techniques", Proceedings of the American Society for Composites, Third Technical Conference, Technomic Publishing Co., Inc., 1988, pp. 313-323.

2. Tam, A.S. and Gutowski, T.G., "Application of Sliding Mode Control to the Cure of a Thermoset Composite", Proceedings of the American Society for Composites, Third Technical Conference, Technomic Publishing Co., Inc., 1988, pp. 255-262.

3. Chapin, C.M. and Joshi, S.P., "Variation of Residual Stresses in Graphite/Epoxy Laminates", ICCM VIII, Hawaii, July 14-18,1991.

4. Shah, D.K., Chan, W.S., Joshi, S.P., and Subramanian, S., "Analysis of Laminates with Embedded Piezoelectric Layers", Recent Developments in Composite Materials Structures, Ed. D. Hui and C.T. Sun, AD-Vol. 19, AMD-Vol. 113, The American Society of Mechanical Engineers, 1990, pp. 19-24.

5. Crawley, E.F. and Anderson, E.H., "Detailed Models of Piezoceramic Actuation of Beams", American Institute of Aeronautics and Astronautics, Inc., 1989, pp. 2000-2010.

Experiments on active vibration control of a thin plate using disc type piezoceramic sensors and actuators.

Authors : Suk-Yoon Hong, Vasundara V. Varadan and Vijay K. Varadan
Penn State University, University Park, PA 16802

For the vibration control of two dimensional structure such as thin plate clamped along all sides, piezoelectric ceramics, PZT (Lead-Titanate-Zirconate) was chosen as sensor, actuator and vibrators because of its large electro-mechanical coupling constant. A series of active vibration control experiments have been performed for the lower several modes with different type of control algorithms for single mode and multi-mode. Although the system structure is stiff enough because of the clamped edges and the material characteristics, we could implement effective vibration control with optimally selected positions and sizes of sensors and actuators.

Coupled mode multi-mode optimal control procedures for the plate have been developed. A general closed-form of mode shape do not exist for the rectangular plate. Therefore, we have used Rayleigh-Ritz procedure to obtain the approximated mode shape constants. Related with modal equation, proportional type damping has been considered to obtain a closed-form solution of the differential eigenvalue problem for the damping systems. Sensing signal conversions have been performed to change curvature signals to displacement signals. Every sensed signal from PZT give us the information for curvatures but, to implement optimal control with displacement sensors, we need the displacement values. In case of single mode, the conversion is straight forward but in multi-mode case, only limited points on the plate satisfy the conversion requirements. Those points could be found using numerical calculation and could be decided as optimum sensor points. Actuating signal also should be converted. PZT actuators can be converted to point force actuator which generate same deflections as those of partially distributed actuator. To do these conversion, finite element method package program, ANSYS has been used considering equivalent loading for the applied PZT moments.

In developing multi-mode velocity feedback control system, we selected specific sensing modes cancellation idea which seems to be pure filters by adding or subtracting each concerned signal and have believed that this idea could improve the control performance in analog multi-mode control.

In this research, the idea for uni-disc type colocated sensors and actuators was proposed and the performance was verified experimentally. The theoretical and experimental work on transduces sizes and positions has been studied to obtain the idea for initial selection of PZT. To take the advantages of digitalized system such as the flexibility of control program and increased logic capability and to simplify experiment system especially for the multi-mode control, digital control system composed of a microcomputer and a data acquisition board has been used.

Robust Performance of an Active Damping System

T.R. Alt, J.T. Harduvel
McDonnell Douglas Space Systems Company
Huntington Beach, California

Abstract

This paper evaluates robust controller synthesis techniques for designing an active damping controller. Five different techniques are used to design controllers for a large space structure. Two of these techniques were then used to design controllers for an experimental truss structure. The performance and stability robustness of these controllers was validated on the experimental truss.

A robust active damping system provides damping for a subset of bending modes in the presence of plant uncertainties. Five controller synthesis techniques were used to design controllers for a dual keel space station. All designs used the same five ideal, colocated linear rate sensors and linear proof-mass actuators. The design techniques were: 1) Low Authority Control (LAC), 2) Positivity (POS), 3) Model Error Sensitivity Suppression (MESS), 4) Maximum Entropy/Optimal Projection (ME/OP), and 5) H-infinity/Mu-synthesis (Mu-Syn). The design goal was to maintain stability of all 30 modes in the evaluation model and at least 5 percent of critical damping on 10 targeted modes in the presence of a 30 percent uniform modal frequency variation. Initially the Mu-Syn controller was the only controller able to meet the robust performance criteria. The insight gained through the Mu-Syn design process motivated a redesign of the other dynamic controllers. The redesigned controllers then achieved nearly the same performance robustness as the Mu-Syn controller.

The H-infinity/ Mu-Synthesis and Low Authority Control techniques were used to design a robust active damping controller for an experimental truss structure. The goal of the controller was to provide at least 5% percent of critical damping (structural damping was 0.22%) for the first two bending modes in the presence of a ±30% uniform modal frequency uncertainty. Robustness to this uncertainty was demonstrated by frequency scaling the controller as though it were designed for a different set of modal frequencies. Frequency responses of the system were performed for the nominal and ±30% frequency scaled controllers. All cases were stable verifying robust stability. The damping ratios of modes 1 and 2, calculated from the frequency responses, were greater than 5% for all cases, verifying nominal and robust performance.

Vibration Characteristics of a Composite Beam with Semi-Active Piezo-Actuators

Sung J. Kim and James D. Jones

Ray W. Herrick Laboratories
School of Mechanical Engineering
Purdue University, West Lafayette, IN 47907
Tel. (317) 494-2146, Fax. (317) 494-0787

ABSTRACT

Low frequency structural noise and vibration is a persistent problem in a variety of lightweight flexible structures such as aircraft, rotorcraft, space structures and automobiles. One promising method of reduction is active noise and vibration control using piezo-actuators. Piezo-actuators have attracted significant attention in recent years because of their distributed character of actuation which allows them to be tailored to selectively reduce structural modes with little control spillover. Piezo-actuators also have several other inherent advantages over conventional actuators including being inexpensive, space efficient, lightweight and easily shaped and bonded-to (or embeded-in) a variety surfaces. Furthermore, in contrast to point actuators, piezo-actuators do not require an inertial back reaction.

Early studies demonstrated the use of piezo-actuators as active dampers in reducing the free-vibration decay time of one-dimensional beams with little spillover to higher order modes of the beam. More recently, piezoelectric passive dampers have been widely investigated, which transform the vibrational energy into dissipating heat energy with piezoelectric materials and resistive electronic components.

Since the active dampers require expensive and complicated control devices and the passive dampers do not provide enough damping, the alternative uses of piezoelectric materials are worth investigating. One promising method is changing the stiffness of the composite structure by applying voltages to the embeded piezoelectric materials. This process is called 'semi-active control'. Shape-memory alloy (SMA) and Electrorheological (ER) fluids have been used for semi-active actuators for composite structures. Activating the embeded SMA fibers in composite materials changes overall stiffness of the SMA hybrid composite structure and consequently changes natural frequencies and mode shapes.

Applying voltages to the ER fluids in hybrid composite result in similar changes in the dynamic characteristics of the composite structure.

In the current work, a semi-active piezo-actuator which changes the characteristics of vibration of a composite beam is investigated. The equations of motion of the composite beam which can incorporate the influence of the initial stresses given by the piezoelectric layers is developed. It is shown that natural frequencies of a composite beam will vary substantially with the voltage applied to the piezoelectric layers. The effects of configuration of the composite beam on the performance of the semi-active piezo-actuators are discussed. Two practical piezoelectric materials, piezo-ceramics and polyvinylidene fluoride, are also discussed. It is shown that the vibration characteristics of the composite beam with semi-active piezo-actuators change significantly as the length of the beam increases.

ADAPTIVE PIEZOELECTRIC SHELL STRUCTURES: THEORY AND EXPERIMENTS†

H. S. Tzou [1] and J. P. Zhong [2]

[1] Department of Mechanical Engineering

[1] Center for Robotics and Manufacturing Systems
University of Kentucky
Lexington, KY 40506–0046

[2] Conmec Inc.
Allentown, PA 18103

ABSTRACT

Active "smart" space and mechanical structures with adaptive dynamic characteristics have long been interested in a variety of high–performance systems, e.g., flexible space structures, flexible robots, "smart" machines, etc. In this paper, an active adaptive structure made of piezoelectric materials is proposed and evaluated. Electromechanical equations of motion and generalized boundary conditions of a generic piezoelectric shell subjected to mechanical and electrical excitations are derived using Hamilton's principle and the linear piezoelectric theory. The structural adaptivity is achieved by a voltage feedback (open or closed loops) utilizing the converse piezoelectric effect. Applications of the theory is demonstrated in a bimorph beam case and a cylindrical shell case. Frequency manipulation of the bimorph beam is studied theoretically and experimentally. Damping control of the cylindrical shell via in–plane membrane forces is also investigated.

---

† Supported by NSF, Army Research Office, and Kentucky EPSCoR.

ENGINEERING OF AN ULTRASTABLE STRUCTURE

by

T. C.Thompson, M. T.Gamble, and J. A.Hanlon

Ultrastable structures, based upon controlled structures concepts, have been developed by the Los Alamos National Laboratory Mechanical Engineering and Electronics Division. A previously designed, deployed, and tested prototype structure, and a new-generation space-qualifiable device evolved from the prototype, are addressed herein. Both systems combine inherent passive stability with a network of motion and acceleration sensors, coupled in a feedback architecture, with force transducers for actively controlling system mechanical stability. The prototype demonstrated precision of >1 μ radian over its 2.5 m length. The current system, based upon the most successful aspects of the prototype and on new, more robust design concepts, is expected to exceed it precursor's performance. The design, deployment, and analysis of these systems, will be discussed detail.

Both systems were designed using advanced composite materials that possess excellent vibration damping, engineered coefficients of thermal expansion (CTE), and high stiffness-to-weight ratios. The systems were fitted with subsystems for stability interrogation and enhancement, including passive vibration attenuation; a 7-beam, heterodyne laser interferometry system configured as a light truss; and a distributed network of accelerometers. The laser metrology system is capable of transforming relative structural motions into spatial degrees of freedom which are introduced into high-authority, fast-feedback piezoelectric force transducers (PZTs) and slow-feedback precision stepping motors as an active control mechanism.

An environmentally stable chamber encompasses the system, allowing accurate performance estimates of critical components and a controlled atmosphere for system operation. Operational environment factors and passive stability were addressed in the design of very low CTE superinvar optical mounts for the metrology system. The current space-qualifiable system will benefit from the wealth of information on the performance of space-based graphite/epoxy composite structures, although more study must be done on environmental influences such as ionizing radiation, vacuum outgassing, and thermal transients in space. The methods applied to these concerns are outlined.

The metrology system for the new detector represents a simplification over the prototype. The metrology system must accurately resolve motion as small as 0.1 μ. The former system transformed component relative motions into 6 spatial degrees of freedom, of which one rotational and two translational degrees were articulated to stabilize the system. The current design will measure only three degrees of freedom with position-sensing detectors and fiber-light sources, enabling a more straight forward quantification of relative motions. This simplified approach will also provide a weight savings for this space-based device. The experiment that will be performed to benchmark the accuracy and bandwidth of this system will provide the most important real-time distortion information used for stability feedback.

Finite element techniques were used to optimize the design of critical components. The space-frame structural members of the current device were engineered to exhibit minimal thermoelastic extension and out-of-plane warping, while demonstrating excellent bending and torsional stiffness. Large, rectangular platens, constructed of Nomex (TM) honeycomb cores with graphite/epoxy facings, were optimized for quasi-isotropic inplane behavior, high bending stiffness, and minimal net CTE. Fundamental natural vibration frequencies and vibration response estimates were obtained for all critical components and assemblies. The engineering tradeoff analyses performed initially were directed toward comparing the costs versus benefits of monolithic materials, such as aluminum and beryllium, to advanced composites. The paper will address the structural and thermal issues associated with designing an ultrastable device.

NANOMETER LEVEL OPTICAL CONTROL ON THE JPL PHASE B TESTBED

by
John T. Spanos and Michael C. O'Neal

Many future space missions will require major advances in the areas of controlling, aligning, and pointing optical instruments mounted on large flexible structures. One of the most challenging applications is optical pathlength control for stellar interferometry[1]. To meet this challenge, the Jet Propulsion Laboratory, in conjunction with a NASA-wide Control Structure Interaction (CSI) program, has developed the Phase B Testbed[2] to explore, develop, and validate emerging technologies and design methodologies. This paper will describe the optical control design methodology and the results achieved.

The experiments described in this paper focus on direct control of the Testbed's optical elements. The objective is to reject disturbances and maintain pathlength control to better than 5 nanometers peak to peak, since this is the level deemed necessary by flight systems[1]. The optical motion compensation system is comprised of three actuators with different strokes and resolutions. The control system designs presented in this paper are based on models created from direct measurements of transfer functions from actuator inputs to changes in optical pathlength. Classical frequency domain loop shaping techniques were used to synthesize multiloop controllers which stabilize the optical pathlength below the 5 nanometer requirement. Once this control objective was attained, the optical configuration was altered to couple more structural motion into the pathlength, creating a more realistic representation of a stellar interferometer. In addition to the environmental disturbances present in the laboratory, representative space disturbances, such as from reaction wheels or tape recorders, were also applied to the structure.

The key to the success of the designs was the identification of transfer function models from frequency response measurements. The measurements were obtained with band limited white noise excitation, windowing, and spectral averaging. Optimal curve fitting algorithms[3] were then employed to obtain the transfer function models from the measured data. In addition to control system design, these models were used in simulation studies for performance assessment both in the linear and nonlinear modes of operation. Details of the technical approaches used and results obtained from the laboratory experiments are provided. A comparison of the different methods and a discussion of the lessons learned from the experience are included as well.

---

[1]Laskin, R.A. and San Martin, M. "Control/Structure System Design of a Spaceborne Optical Interferometer", AAS/AIAA Astrodynamics Specialist Conference, Stowe, Vt., August 1989.

[2]O'Neal, M., Eldred, D., Liu, D., and Redding, D., "Experimental Verification of Nanometer Level Optical Pathlength Control on a Flexible Structure", 14th AAS Rocky Mountain Guidance and Control Conference, Keystone, CO, Feb. 1991.

[3]Spanos, J.T., "Algorithms for l_2 and l_∞ Transfer Function Curve Fitting", AIAA Guidance, Navigation, and Control Conference, New Orleans, LA. August 1991.

Vibration Isolation For Micro-Gravity Applications

Y.T. Chung and J.J. Tracy[1]
McDonnell Douglas Space Systems Company

A. H. von Flotow
Massachusetts Institute of Technology

A study by the Space Station Freedom Micro-gravity Committee showed that the dynamic responses to un-isolated disturbances for the Space Station Assembly Complete (AC) configuration exceeded the allowable levels by as much as 200 times at low frequencies. The micro-gravity environment required for material processing, life science research, and to maintain payload pointing precision may not be met unless an effective mechanism that can reduce the vibration level is developed. This paper describes the development of an actively controlled micro-gravity vibration isolation mount which provides broadband isolation using piezoelectric actuators. The goal of the isolation system will be to provide at least a factor of 200 reduction in the acceleration levels experienced by a typical payload subjected to a realistic disturbance environment.

We have selected a soft mount isolation concept based on the review of the anticipated disturbance environment, isolation and pointing requirements, and past and potential hardware and software approaches to the vibration isolation problem. To determine the values of the isolator design parameters for the micro-gravity environment, modal transient analyses were performed using a Space Station Assembly Complete finite element model. A simple mass-spring-damper model representing the isolation system was used in the parametric study to narrow down the range of the design parameters. An isolator frequency of 0.1 Hz was determined to best meet the micro-gravity requirement for the most severe disturbance, a 196 lb crew member jogging at 6 mph. Since the space station will experience random disturbances as well as harmonic excitations, a tuned passive isolation system will not be adequate to maintain a micro-gravity environment at all times. This leads to the design of an active control system. Based on the analytically predicted accelerations and the micro-gravity requirement, a disturbance rejection profile was defined. An initial piezoelectric actuator was designed to meet this rejection profile. . A heuristic

[1] Please address all correspondence to J. Tracy at MDSSC, 5301 Bolsa Ave, Huntington Beach, CA 92647 or phone (714) 896-5169.

comparison between a piezo-polymer film actuator approach and a magnetic field based approach is also presented.

Through M.I.T, we have developed and built an actuator made of piezoelectric PVDF film and established gluing and electrical connection techniques for fabricating the actuators. We have verified the actuator gross operational parameters such as stiffness, deflection, and internal dynamics. A prototype mount assembly, controller, and experiment configurations for verifying and validating the design of the piezoelectric actuators have been developed.

A three degree of freedom experiment was designed and assembled for an experimental verification to validate the isolator performance by measuring the transmissibility of a system with piezoelectric actuators subjected to simulated Space Station excitations. Because of the presence of 1-g, the motion of the experiment was measured in the horizontal plane to minimize the effect of the gravity bias. The test setup consisted of an inner box representing a micro-gravity processing facility, an outer box representing the Space Station, two piezoelectric actuators in each translational axis, and three Ling model 420 shakers that provided in-plane translational and rotational disturbances. The actuators provide the only load path between the inner box and outer box. Three Sundstrand QA-1400 accelerometers were mounted on the top of the inner box for measuring the translational and rotational displacements.

The transfer functions measured from the open loop experiment indicate that a closed loop actuator system is required to achieve the micro-gravity environment. A control law which produces a control force based on the acceleration of the payload was developed and incorporated with the laboratory test set up. A series of tests for both single-axis and multi-axis excitations were performed to verify the active control actuator design and validate its effectiveness for the micro-gravity environment.

ABSTRACT

The Dial-a-Strut Controller for Structural Damping

B. Lurie, J. O'Brien, S. Sirlin, J. Fanson

Jet Propulsion Laboratory
California Institute of Technology

At the Jet Propulsion Laboratory (JPL), we have been experimenting with the use of active structural members as part of the Control/Structure Interaction (CSI) program. The program goal is to develop and demonstrate the technology necessary for future large space structures, such as interferometers, which require stabilization of an optical pathlength to the nanometer level. Such precise control will require an hierarchical approach including suppression of structural vibration as well as high bandwidth articulation of optical elements.

A testbed has been developed on the ground (the JPL CSI Phase B testbed) consisting of a 3m flexible truss structure with active structural members as well as articulating optical elements for control of a laser pathlength. Certain passive structural truss members on the testbed have been replaced with active piezoelectric members. These active members have embedded force and displacement sensors, and may be used for a variety of purposes in a flight system, including compensation for thermal distortion as well as reduction of structural vibration. This work addresses a robust local feedback design that increases the damping in the structure, and hence reduces vibration levels. It is a continuation of the work presented in [1].

The Dial-a-Strut controller is designed to emulate an elastically coupled viscous damper over the frequency band of interest while maintaining a high stiffness at low frequency. The frequency range, damping and stiffness parameters are tuneable. The installation procedure of the controller consists of one structural impedance measurement, some simple algebraic calculations and then "dialing in" three resulting parameters that govern the controller behavior. These parameters are physically changed by moving a dial that changes certain capacitors and resistors in the circuit. In the experiment described, 8 piezoelectric active members with Dial-a-Strut loops closed around them will be embedded in the testbed. Addition of Dial-a-Strut feedback produces a significant increase in the structural damping.

The control uses both position and force sensors in a multiloop feedback design. The feedback loops of the controllers were designed specifically using Bode design methodology to have sufficiently wide stability margins to make them applicable to almost any flexible structure. In addition, the tuning of the Dial-a-Strut parameters is simple since the resulting performance is not highly sensitive to these parameters. In the experiment described, the system will be globally stable.

The paper will present experimental frequency responses and comparison to theoretical predictions. The simulations will use a Pro-Matlab model based upon a NASTRAN model and an experimentally identified input-output plant model. In addition, simulations will be presented demonstrating the robustness of the feedback design to mechanical changes in the plant.

ACKNOWLEDGEMENT

This work was performed at the Jet Propulsion Laboratory, California Institute of Technology under a contract with the National Aeronautics and Space Administration.

REFERENCES

1. J.L. Fanson, B.J. Lurie, J.F. O'Brien, C-C Chu, R.S. Smith, "System Identification and Control of the JPL Active Structure," AIAA/ASME/ASCI/AHS/ASC Structures, Structural Dynamics and Materials Conference, April 1991.

ELECTRO-RHEOLOGICAL FLUID TORSIONAL DAMPER FOR AN AUTOMOBILE STEERING SYSTEM

James R. Salois Project Engineer General Motors Corporation
Saginaw Division, Saginaw, Michigan

Mechanisms, machinery, and systems have continuously improved to operate at higher speeds, more efficiently, and more economically than their comparatively slower and cumbersome predecessors. This has led to great advances in productivity and manufacturing cost reductions. However, as these improvements have been accomplished the systems have become more specialized, operating within specific parameters and conditions. When perturbations are introduced into these specialized systems undesirable responses may appear even though many were considered in the initial design. The predecessor systems, on the other hand were able to forge through these perturbations mainly because of their higher mass and slower operating speeds.

Systems are now required to control these unwanted effects while remaining transparent to the operation of the original system. The transparent technology investigated in this work is electro-rheological fluids. These fluids change viscosity with the application of an electrical field across them. For this application, in a torsional damper, this type of fluid is well suited to provide adequate and predictable control of unwanted effects.

A torsional damper was developed to attach to a steering column steering shaft. The unwanted effects being experienced consisted of excessive road input feedback through the steering wheel to the driver, and a soft "on-center" feel at higher speeds. These unwanted effects are the consequence of providing a system that performs extremely well during slow-speed maneuvers, which normally would require a compromise.

The response of the vehicle to step inputs was modeled using a single-degree-of-freedom system to simulate the steering wheel motion at increasing linear vehicle speeds. (There was a correlation that as vehicular linear speed increased so did the unwanted feedback.) To counteract this effect a computer model was developed to simulate the motion of the steering wheel at the increasing speeds. The damping constant was varied linearly as the vehicle speed increased to yield a satisfactory response. The damping constant was then related to the viscosity of the fluid required in the torsional damper. Thus, using a linear velocity feedback from the vehicle, the viscosity in the damper was varied to provide the desired damping.

In the above application a very simple control system was utilized to control the undesirable effects while not adversely affecting the system during slow speed maneuvers. Thus, the system was not compromised in any way, only enhanced through the use of the electro-rheological fluid torsional damper.

AN INNOVATIVE CLASS OF SMART MATERIALS AND STRUCTURES INCORPORATING HYBRID ACTUATOR AND SENSING SYSTEMS

M.V. Gandhi, B.S. Thompson, S.R. Kasiviswanathan,
S.B. Choi[†], B. Hansknecht[††], M. Soomar, X. Huang
Intelligent Materials and Structures Laboratory
and

C. Chmielewski
Department of Chemical Engineering

C. Foiles
Department of Physics

Michigan State University
East Lansing, MI 48824-1326

A new generation of smart materials and structures incorporating *embedded hybrid multiple actuation systems* which capitalize on the diverse strengths of both electro-rheological fluids and piezoelectric materials, and operate in conjunction with fiber optic, and/or conventional sensing systems is proposed herein for the active continuum vibration control of structural and mechanical systems. By judicious selection, the smart-materials designer can synthesize numerous classes of hybrid actuation systems from a variety of actuator systems to satisfy a broad range of performance specifications that cannot be satisfied by deploying a single class of actuator systems alone. This paper is focused on tailoring the elastodynamic characteristics of beam and plate-like structures,

† currently at Korea Institute of Machinery and Metals

†† currently at Ford Motor Company, Plastic and Trim Products Division

and a variety of mechanical systems in real time. Therefore, the hybridization of ER fluids and piezoelectric actuator systems is proposed. This hybridization philosophy enables vibration-tailoring capabilities to be accomplished in real-time with minimal energy consumption and high reliability due to the inherent characteristics of ER fluids and piezoelectric materials. Clearly actuator efforts resulting in large changes in the geometry of the structure are precluded from this investigation.

This paper presents a summary of analytical and experimental investigations on characterizing the constitutive characteristics of hydrous and anhydrous electrorheological (ER) fluids. The experimental work was undertaken on a Rheometrics RMS 800 retrofitted in order to impose a variety of electrical fields on ER fluid specimens. Typical results from these investigations are presented in Figures 1 and 2. Similar studies have also been undertaken on characterizing both piezoceramics and polymeric piezoelectric materials.

The above investigations on characterizing ER fluids and piezoelectric materials were instrumental in the development of smart structural systems featuring composite and monolithic materials with embedded ER fluids and surface bonded piezoelectric actuator elements. Subsequently beams and plates were fabricated featuring a variety of ER fluids and piezoelectric materials prior to undertaking experimental investigations involving both free vibration and also forced vibration. Figures 3 through 5 present some results from these investigations.

Control strategies and methodologies have also been developed to actively control the elastodynamic response on these structures in an autonomous manner. Fiber-optic and/or conventional sensors have been employed in developing these control strategies. Investigations have also been undertaken to control the elastodynamic response characteristics of structural systems which exhibit a chaotic behavior. Typical waterfall plots from these investigations are shown in Figure 6.

The above work on developing smart materials featuring ER fluids and piezoelectric materials was subsequently extended to the development of smart mechanisms and also robotic systems. The objective of these investigations was to actively control the elastodynamic response of these complex mechanical systems by changing the natural frequencies and stiffnesses of the articulating members. Figure 7 shows typical results of investigations focused on the control of a single-link robotic system featuring an embedded ER fluid. Investigations of linkage mechanisms have also been undertaken and Figure 8 presents the elastodynamic response characteristics of the midspan of a flexible connecting rod of a slider crank mechanism. The behavior of the link is clearly a function of the applied voltage.

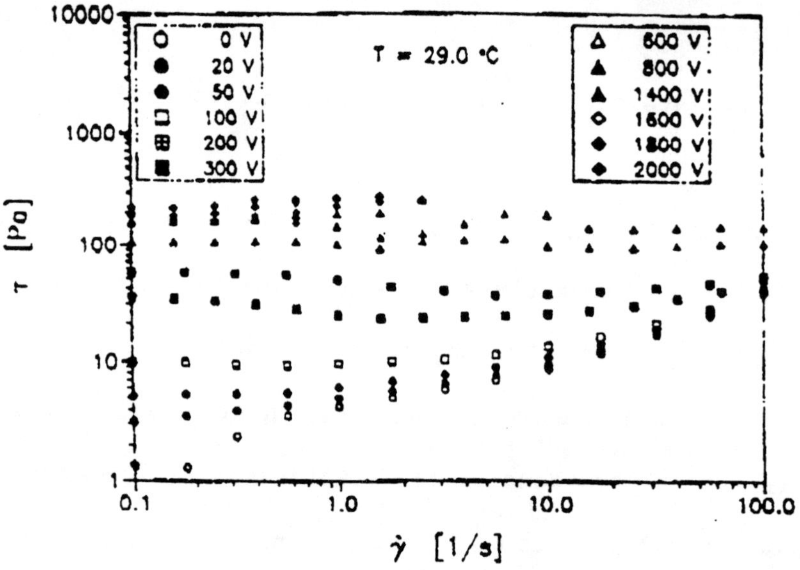

Figure 1

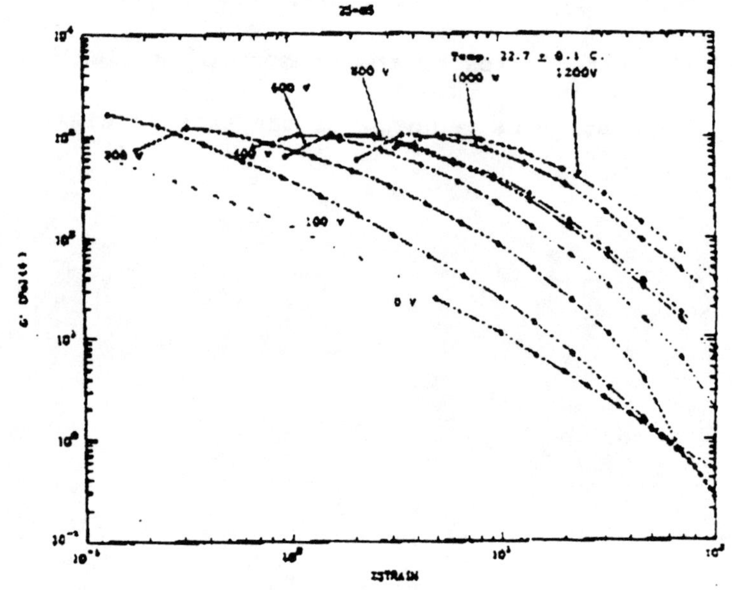

Figure 2

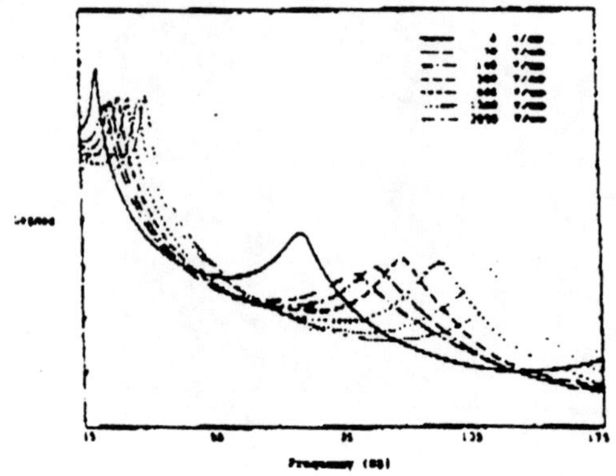

Figure 3: Frequency Response of Smart Cantilevered Beam Containing an ER Fluid.

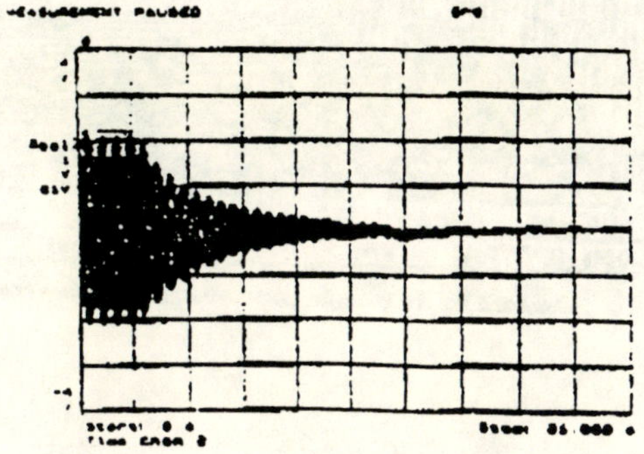

Figure 4: Transient Response of a Smart Beam Featuring Piezoelectric Actuator: Zero Volts.

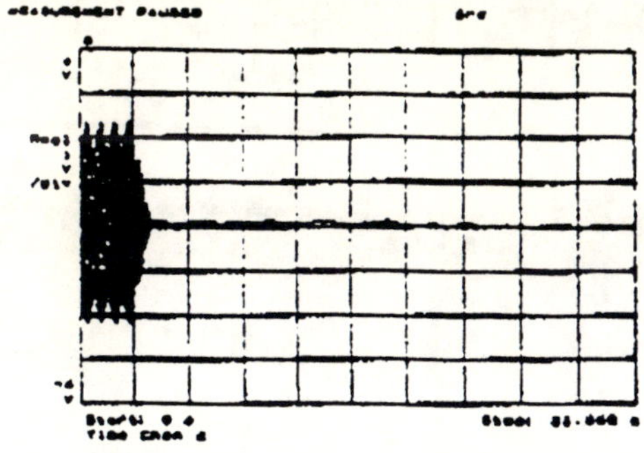

Figure 5: Transient Response of a Smart Beam Featuring Piezoelectric Actuator with Finite Voltage.

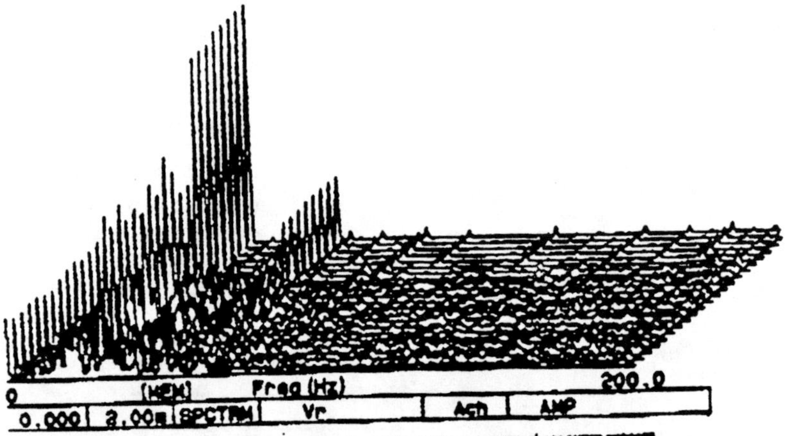

Figure 6 waterfall plot of response for [illegible] process

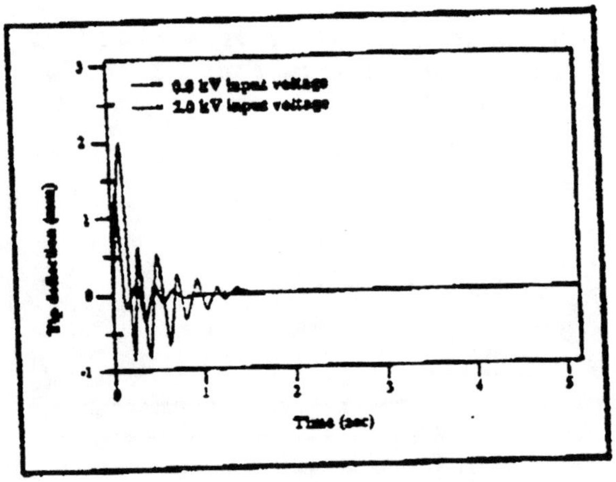

Figure 7 Measured step response with feedback gains of $K_p = 0.5$ and $K_v = 0.25$.

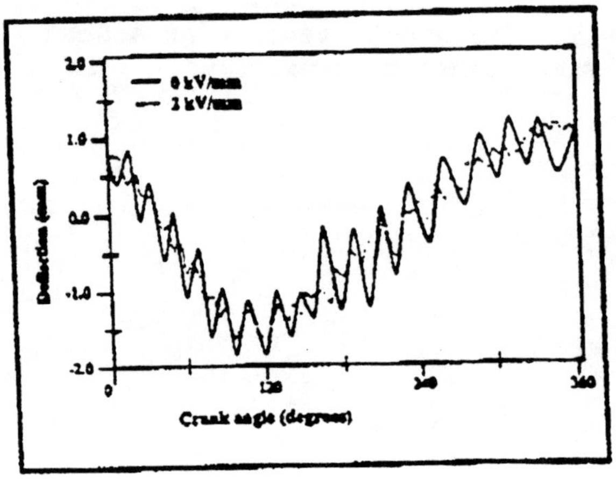

Figure 8 Midspan transverse deflections of the dynamically tunable connecting rod; mechanism operating speed 95 rpm.

Design of Anhydrous Electro-Rheological (ER) Suspensions and Mechanism Study

Wei-Ching Yu, Rex C. Kanu and Montgomery T. Shaw*
Institute of Materials Science, U-136
The University of Connecticut
Storrs, CT 06269-3136

Electrorheological (ER) fluids are suspensions of highly-polarizable fine particles dispersed in an insulating oil. The attractive features of the fluids for practical application are the speed and reversibility of the liquid-to-solid and solid-to-liquid transition when an electric field is turned on and off. These make possible feed-back control for robotic, automotive, and other related applications. However, the many conceivable applications of ER technology have yet to be realized because the performance of available fluids is deficient. Our efforts in this study are directed toward (1) understanding the electronic polarization in particles, (2) design of anhydrous ER suspensions, and (3) development of a new method for the measurement of dielectric interaction force under high fields.

First, a two-component model for dielectric properties of suspensions is proposed to describe the fluids that contain dielectrically anisotropic particles or nonspherical particles oriented with a certain angle to the applied electric field. The roles of the shape and electrical conductivity of particles to the dielectric properties of the ER fluids are elucidated by using the model for Maxwell composites. The role of particle conductivity is to expedite the particle polarization in the ER suspensions. At the same time, high particle conductivity causes interparticle conduction, which reduces the performance of the fluid.

From these findings, we designed anhydrous ER suspensions based on semiconductive materials, I_2-doped poly(pyridinium salt) and poly(p-phenylene-2,6-benzobisthiazole) (PBZT) (Figure 1). The I_2-doped particles were further processed into the ones with an insulating skin, which reduces interparticle conduction. The designed anhydrous suspension showed a strong ER effect (Figure 2) and low current density (Figure 3), demonstrating the importance of conductivity and the insulating skin. PBZT, a robust liquid crystalline polymer featuring a highly aromatic, ladder-like structure is also being studied as the dispersed phase. Rheological measurements have revealed that PBZT is a good candidate for the dispersed phase of an ER fluid. In steady shear measurements for 10 v% of PBZT particles in mineral oil, the shear stress increased by 40% for an electric field of 4 kV/mm. The dynamic elastic modulus G' for the fluid comprising 20 v% of PBZT particles in silicone oil showed a positive dependency on electric field in the range of 0.7 to 3.5 kV/mm (Figure 4). Further evidence of ER behavior of the PBZT suspension by optical microscopy showed that the stiffening of the ER fluid, portrayed by the increase in G' with electric field, is due to particle chaining.

The dielectric properties of the two suspensions are also studied by dielectrometry. A new method was developed to evaluate the dielectric interaction force under high fields. Conventional methods for the measurement of dielectric constant are to monitor the charge accumulation on the capacitor that contains the test sample. Because ER suspensions are used at high fields, the conventional methods become unrealistic and difficult. To overcome this problem, we monitor the attractive force instead. The effective dielectric constant under high field can be extracted from the measured attractive force. To demonstrate the validity of the new method, the effective dielectric constant of mineral oil by this method is shown to be consistent with the conventional measurement.

Numerous experiments have been done using applied fields ranging from 0.5 to 3.0 kV/mm and shear rates from 0.1 to 100 s^{-1}. It was found that a higher shear rate gives a smaller value of the effective dielectric constant of the designed suspensions. This implies that the ordered structure (chaining structure) of the dispersed phase was destroyed at high shear rate and gave a smaller value of effective dielectric constant of the suspension. These are consistent with the measurements by dielectrometry before and after the dispersed phase in the suspension was aligned with a high field.

The excess normal attractive stress (ΔN) is defined as the difference of the normal attractive stresses due to the suspension and the suspending medium. The excess normal attractive stress (ΔN) represents the sum of dielectric interaction force of the suspension. The shear stress arising from ER effect is linearly proportional to the dielectric interaction force (ΔN) for the designed ER suspension in a system containing 15 v% of I_2-doped particles (Figure 5). The universal response is a demonstration, in essence, of the appropriateness of the chaining theory for the ER effect.

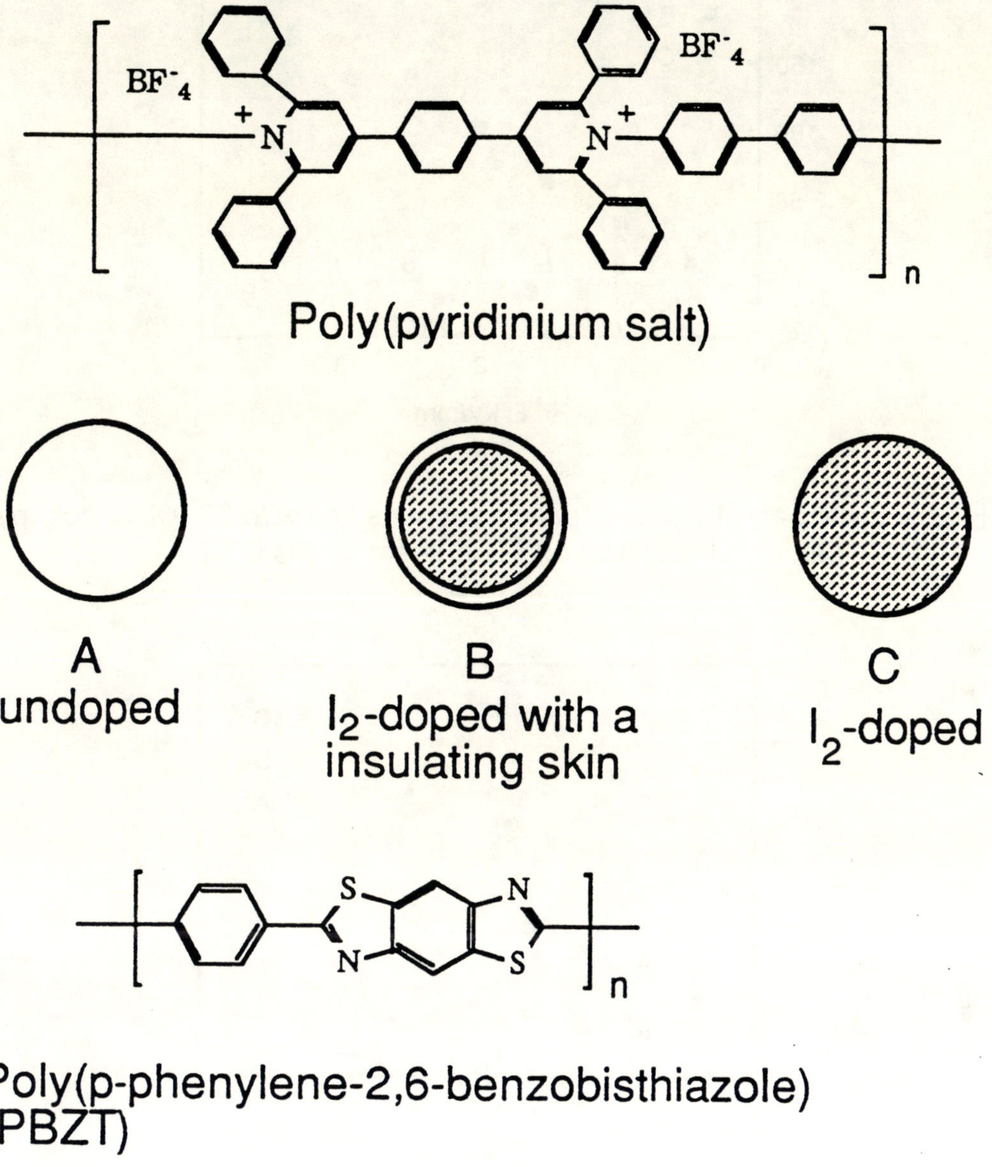

Figure 1 Structures of I2-doped poly(pyridinium salt) and poly(p-phenylene-2,6-benzobisthiazole) (PBZT)

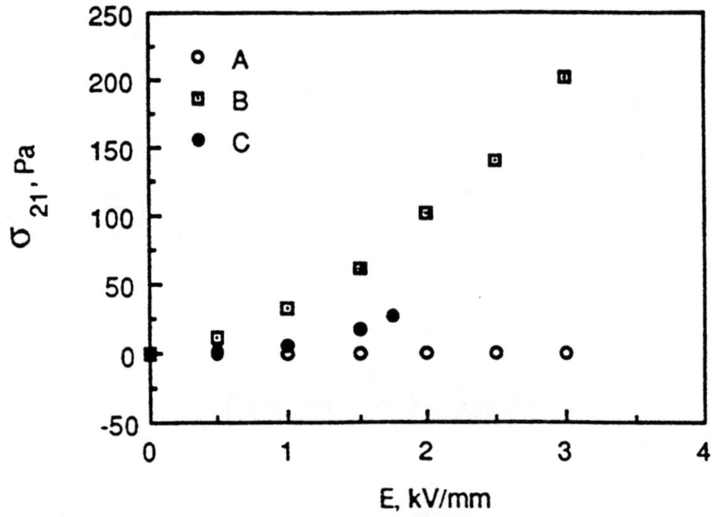

Figure 2 Apparent shear stress of suspensions with 20 v% of poly(pyridinium salt) particles in mineral oil. The shear rate is 0.5 s^{-1}.

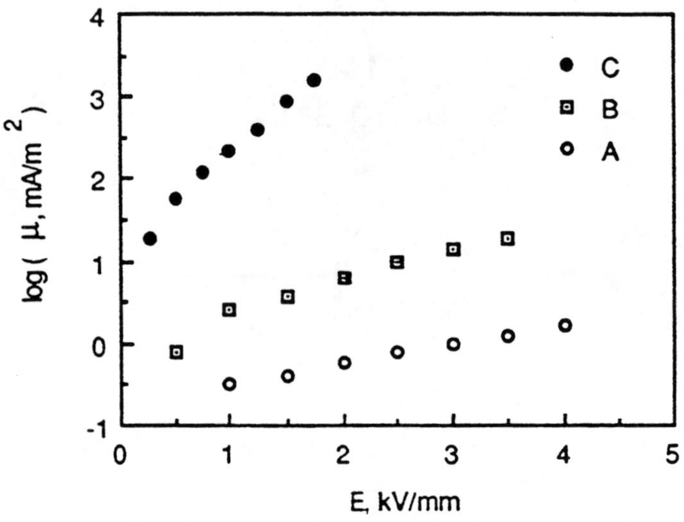

Figure 3 Current density of suspensions with 20 v% of poly(pyridinium salt) particles in mineral oil.

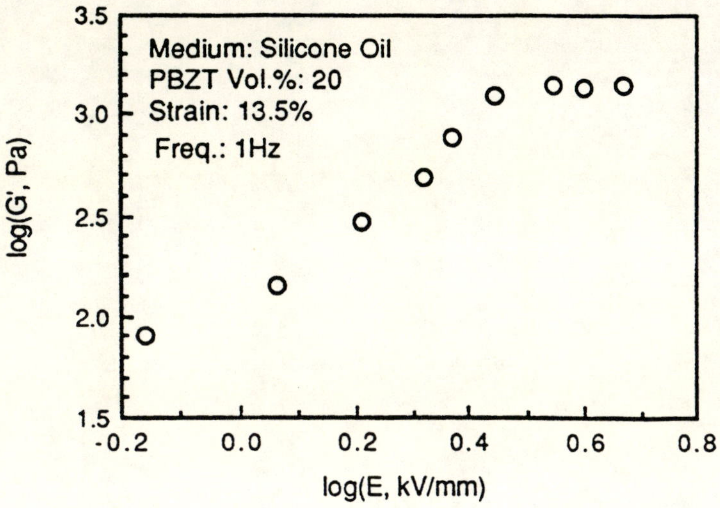

Figure 4 Stiffness dependence of PBZT-based ER fluid on electric field

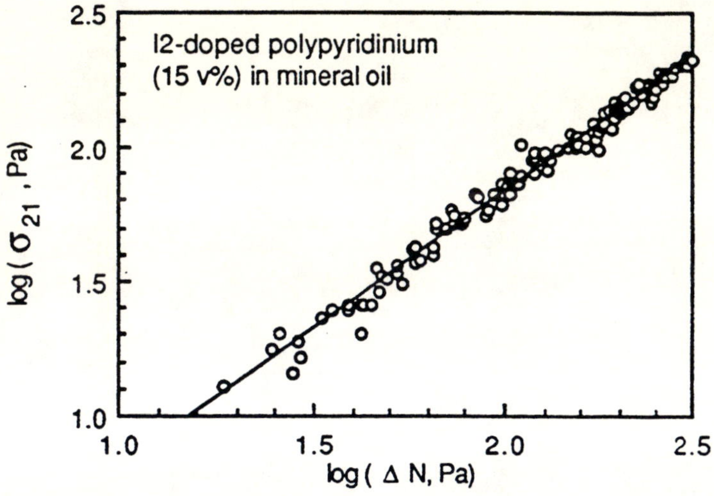

Figure 5 Relationship of apparent shear stress (σ_{21}) and dielectric interaction force (ΔN)

AN ANALYTICAL AND EXPERIMENTAL INVESTIGATIONS OF ELECTRORHEOLOGICAL FLUIDS

by

S.R.Kasiviswanathan, B.S. Thompson, M.V. Gandhi
Intelligent Materials and Structures Laboratory
Machinery Elastodynamics Laboratory
Department of Mechanical Engineering

and

C. Chmielewski
Department of Chemical Engineering
Michigan State University, East Lansing, MI 48824, U.S.A.

This paper presents a summary of analytical and experimental investigations on characterizing the constitutive characteristics of electrorheological(ER) fluids. The steady-state rheological properties of an ER fluid are investigated, and a constitutive rheological equation is proposed for the ER fluid. A variety of ER fluids are considered here with different particulate suspensions in a dielectric medium. In this paper, the nature of an ER fluid, the influence of important variables such as shear stress, shear rate, electric field strength, field frequency, temperature and fluid composition are presented. The results obtained from the experimental investigations are employed in the mathematical modeling of the constitutive rheological equation of the ER fluid.

The rheological properties of the ER fluid were studied while subjected to electric fields ranging from 0 to 2200 volts.The experimental work was undertaken on a Rheometrics RMS 800 mechanical spectrometer, upgraded with a high voltage capability (5000V). Using a double walled couette fixture, steady shear measurements were made at rates ranging from 0.1 to 100 1/s. All measurements were made at temperatures ranging from 24° C to 29° C. These studies will not only illustrate the rheological effect of an electric field on ER fluids, but also will provide an insight for the development of future experimental programs and procedures employing them. The rheological investigations on these fluids indicate that the higher imposed voltages on the fluid apparently lead to more rigid fluid structures causing an observed increase in the fluid's shear viscosity. Typical results from these investigations are presented in Figures 1-4.

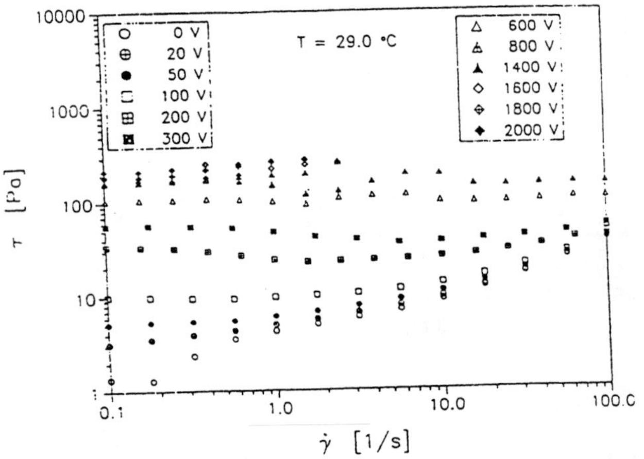

Figure 1: Shear Stress vs. Strain Rate

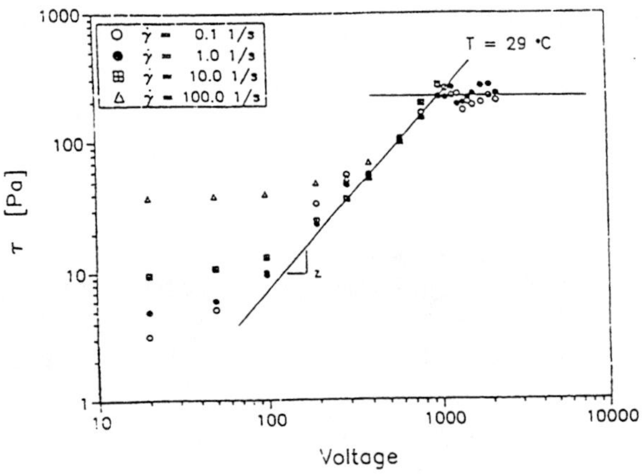

Figure 2: Shear Stress vs. Electric Field

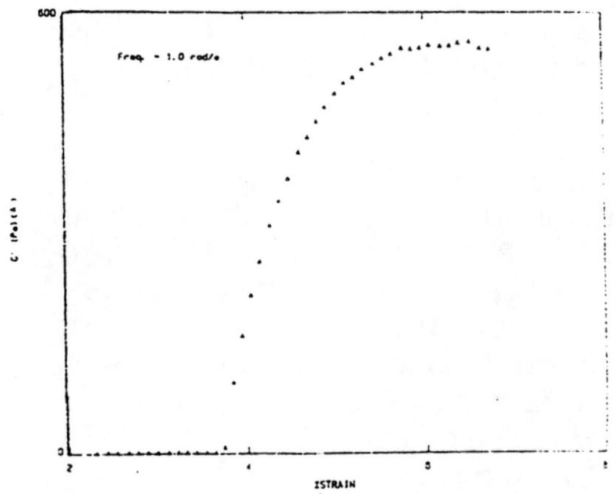

Figure 3: Storage Modulus vs. Percent Strain

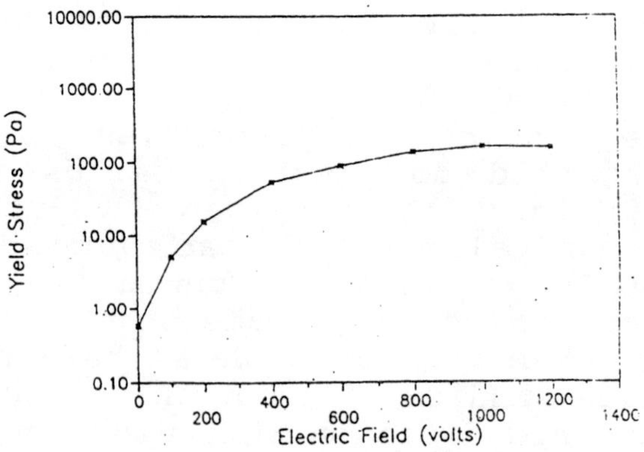

Figure 4: Apparent Yield Stress vs. Electric Field

Simultaneous Single Optical Fiber Communications and Sensing for Smart Structures Applications

P. L. Fuhr
University of Vermont
College of Engineering
Burlington, Vermont 05405

W. B. Spillman, Jr.
BFGoodrich Aerospace Division
Simmonds Precision Aircraft Systems
Vergennes, Vermont 05491

Abstract

Fiber optic technology will probably be used first in operational smart structures for very high speed telecommunication links between sensing nodes and processing locations. If analysis of the operational parameters of the data links can also provide information about the physical link environment, e.g. vibration levels, then the smart structure can utilize that information to adapt itself into the most effective configuration for its environment. The implications of simultaneous sensing and telecommunication on a single optical fiber have been discussed for some time but surprisingly little experimental work has been carried out. Initial studies using a single analog tone frequency and vibration sensing were first described within the last few years. We have extended that earlier work to the domain of high data rate digital optical communications and speckle pattern-based vibration sensing along a single multimode optical fiber. The data link parameter used as an independent variable for the determination of vibration levels was the bit error rate (BER). The ability to transmit a digital pseudo-random-bit-sequence (PRBS) data stream at the CEPT 34 Mb/sec telecommunications interface rate while simultaneously monitoring fiber vibrations ranging from DC to 100 Hz was studied. In addition, by adding a frequency subcarrier to the laser diode light source, the ability to shift the vibration signal away from low frequency noise sources (e.g., line noise, 1/f noise) and obtain a significantly higher SNR and thus improved vibration signal detection. Measurements of BER for this on-off keyed digital bit stream showed that the system performance varied from 10^{-8} to 0.5 depending on the magnitude of the fiber vibration. Experimental results for alternate system configurations have been investigated as have the overall cross-talk effects of the communication channel and the vibration sensing signal. Finally, techniques for minimizing the degradation effects of the communications channel and sensing system have been examined. The results of this work will be presented and recommendations will be made for practical implementations.

Non-Destructive Evaluation of PMN Actuator Elements for Adaptive Structures

John A. Wellman
Litton/Itek Optical Systems
10 Maguire Road
Lexington, Massachusetts 02173-3199

Actuator elements are incorporated into adaptive structures as critical structural members to apply forces that distort the structure in a precise manner. The structural properties and integrity of the actuators are critical to the design and performance of the adaptive structure. Ultrasonic measurement techniques are presented for non-destructive evaluation of Lead Magnesium Niobate (PMN) multi layer stacked actuators for use in adaptive structures. An ultrasonic velocimeter is used to measure the acoustic velocity and attenuation of high frequency sound waves transmitted through actuator segments. Experimental results are presented from which elastic properties of PMN actuator segments are deduced. Variations of acoustic velocity were found to correlate with differences in electrostrictive strain field sensitivity between actuators. Acoustic velocity fluctuates under an applied electric field mapping the elastic property changes with the electrostrictive effect. Echo patterns of sound waves transmitted through the segment can be used to detect defects or delaminations with in the multi layer actuator structure. Ultrasonic velocimetry techniques presented offer a non-destructive means to evaluate the structural properties of PMN actuators to improve the reliability of adaptive structures.

Composite Embedded Fiber Optic Data Links and Related Material/Connector Issues

The objective of this effort is to explore the 'Smart Skins/Structures' concept of composite-embedded optical fibers and utilize it in conjunction with composite material developments in airborne electronic packaging. Both Polymer Matrix Composites and Metal Matrix Composites (MMC) are being developed.

In order to achieve signal distribution, various coupler methods are being investigated. Both fused or biconical couplers and evanescent couplers are being considered. Specifically, an embedded spiral is being tested for signal distribution by means of evanescent coupling from a motherboard to individual circuit cards.

To achieve interconnection between individual panels, a design for an embedded connector is being developed and tested.

High temperature optical fibers are being investigated for embedment in MMC. Hollow sapphire waveguides have been embedded in aluminum. Testing, both optical and material, is in progress.

Authors: R. E. Morgan, S. L. Ehlers, and K. J. Jones
Naval Avionics Center
6000 East 21st Street
Indianapolis, IN 46219-2189
Telephone (317)353-3826
FAX (317)353-3583

Active and Adaptive Optical Components: A General Overview

Mark A. Ealey

Litton/Itek Optical Systems
10 Maguire Road, Lexington, Massachusetts 02173-3199

ABSTRACT

Since the revolutionary development of the laser in the late 1950's, the optics industry has strived to achieve optical components and systems with near diffraction limited performance. Optical components were designed and manufactured with precision tolerances to minimize fixed distortions. As the laser device improved in terms of power and beam quality, time-varying distortions induced by laser gain medium nonlinearities, atmospheric turbulence, and thermal blooming impeded system performance. A new generation of active optical components and adaptive techniques were realized to correct the time-varying distortions in real time. Active optics is a generic term applied to those systems and components whose characteristics are controlled during actual operation to modify the optical wavefront. Adaptive optics is a term which includes the addition of a wavefront sensor, a wavefront processor, and a servo control mechanism to implement real time closed loop control of the optical wavefront.

Since the advent of adaptive optics over two decades ago, a number of devices have been conceived and studied for wavefront correction. All adaptive optical systems require novel devices to implement the phase shift operation necessary for wavefront control. The phase of the wavefront can be controlled by either changing the propagation velocity or the optical path length. Refractive index varying devices such as spatial light modulators and other ferroelectric or electro-optical crystal devices have been used with limited success in using velocity control to implement phase change. Frequency response and amplitude limitations have been limiting factors for the crystal devices. Reflective surface modifying devices such as segmented and deformable mirrors have been successfully demonstrated in several systems and applications. High bandwidth, large amplitude control with several hundred degrees of freedom are routinely manufactured for use in adaptive systems. These structures may be grouped into two categories, 1) segmented facesheet and 2) continuous facesheet. In addition there are two methods of closed loop operation, 1) zonal control and 2) modal control. A general overview of the technology of active and adaptive optical components will be described below in terms of their ability to implement control with precision equal to a fraction of the wavelength of visible light.

Neural Network Applications in Structural Dynamics

M. E. Regelbrugge and R. Calalo

Structures Laboratory
Lockheed Palo Alto Research Laboratories
Palo Alto, CA 94304

Introduction

Effective utilization of active materials in adaptive structures depends on the proper synthesis of structural configurations, sensors, actuators and the means to direct these elements in harmony to achieve desired performance. Active materials promise a structure-integrated capability for large-scale, distributed sensing and actuation. Kinematically designed structures offer the necessary freedom to reconfigure and adapt load paths to changing conditions. Distributed and parallel-processing networks offer the needed capability to direct appropriate actuation and adaptation in response to external stimuli and changing conditions. One promising architecture for large-scale, distributed processing that appears ideal for adaptation is the neural network.

As in biological systems, artificial neural networks offer the means to quickly recognize a diversity of complicated patterns evident in external stimuli. Present neural network models attribute this capability to distributed storage of pattern information among a large number of highly interconnected yet simple elements. The network elements and interconnections typically have a limited number of parameters and weights that are adjusted through a learning process to yield repeatably correct classifications of input patterns. The highly parallel topology of the network admits fast response to a stimulus, while the distributed representation of learned pattern data has been shown in many cases to enable correct classification of input datasets whose contents are noisy or imprecise. These two properties have raised significant interest in applications of artificial neural networks to real-time recognition, classification and identification of real-world operational patterns of complicated systems.

As a first step, this paper explores issues that arise in applying artificial neural networks to the identification of vibratory characteristics of large structures. Important aspects of network integration with a dynamic structure are explored. Of particular interest are the synthesis of network input patterns appropriate to identify structural motions, and of network output quantities appropriate to direct control or adaptation. With respect to these issues, schemes are sought which preserve a robust autonomy in operation. Autonomy is critical for the practical use of neural networks to monitor dynamic characteristics of structures beyond immediate human accessibility, e.g. structures deployed in space. Furthermore, fast-learning, autonomous networks may be employed to identify or estimate the dynamic characteristics of evolving structures (i.e. those with time-varying configurations or parameters).

The issue of autonomy in neural-network learning and operation hinges on the reliability of the learning algorithm to produce accurate identifications of observed patterns, where identification can be viewed as the expression of stimuli in terms of a limited set of useful (output) parameters. Inaccurate identifications can result in two ways:

1) the learning algorithm can fail to adjust network parameters appropriately to identify a familiar stimulus, or
2) the network may be faced with a stimulus sufficiently foreign as to preclude identification with any confidence.

In the first instance, the learning algorithm is not capable of configuring the network to respond to the observed system. In the second instance, the learning algorithm is not able to determine whether or not a stimulus is represented within the domain of the observed system. Naturally, one should expect some practical bounds on the degree to which any network, however sophisticated, can confidently identify unfamiliar stimuli. Autonomous learning requires the network to be able to differentiate between stimuli falling on either side of these bounds. One should note the complexity of such differentiation depends to a large degree on the constitution of the stimulus. For simple networks, stimuli should be simply constituted to avoid ambiguous or pathological identifications.

Network Architectures

Several network architectures have been proposed for application to dynamics and signal identification tasks. Among these are perceptrons and other back-propagation (BP) networks [1,2], Hopfield networks [3,4], self-organizing networks (SONN) [5], Bayesian probability networks [6,7], the so-called CMAC networks [8] and Widrow's Adeline networks [9]. Many of these networks have been applied to identification strategies resulting in adaptive active control.

Of these types of networks, perceptron, BP, SONN, CMAC and Adeline networks learn by iteration. The topology of all of these network types, save SONN, is fixed initially and network-element interconnection weights are adjusted by a suitable learning algorithm to yield desired input-output behavior of the network. The learning algorithms are usually based on gradient descent methods such as steepest descent, conjugate gradient or, in certain special cases, Newton methods. In CMAC networks, typically only the output of a single stage of the network is weighted. The SONN topology and interconnection weights are determined by an heuristic optimization procedure known as simulated annealing.

Unfortunate aspects of iterative learning schemes are that they may converge slowly and, during the process of convergence, they may produce incorrect identifications of previously learned input patterns. Furthermore, when converged, these networks will always yield identifications in terms of the learned pattern set, i.e. they have no capacity to recognize when stimuli fall outside the domain of their experience. This places a premium on adjusting network parameters over the entire expected domain of system performance. It also reduces the degree of autonomy in network learning and precludes adaptability to unforeseen evolutions of the system without explicit redirection to undergo additional learning cycles. Finally, the establishment of robust network convergence criteria is absolutely critical to determine when the network becomes useful as an accurate identifier. This task is considerably complicated by uncertainties in system characteristics.

Hopfield networks are analog networks whose response to an initial stimulus converges to a stable, steady-state value which minimizes the energy state of the network. For the network to provide physically meaningful outputs, an energy functional must be synthesized in terms of the output quantities, and the network topology configured to mimic the synthesis of the functional. Hopfield networks can be made as fast as analog electronics allow for any given application.

Bayesian, or probabilistic neural networks (PNN) [7], alleviate the difficulties associated with fixed network topologies and iterative learning algorithms by instantly learning classifications based entirely on observations. Contrary to networks with fixed topologies that learn over time, PNNs learn instantly by increasing their order in neuron space. This is done by allocating discrete neural elements for each observed input pattern.

An extension of the probabilistic neural network, the Gaussian Potential Function Network (GFPN) [10] employs similar concepts to the PNN but allows iteration over vertain network parameters to improve identifications over time. The GPFN offers a practical a tradeoff between network order and learning time for a required accuracy.

Application to Signal Identification

The PNN or GPFN architectures may be employed to identify structural dynamic motions by configuring a network to construct mappings between local histories of observed motions and the motions which follow a history (the "future" motions). Toward this end, numerical simulations have been conducted which illustrate the relative performances of PNN and GPFN networks for a multi-degree-of-freedom system. Results of these simulations are depicted in Figures 1 and 2, which illustrate observed and predicted motion for PNN-based and GPFN-based identification networks, respectively. Each network was allocated 200 neurons, and the GPFN network was further allowed some 10,000 sample-iterations to converge to a predictive accuracy of 0.10%.

As the figures show, a better predictive accuracy is obtained by the GPFN network albeit at a high cost in iteration. Thus, a central issue in the application of such networks is the inherent tradeoff between how accurately they must identify observations and how fast they must respond to changing characteristics. These characteristics need to be defined for specific systems on an individual basis. Finally, one should note that networks of the type discussed can be applied to identify information locally within a large structure irrespective of global variations of response. Such a scenario may be especially useful to monitor integrity and pinpoint damage locations in very large structures.

References

[1] Narendra, K. S. and Parthasarathy, K., "Identification and Control of Dynamical Systems using Neural Networks," *IEEE Trans. Neural Networks*, **1**, March 1990, pp. 4–27.

[2] Cabell, R. H. and Fuller, C. R., "Pattern Recognition System for Automatic Identification of Acoustic Sources," *AIAA J.*, **29**, February 1991, pp. 180–186.

[3] Tank, D. W. and Hopfield, J. J., "Simple "Neural" Optimization Networks: An A/D Converter, Signal Decision Circuit, and a Linear Programming Circuit," *IEEE Trans. Circuits and Systems*, **33**, May 1986, pp. 533–541.

[4] Chu, S. R., Shoureshi, R., Tenorio, M., "Neural Networks for System Identification," *IEEE Control Systems Magazine*, April 1990, pp. 31–35.

[5] Tenorio, M. F., Lee, W.-T., "Self Organizing Neural Networks for the Identification Problem," in *Advances in Neural Information Processing Systems*, **1**, D. S. Touretzky, ed., Morgan Kaufmann Publishers, Palo Alto, CA, 1989, pp. 57–64.

[6] Malkoff, D. B., "A Neural Network for Real-Time Signal Processing," in *Advances in Neural Information Processing Systems*, **2**, D. S. Touretzky, ed., Morgan Kaufmann Publishers, Palo Alto, CA, 1990, pp. 248–255.

[7] Specht, D. F., "Probabilistic Neural Networks," *Neural Networks*, **3**, 1990, pp. 109-118.

[8] Albus, J. "A New Approach to Manipulator Control: The Cerebellar Model Articulation Controller (CMAC)," *J. Dyn. Sys., Meas., Contr.*, **97**, 1975, pp. 270–277.

[9] Widrow, B. and Stearns, S. D., *Adaptive Signal Processing*, Prentice-Hall, Englewood Cliffs, NJ, 1985.

[10] Lee, S. and Kil, R. M., "A Gaussian Potential Function Network with Hierarchically Self-Organizing Learning," *Neural Networks*, **4**, 1991, pp. 207–224.

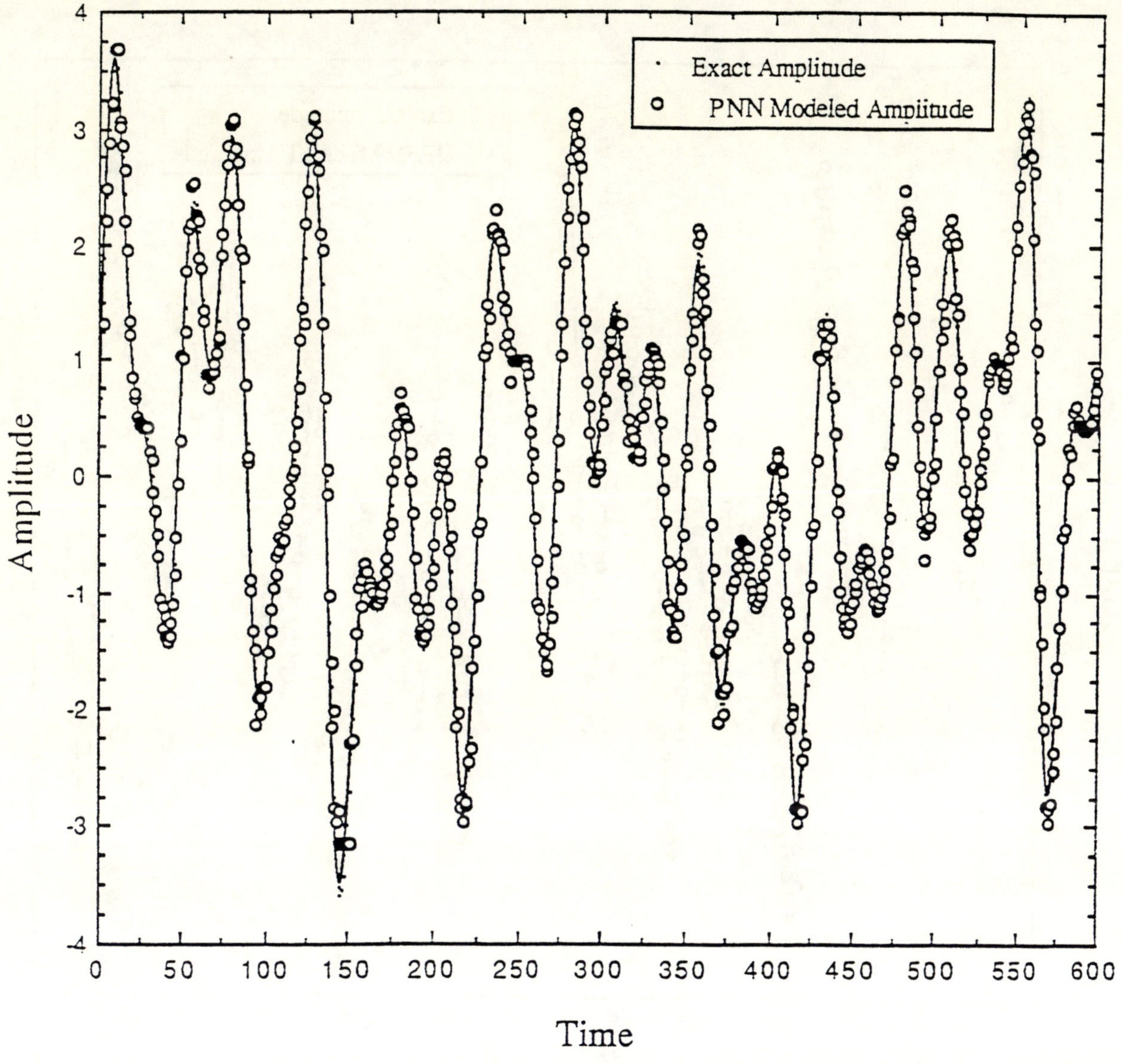

Figure 1. MDOF Oscillator Identified by PNN

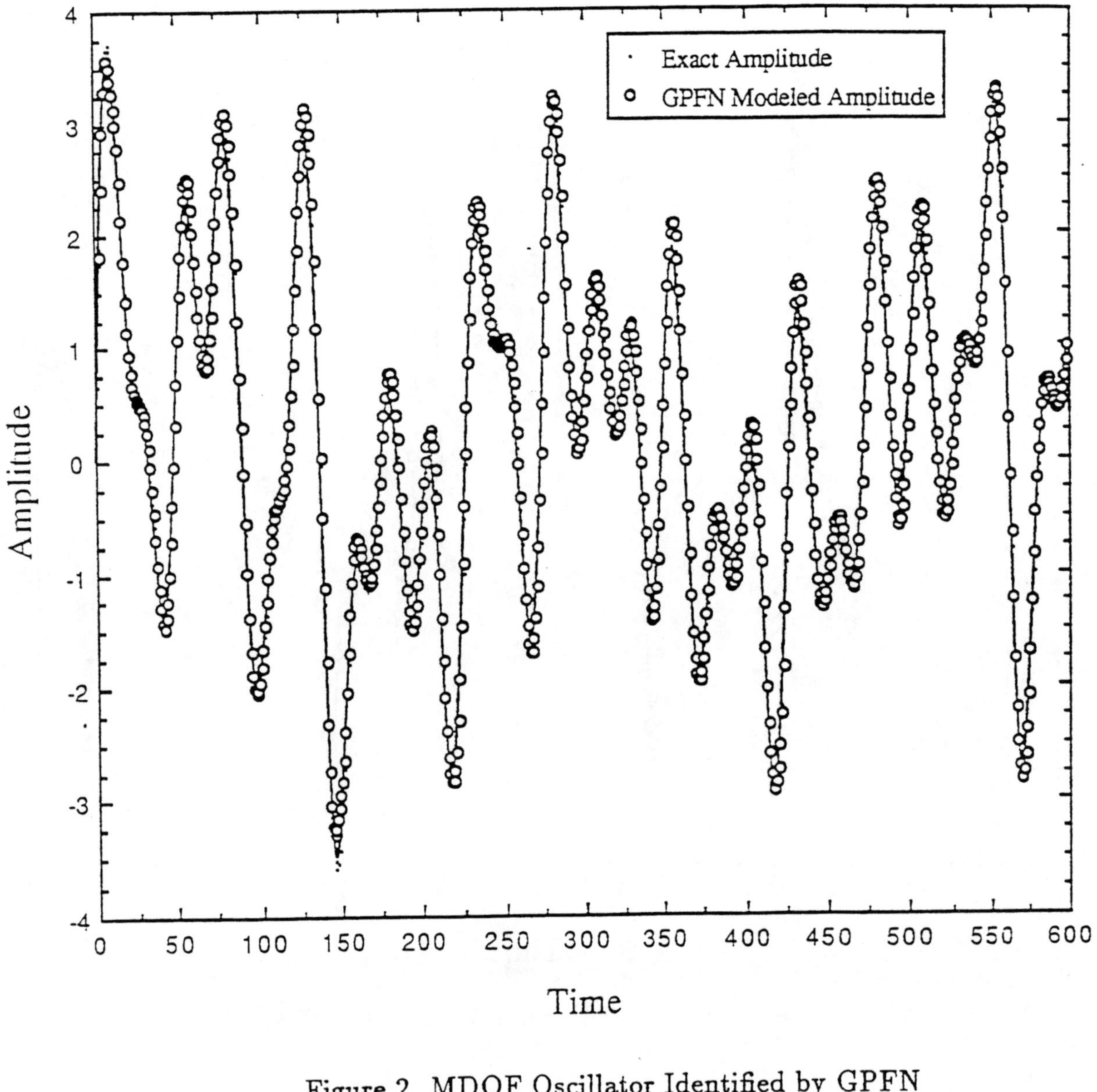

Figure 2. MDOF Oscillator Identified by GPFN

Low Level Damping and Hysteresis of Damped Structures

B. Tse, and D. Werner, Lockheed Missiles and Space Division,
Sunnyvale, CA

ABSTRACT

Built-up precision structures from composites are being used in spacecraft to achieve better dimensional tolerance and fastener weight saving. The reduction of connections results in a low damped structure. To improve the performance of the system, constrained layer damping can be designed into the structure. There are concerns that the damping may be amplitude dependent, and the prediction from large amplitude test results may not be applicable to sub-micron vibrations. Another concern for precision structure is the possible hysteresis that may exist at sub-micron levels, and the damping treatment on the structure may also introduce additional hysteresis.

These concerns has been investigated on a damped honeycomb structure using constrained layer damping. Damping for displacement amplitude has been measured from .2 milli-meter down to 1 nano-meter. The measurement indicates little variation in damping between these vibration levels. Hysteresis measurements for the bare structure and damped structure were made with displacement sensors. The micro hysteresis time history indicates that a relaxation time is required for the precision built-up structures to recover from displacement due to application and removal of loading. Sub-microns residual hysteresis does remain after load removal, depending on the load magnitude and duration. The equivalent hysteresis for the damped structure is about 3 times higher than the bare structure. These factors must be taken into account in the design performance study of precision built-up structure.

Abstract
for the
ADPA/AIAA/ASME/SPIE Conference
on
Active Materials and Adaptive Structures

Structural Control Sensors for the CASES GTF

H.W. Davis (*) and A.P. Bukley (**)

Marshall Space Flight Center (MSFC) has nearly completed development of its ground test facility (GTF) for the Controls, Astrophysics and Structures Experiment in Space (CASES). CASES is a proposed Shuttle-based experiment designed to produce x-ray images of the galactic center and solar disk with unprecedented resolution. This requires precision pointing and suppression of the vibrations in the long flexible structure that comprises the 32-m x-ray telescope optical bench. The CASES experiment has undergone both Phase A and Phase B studies; therefore, a detailed design exists. Another primary objective of the experiment is to conduct controls-structures interaction (CSI) experiments on-orbit and evaluate structural response to several different control system methodologies. The GTF was created to support CSI technology development in the areas of controller design and implementation methodologies, system identification techniques, actuators, sensors, and associated interfaces.

Two separate electro-optical sensor systems are provided for the GTF. The sensors are based on the Remote Attitude Measurement Sensor (RAMS), designed by the Electro-Optics/Cryogenics Division of Ball Aerospace. The two sensor systems include the Tip Displacement Sensor (TDS) and the Boom Motion Tracker (BMT). The TDS measures lateral motions of the boom tip and provides feedback for the closed-loop control system that maintains tip position. The BMT measures three-axis translations of 42 reflective targets attached along the length of the boom. The BMT data is processed post-mission to determine the mode shapes of the boom and to aid in evaluating the effectiveness of various control system designs. The TDS updates the position of each of four targets at rates up to 500 Hz and to an accuracy of 0.008 inches. The BMT updates each of 42 targets at a rate of 100 Hz and to a position accuracy of 0.01 inches.

This paper describes the design features of RAMS and how they satisfy the requirements of the TDS and BMT. A discussion of the design factors will include geometry, radiometry, optics, interfaces, alignment, and verificattion. Predicted performance and preliminary characterization results, obtained from acceptance tests performed at Ball and from tests performed after installation into the CASES facility at MSFC, will be presented and discussed.

* Ball Electro-Optics/Cryogenics Division
** NASA/MSFC

Evaluation of Acrylate and Polyimide Coated Optical Fibers as Strain Sensors in Polymer Composites

L. D. Melvin, R. S. Rogowski, M. S. Holben, J. S. Namkung

NASA Langley Research Center
Mail Stop 231
Hampton, Virginia 23665-5225

Abstract

Optical fibers have been used as strain, pressure, temperature, and chemical sensors. These fibers have been attached or embedded in configurations for health monitoring of structures over their predicted lifetime. However, to make such measurements over the structure's service life will require the fiber sensing arm to remain embedded or attached to accurately report the condition of the component especially before failure. To compare the effect of different fiber coatings on sensor performance, acrylate and polyimide coated optical fibers were embedded in 8 ply graphite/epoxy composite samples. Strain was measured under tensile and cyclic loads with both surface mounted resistive strain gages and the embedded optical fibers using an Optical Phase Locked Loop (OPLL) system. Investigation of the optical fiber coating/matrix interfacial condition after mechanical cycling was performed using optical, electron and acoustic microscopy.

Embedded Optical Fiber Sensors for Monitoring Cure Cycles of Composites

Mark A. Druy, Paul J. Glatkowski, and W. A. Stevenson

Foster-Miller, Inc.
350 Second Avenue
Waltham, MA 02154

The repeatable processing and manufacture of advanced composite materials is perhaps the major obstacle inhibiting the widest possible acceptance of resin matrix composites as aircraft structural elements. Considerable research and development has been expended in government, university and aerospace laboratories in efforts to improve the quality and reliability of this class of materials. We address this critical issue from a uniquely fundamental standpoint. The experimental technique involves the use of infrared spectroscopy and the development of infrared transmitting optical fibers as sensors for monitoring the cure of graphite fiber/resin matrix materials. In this paper we report the use of this experimental technique for monitoring the extent of cure in both the autoclave and high temperature press environment.

The real-time in situ monitoring of the chemical states of epoxy and polyimide resins were investigated during cure using an embedded fiber optic sensor and a Fourier transform infrared spectrometer (FTIR). In this work a short length of sapphire fiber is used as the sensor for monitoring the cure of the epoxy, while for the polyimide resin, we use a chalcogenide fiber as the sensor. The cure of the epoxy resin/graphite fiber composite is monitored in an autoclave, while the cure of the polyimide resin/graphite fiber composite is monitored in a high temperature press. The sapphire sensor is connected to infrared transmitting zirconium fluoride optical fiber cables which penetrate the wall of the autoclave and interface to the FTIR spectrometer. The chalcogenide sensor connects to other chalcogenide fibers which act as a transmission link to the FTIR spectrometer. The results indicate that this equipment and sensors are suitable for monitoring the degree of cure of the laminates throughout the entire cure cycle.

In previous work, we have reported on the use of this technique in the laboratory for materials such as polyimides and epoxies[1-3]. The earlier results were obtained using a prototype equipment and optical fibers. In the current study which we are reporting on here, we are using a spectrometer and optical fiber arrangement which has been further developed to enable its use either in a laboratory autoclave or laboratory press environment. The results which are discussed in this paper were obtained on this new system.

This technique, particularly in combination with other types of sensors, holds promise for alleviating some of the problems associated with the manufacture of advanced materials.

ACKNOWLEDGMENTS

The research and development presented in this paper is supported by the NASA Langley Research Center through a Small Business Innovative Research (SBIR) contract. The authors also wish to thank Philip R. Young for his continued interest in this effort.

REFERENCES

1. D. A. C. Compton., S. L. Hill, N. A. Wright, M. A. Druy, J. Piche, W. A. Stevenson and W. Vidrine, Appl. Spec., August 1988.

2. M. A Druy, L. Elandjian and W. A. Stevenson, SPIE Proceedings, Vol. 986, pp.130-134, 1988.

3. P. R. Young, M. A. Druy, W. A. Stevenson and D. A. C. Compton, SAMPE J., Vol. 25, No.2, pp. 11-16, 1989.

Simultaneous Measurement of Strain and Temperature Variations in Composite Materials.

W.Craig Michie, B. Culshaw, S.S.J. Roberts, R.Davidson.

The discrimination between the effects of temperature and strain in a single integrated measurement has been investigated using an optical fibre sensor. A sensitivity to strain and temperature variations to within 20 $\mu\epsilon$ and 1 K over a strain and temperature excursion of 2 mϵ and 45 K has been achieved over an 80 cm sensing length. This high degree of resolution is obtained through the use of single sensing fibre interrogated simultaneously with two different sensing schemes: a polarimeter combined with a dual moded interferometer.

Dual moded and polarimetric sensing techniques are well understood. This present approach employs both schemes in order that the relative difference in sensitivities of the two approaches can be used to recover strain and temperature information from a single integrated measurement. A polarisation maintaining fibre is interrogated as a polarimeter with a laser diode of centre wavelength 830 nm while operated simultaneously as a dual moded interferometer with a laser source of 633 nm wavelength. In principle, the use of two separate wavelengths and two separate sensing schemes allows the sensitivity of the system to be optimised by using the dispersive properties of the fibre and the relative differences in the sensitivities of the two methods. The sensing fibre is first characterised in terms of the relative sensitivities of the dual moded and polarimetric interferometers to temperature and strain and represented in a two dimensional matrix form. Strain and temperature recovery is performed through measurement of the total phase change in the two measurements systems and matrix inversion. The choice of fibre determines the relative sensitivities of the two sensing schemes thus controlling the conditioning of the matrix and hence the overall accuracy of the measurement. In the present experiment a number of different fibre types have been investigated. The elliptical core fibre from the Andrews

Corporation has been found to be the better suited to the task than stress birefringent fibres. In an elliptical core fibre the contribution to the birefringence from the internal stress variation across the fibre core diameter is small, therefore the dominant influence in determining the strain sensitivity is the fibre elongation. Consequently the dual moded signal is an order of magnitude more sensitive to strain than the polarimeter. Both sensing schemes have however a similar sensitivity to temperature thus the relative difference in sensitivities allows the temperature and strain to be recovered in a single integrated measurement.

Experimental evaluation of the measurement method was carried out over an 80cm sensing length of fibre encapsulated in a chamber which allowed the fibre to be strained in a temperature controlled environment. The temperature of the fibre was obtained from averaging the measurements of three thermocouples placed at the top, middle and bottom of the fibre length. The sensor was heated and allowed to cool by approximately 45 K while being simultaneously extended in length by 1.9 mm. During this process the recorded changes in the amplitude of the polarimeter and the dual moded signal indicated changes in respective phases of -8.16 rads (approximately 1.3 cycles) 22.5 cycles (141.4 rads). This corresponds to a measured change in length of 1690 μm and a change in temperature of - 45 K. A rapid relaxation of the applied strain was carried out during which the fibre experienced a drop in temperature of approximately 3.8 K. The measured changes during this process were -1689 μm and -4.2 K. These measured changes in temperature and strain are in excellent agreement with those applied.

Further to the investigation of the suitability of the sensing scheme on a general basis, a length of elliptical core fibre has been embedded in an 8 ply unidirectional Carbon Fibre Reinforced Plastic composite sample. This sample has bee subjected to similar temperature and strain cycling. A comparative analysis between the performance of the embedded and the unembedded fibre sensor will be presented.

Bend-Insensitive Single Mode Fiber for Embedding in Composite Materials

Gérard Orcel

SpecTran Corporation, 50 Hall Road, Sturbridge, MA 01566

and

Russell May, Jonathan Green, Richard O. Claus

Fiber & Electro•Optics Research Center, Bradley Department of Electrical Engineering, Virginia Tech, Blacksburg, VA 24061-0111

Optical fibers are being applied or contemplated for data transmission, and sensing and diagnostic applications in advanced aircraft. For aircraft is constructed of composite materials, the optical fibers may be embedded directly in the material. Possible uses for embedded fibers include monolithic backplanes for avionics computers, or strain, temperature, and pressure sensors for smart structures.

One difficulty in the use of embedded optical fibers is that structural reinforcing fibers often generate microbends in the optical fibers, inducing unacceptably large losses. The magnitude of these losses depends on fiber refractive index profile, fiber coating modulus, and the type and orientation of the structural fibers.

We report the development and evaluation of singlemode bend-insensitive (SMBI) optical fibers with high temperature polymeric coatings for embedding in composite materials. An SMBI fiber was fabricated for use at 1300 nm by designing the fiber's refractive index profile to increase the numerical aperture to a value higher than that of a standard telecommunications fiber. This design reduces the tendency of the mode field to radiate when the fiber is bent. A peel point attenuation test, which determines the loss induced in a fiber by a single quarter turn around a round mandrel, demonstrated that the SMBI fiber suffered 4.5 dB less loss than a singlemode matched clad fiber when bent around a 3 mm diameter mandrel. The fiber was coated with a high temperature polyimide polymer coating capable of withstanding continuous exposure of temperatures up to 300° C with no reduction in fiber strength.

The attenuation of the SMBI fiber when embedded in graphite/epoxy coupons was tested under varying temperature and tensile strain extremes. Standard telecommunications fiber with identical polyimide coating was also embedded and submitted to the same test. In addition, SMBI and standard fibers with UV-cured acrylate coatings were also embedded and tested.

For the temperature tests, the fibers were embedded in the middle of a twelve-ply Hercules AS4/3501–6 Gr/Ep laminate. The dimensions of the finished coupons were 2"x6"x0.072". In half of the specimens, the optical fiber was laid parallel to adjacent plies, and the remainder were fabricated with the fiber perpendicular to adjacent plies. The optical attenuation of the fibers was monitored as the temperature of the coupons was varied between -40°C and +85°C.

For the tensile strain tests, the fibers were embedded in the middle of six-ply Gr/Ep with dimensions of 1"x6"x0.036". The optical attenuation of the fibers was monitored as the coupons were loaded up to 20 KN force in a load frame.

In addition, the interfacial shear strengths of the fiber/coating/matrix interfaces was evaluated for both the polyimide coated SMBI fiber and the acrylate coated SMBI fiber. Single fibers were embedded in dogbone shaped specimens of neat Hercules 3501-6 resin, and pullout tests were performed using tensile test instruments.

Decentralized Control: Distributed Intelligence for Smart Structures

K. David Young
Lawrence Livermore National Laboratory
University of California
Livermore, California 94550

Umit Ozguner
Ohio State University
Columbus, Ohio 43210

Abstract

Decentralized Control is intuitively appealing for vibration control of large flexible structures: It offers simplified control system implementations which only require the feedback of local measurements to close the control loop for systems that may have a large number of control loops. In Smart Structures, it is likely that actuators, sensors, and control computations are widely distributed throughout the structure. While the theory and practice of Decentralized Control in the control of large structures have made substantial progress in the past decade, the advent of Smart Structure concepts requires an assessment of the current status and future needs of Decentralized Control and its applicability to Smart Structures. In this paper, we present an overview of the current status of Decentralized Control and examine the modeling and control system design issues in Smart Structures from a distributed intelligence point of view.

PIEZOCERAMIC/DSP-BASED INTEGRATED WORKSTATION FOR MODAL IDENTIFICATION AND VIBRATION CONTROL

J. Su, M. Rossi, G. Knowles, F. Austin
Grumman Corporation
Corporate Research Center
Bethpage, NY 11714
Mail Stop A08-35

ABSTRACT

A piezoceramic/DSP-based integrated workstation was developed for modal parameter identification and vibration control of flexible structures. Utilizing a novel technique, the workstation was configured to perform very accurate estimation of the modal parameters of a cantilever beam. Active vibration control of a cantilever beam also was accomplished by configuring the workstation as a simple gain-controlled feedback loop. Although all preliminary usage involved a cantilever beam, the workstation proved to be an effective data acquisition, processing, and waveform generation system that can be configured for the modal identification and vibration control of all structures.

The workstation is an integrated system of an IBM 80486 computer and a Texas Instrument TMS320C30 DSP with analog-to-digital (AD) and digital-to-analog (DA) converters, piezoceramic sensors and actuators, power amplifiers, and wide-dynamic-range instrumentation amplifiers. The IBM 80486 computer is the host to the DSP, and provides the interface between the user and the DSP. DSP software development, loading and execution, as well as DSP data visualization and postprocessing all are performed there. The DSP is the main processor. It analyzes the sensor data, and generates control signals to the actuators. The workstation can process structural acceleration and strain data from 34 piezoceramic sensors and generate control signals for 18 actuators. The 200 W power amplifiers provide the power to drive the piezoceramic actuators; its instrumentation amplifiers provide 1 to 8000 gain to provide full AD resolution to piezoceramic sensor signals from a wide dynamic range.

Modal parameter estimation of a cantilever beam was accomplished with the workstation by measuring the steady-state response of the beam under sinusoidal excitation. The modes were first detected using stochastic methods. Frequency response of the beam near each mode was then determined using a novel technique that measured the steady-state response of sinusoidally excited structures with high accuracy and precision. The modal parameters, which consisted of the natural frequency and the damping were then calculated from the response by locating the 90 degree phase shift, the peak amplitude, and the 3 dB bandwidth. Very accurate estimates were obtained even for the low frequency modes.

The novel technique involved exciting the structure under test with a high-precision sine wave generator, measuring its steady-state response using an adaptive noise canceller to cancel the sinusoidal components of the measurement with the same frequency, and using the filter's impulse response as an estimate of the response. This corresponds to exciting the structure with a pure sine wave and then measuring the output with a very narrow bandpass filter tuned to the frequency of the sine wave. Very precise and accurate frequency response measurements can be made this way even when the steady-state response is buried in the noise. Implemented on the workstation, this technique can determine the frequency response from 0.01 Hz to 4000 Hz.

Active vibration control of the cantilever beam was performed by configuring the workstation in a feedback loop. Acceleration data from the cantilever beam were converted to velocity and filtered to pass only the first mode. This signal was then multiplied by a gain and fed back to the piezoceramic bending motor at the base of the cantilever beam. Preliminary results are encouraging.

The Intelligence Between Sensing and Actuation for Smart Structures

Ü. Özgüner and L. Lenning

The Ohio State University
Department of Electrical Engineering
2015 Neil Avenue
Columbus, Ohio 43210

Most of the recent effort in the Smart Structures area has been concentrated in developing imbedded sensors and actuators for the structures. It is clear, however, that pure sensing and actuation is not what makes a system, a structure, or even a living organism "smart". In this talk, we address the basic intelligence required to connect the distributed sensors and actuators, so that the structure acts in a desired way. We claim that a knowledge of "self" is required as a building step for any intelligence or smartness to be claimed. To this end, a smart structure will need to have a model of its own dynamics, and if the intelligence is to be distributed, such models have to be distributed throughout the structure. Based on such models, a truly smart structure can control its own actions.

With the above viewpoint, we shall consider models of flexible mechanical structures realized as analog circuits distributed and embedded on the structure itself. The circuits are developed based on analogies betwean voltages/currents in electrical circuits, waves along wave-guides and vibrations in mechanical structures that can be exploited in both modeling and controller design. Such analogies have been mentioned in classical texts, and even used to a certain extent in the design of mechanical filters. In this talk, we shall claim that it is time for a new look at insights provided by such analogies.

We are basically interested in active vibration damping in large flexible structures, especially for utilization in space. We shall consider the coupling of substructures, overlapping controller design and the effect of proof–mass and piezoelectric actuators from a circuit equivalent viewpoint. We shall also show how the circuit analogies can set the stage for sensitivity minimization and tie in with standard decentralized controller design.

Decentralized Control Experiments: Implications for Smart Structures

D. C. Hyland, E. G. Collins, Jr.,
D. J. Phillips, and J. A. King

Harris Corporation
Government Aerospace Systems Division
MS 22/4847
Melbourne, FL 32902

Abstract

This paper considers the implications of using decentralized control for active vibration suppression of flexible structures. The discussion is based on results obtained from two experiments conducted as part of the NASA CSI Guest Investigator Program. The first experiment was conducted using the ACES structure at NASA Marshall Space Flight Center while the second expeririment involved the Mini-MAST structure at NASA Langley Research Center.

The basic test article of the ACES testbed is a spare Voyager Astromast, a deployable, lightweight (about 5 lb.), lightly damped beam, approximately 45 ft. in length. The ACES configuration consists of an antenna and conterweight legs appended to the Astromast tip and pointing gimbal arms near the Astromast base. Overall, the structure is very flexible and lightly damped. It contains many closely spaced, low frequency modes (more than 40 modes under 10 Hz.).

The goal of control design for the ACES experiment was to position the laser beam in the center of the detector. In our control design and implementation we used 8 control inputs and 8 measurement outputs. A decentralized configuration was chosen for the controller. This paper details the design procedure and describes the substantial performance improvement achieved with a decentralized structure. The paper also discusses some of the reasons that a decentralized controller was so effective with this controller.

The basic test article of the Mini-MAST testbed is a generic space truss designed and manufactured by Astro Aerospace Corporation. The truss beam is deployable and retractable and has a triangular cross section. The total height of the truss is 20.16 meters and the truss consists of 18 bays, each of which is 1.12 meters in height.

The basic control objective for the Mini-MAST experiment was to minimize the displacements at the tip of the Mini-MAST. The control designs used 3 control inputs and up to 5 sensor outputs. Both centralized and decentralized controllers were designed and implemented. This paper presents some of the highlights of this design process and compares the performance obtained with decentralized control with the performance achieved with centralized control.

Active Materials and Adaptive Structures Conference
Arlington, VA

Nov. 5–7, 1991

A Workstation Environment for Design of Vibration Control for Flexible Structures Using Digital Signal Processors*

William H. Bennett, Ph.D.
TECHNO-SCIENCES, INC.
7833 Walker Dr., Suite 620
Greenbelt, MD 20770
(301)345-0375

ABSTRACT

Design of multivariable control laws using modern methods of optimization (e.g. H_∞, H_2) requires extensive numerical computational algorithms to evaluate design tradeoffs. Modern computational algorithms for standard optimal control design utilize state space constructions for computations and control law realization. Such constructions are often awkward for identification of optimal control laws for distributed parameter systems arising in control of vibrations in structures with multiple actuators and sensors and must rely on model order reduction *prior to control law optimization*. One alternate approach under development at TSI is to utilize explicit frequency domain constructions for modeling of distributed parameter dynamics arising in control of vibrations in structures. Optimal control design and implementation of control laws is achieved using computational algorithms based on data sampling of the frequency response. Optimal Wiener-Hopf control laws are obtained by solving a spectral factorization based on frequency response data. *No model order reduction is used in the optimal control computations.*

In this presentation we demonstrate a special purpose workstation configuration and software environment for frequency sampled modeling and computations for design of vibration control for flexible structures. The workstation configuration utilizes high speed Digital Signal Processors (AT&T DSP-32C) to emulate and implement the optimal control law using a MIMO FIR construction which interpolates the frequency samples obtained by optimal control computations. A computational example will be illustrated using the workstation environment.

*Research supported by SDIO/SBIR and managed by WRDC/FIBRA, Wright-Patterson AFB

An Abstract of a paper submitted
for presentation at the
AMAS Conference
November 5-7, 1991, Virginia

Modal Survey and Test-Analysis Correlation of a Multiply-Configured Three-Stage Booster [1]

Edward L. Marek [2]
Linda J. Branstetter, Member ASCE
Thomas G. Carne, Member AIAA
Randall L. Mayes
Sandia National Laboratories
Albuquerque, New Mexico, USA, 87185

A three-stage solid propellant booster system has been developed to launch a variety of payloads. A critical issue to the success of any flight is the possible interaction of the elastic vibrational modes of the booster with the control system. It is important to have high confidence in the accuracy of the mathematical model of the booster not only at launch, but also at certain other critical flight times when its configuration has changed drastically.

A three-dimensional continuum finite element model (FEM) of the booster was assembled using superelements and component mode synthesis (CMS) reduction. The MSC/NASTRAN [1] model appears on the lefthand side of Figure 1. A subset of selected grid locations from the FEM was used to define a structural display model, useful for clear and computationally efficient visualization of the booster mode shapes. The display model is shown on the righthand side of Figure 1. The first two free-free bending modes of the booster at launch are shown schematically with the display model in Figure 2. It was extremely important to correlate the FEM with modal test data. This was due to the composite material used in the construction of the booster motor cases,

[1]This work was performed at Sandia National Laboratories and supported by the U.S. Department of Energy under contract DE-AC04-76DP00789.

[2]Address Correspondence to Edward L. Marek, Sandia National Laboratories, Division 1545, P.O. Box 5800, Albuquerque, NM, 87185.

the complex internal propellant grain geometries of the first and second stage motors, uncertainty in the propellant dynamic characteristics, and several other factors.

A series of nominally free-free modal tests was performed using both inert and live propellants with several booster configurations. For each of the tests, a single payload simulator was mounted to the payload adapter plate. Three tests were performed with available hardware and inert propellants, to obtain a maximum amount of structural response information early in the development program. Closer to first launch, two tests were performed using live propellants. Several supporting tests of subassemblies were also performed. Schematic diagrams of the tested configurations appear in Figure 3. In addition to the system-level tests, a separate test of the payload simulator alone was performed. For the all-inert system-level tests, the free-free condition was closely approximated by suspending the booster vertically from an overhead bridge crane with a long flexible strap. A photograph of the inert system-level test with an empty first stage motor case (see Figure 3(c)) is shown in Figure 4. For the test configurations involving live propellant, the booster was suspended by its nose using an assembly of straps, rigid links, pulleys, cables, and a hydraset. This suspension system was attached to the boom of a truck-mounted hydraulic crane.

It was necessary to assemble an accurate model of the crane and suspension system, for the purpose of test-analysis correlation for the live-propellant tests. During pre-test calibrations with a rigid booster mockup, the initial suspension system design was found to have significant coupling with the booster. This indicated that the suspension would not provide ideal free-free isolation of the test article. A finite element model of the preliminary suspension system was developed, correlated to the modal data using the booster mockup, and then altered to reflect needed design changes. A differential stiffness formulation captured the effects of tensioning of the suspension system elements, which behaved much like a tensioned cable of varying weight per unit length. The suspension system was designed so that modes involving a significant amount of booster/suspension coupling occurred well below the first bending frequency of the booster. The overall length of the suspension system was constrained by the height of the test facility. Results of the correlation of the preliminary suspension system appear in Table 1. Inclusion of accelerometers at various positions on the suspension system during the test allowed reliable correlation of the final booster/suspension/crane model to the test results.

Test-Analysis Models (TAMs) were derived from the FEM. A TAM is a reduced representation of the FEM, where the TAM degrees of freedom (DOF) correspond one-to-one with the test accelerometers. An accurate TAM will represent the same stiffness and mass properties as the FEM. The TAM is a mathematical link between the FEM and the test results. This allows for optimal selection of accelerometer quantities, locations, and orientations, and for on-site data evaluation. TAMs were derived from the FEM using three different methods: static (Guyan) reduction [2]; the Improved Reduced System (IRS) method [3]; and the Hybrid TAM method [4]. It was found that for this system, the static TAM was inadequate for the selected number of measurement DOF, because static reduction ignores the inertia effects of mass at non-instrumented DOF. The approximate dynamic mass matrix formed with the IRS approach allowed dramatic improvement in the TAM when numerous internal propellant grids were included in the reduced order CMS model. The Hybrid TAM, which provides an exact reduction of the FEM mass matrix, was very accurate.

Table 1. Correlation results for a preliminary suspension system design.

Mode Shape Description	Test Frequency (Hz)	Correlated Suspension System FEM Frequency (Hz)
Pulley/cable 1st bending	4.13	4.11
Pulley/cable 1st bending	4.25	4.27
Pulley/cable 2nd bending	11.75	11.71
Pulley/cable 2nd bending	12.88	12.85
Crane lateral bending	1.74	1.73

Exciter locations and combinations were identified during pretest analyses, which were successful in adequately exciting all booster modes of interest. Using the FEM, frequency response functions (FRFs) were computed for every TAM accelerometer channel, using several potential exciter locations for the live-propellant configurations. The FRFs were used to compute mode indicator functions (MIFs) and multivariate mode indicator functions (MMIFs). Review of this data allowed each exciter location to be evaluated prior to the test.

Numerous researchers have used identification procedures to correlate structural models to the results of modal tests [5,6,7,8]. Recently, test-analysis correlation using design sensitivities has been applied to numerous aerospace structures, including the STS Centaur, the Space Shuttle solid rocket motor, and the commercial Titan dual payload carrier [9,10,11]. Design sensitivity analysis was used in conjunction with an optimization algorithm to correlate the FEM of the booster and the suspension system to the modal test data. MSC/NASTRAN design sensitivity coefficients were utilized to assess the effects of changes in various system parameters on the modal frequencies. Simultaneous use of data from the multiple test configurations allowed a unique and consistent parameter identification to be accomplished which provided confidence in the resulting FEM at important flight times. The resulting test-verified FEM components were assembled into a set of NASTRAN databases. When coupled with separate test-verified superelement models for the flight-dependent booster payloads, this data will be

used for timely and efficient calculation of the system structural dynamic characteristics (modal properties and transient responses) for each mission.

References

[1] MSC/NASTRAN User's Manual, Version 65, Los Angeles, California, MacNeal Schwendler Corporation, November, 1985.

[2] Guyan, R. J., "Reduction of Stiffness and Mass Matrices", *AIAA Journal*, Vol. 3, February 1965.

[3] O'Callahan, J., "A Procedure for an Improved Reduced System (IRS) Model", *Proceedings of the 7th International Modal Analysis Conference*, Las Vegas, Nevada, January 1989.

[4] Kammer, D. C., "A Hybrid Approach to Test Analysis Model Development for Large Space Structures", Submitted for publication in *Journal of Spacecraft and Rockets*, November 1989.

[5] Collins, J. D., Hart, G. C., Hasselman, T. K., and Kennedy, B., "Statistical Identification of Structures", *AIAA Journal*, Vol. 12, February 1974.

[6] Hart, G. C., and Martinez, D. R., "Improving Analytical Dynamic Models Using Frequency Response Data - Application", *Proceedings of the 1982 Structures, Structural Dynamics, and Materials Conference*, Paper 82-0637.

[7] Carne, T. G., and Martinez, D. R., "Identification of Material Constants for a Composite Shell", *Proceedings of the 5th International Modal Analysis Conference*, London, England, April 1987.

[8] Allen, J. J., and Martinez, D. R., "Techniques for Implementing Structural Model Identification Using Test Data", *SAND90-1185*, Sandia National Laboratories, Albuquerque, New Mexico, June 1990.

[9] Flanigan, C. C., "Test/Analysis Correlation of the STS Centaur Using Design Sensitivity and Optimization Methods", *Proceedings of the 5th International Modal Analysis Conference*, London, England, April 1987.

[10] Brillhart, R. D., Hunt, D. L., and Kammer, D. C., "Modal Survey and Test-Analysis Correlation of the Space Shuttle SRM", *Proceedings of the 6th International Modal Analysis Conference*, Orlando, Florida, February 1988.

[11] Brillhart, R. D., Freed, A. M., Hunt, D. L., and Chism, T. L., "Modal Test and Correlation of the Commercial Titan Dual Payload Carrier", *Proceedings of the 8th International Modal Analysis Conference*, Kissimmee, Florida, January 1990.

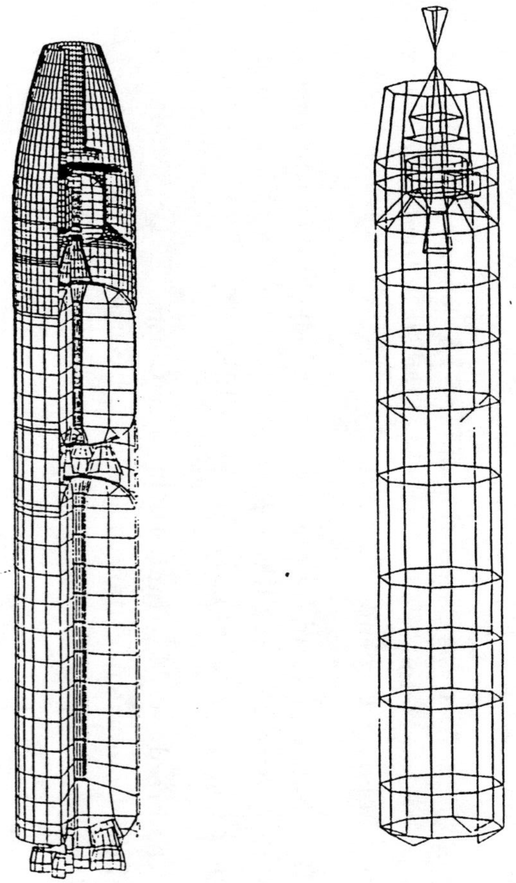

Figure 1. MSC/NASTRAN finite element model, and a corresponding display model, of a three-stage solid propellant booster system.

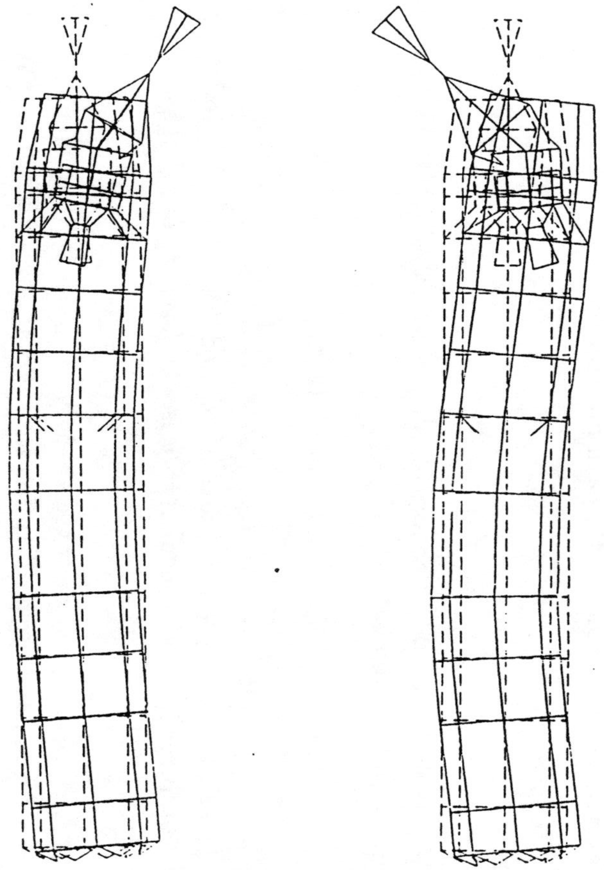

Figure 2. First two free-free bending modes of the booster in its launch configuration.

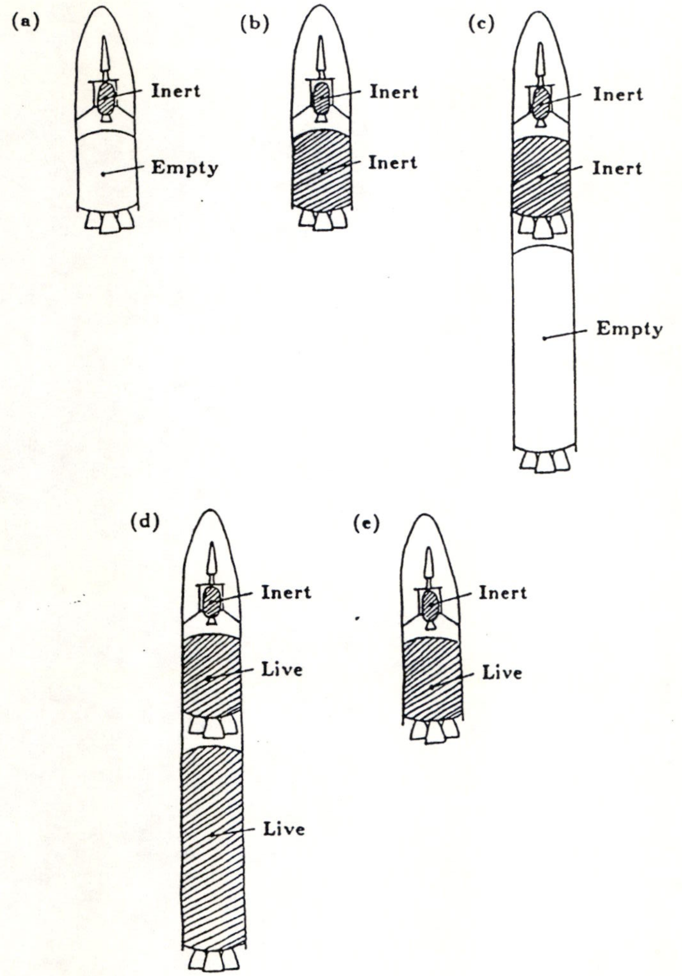

Figure 3. Booster configurations used for system-level modal tests involving both inert and live propellants.

Figure 4. Inert system-level test with an empty first stage motor case.

Comparison of Four Methods for Calculating Vibration Mode Shape Sensitivities

Farhang Aslani[1] , Nickolas Vlahopoulos[1] ,
Ichiro Hagiwara[2]

Frequency and mode shape sensitivities are often needed in a process involving calibration or design optimization of a dynamic structural model. Either process uses an optimization scheme in which the sensitivity coefficients can be used to reduce the number of iterations needed to reach a solution giving optimum changes in design variables. Unlike mode shape sensitivities, frequency sensitivities involve minor computational effort. The earlier work by Fox and Kapoor [1] provided the formulation for calculation of frequency sensitivities while the formulation given for mode shape sensitivities suffered from inaccuracy. Nelson suggested an exact method for mode shape sensitivities [2] which was computationally intensive since it required one decomposition per mode of interest. The Nelson's Simplified Method (NSM) has been installed on MSC/NASTRAN [3] and it proved to be very expensive for working environment structures having several degrees of freedom.

In the search for a method more computationally efficient than the NSM, other methods have been suggested by some researchers [4-7]. These methods are based on finite-difference method or modal method. Sutter [8] has compared the efficiency of these methods on two small structures. However, the results of comparisons are not valid for working-environment structures having thousands of degrees of freedom. The purpose of the study presented in this paper is :

1. To develop a new method for calculation of mode shape sensitivities based on finite difference method. It indirectly calculates the modes shapes of perturbed structure in a reduced space. The method will be called Indirect Reduced Basis Method (IRBM).

---

[1] Staff Engineer, Automated Analysis Corporation, Ann Arbor , Michigan.
[2] Research Scientist, NISSAN MOTOR Co., Yokosuka, Japan

2. To implement the three methods in MSC/NASTRAN using Direct Matrix Abstraction Program (DMAP). Thease methods are :

   a. Implicit Modal Superposition Method (IMSM) by Wang [7]

   b. Direct Reduced Basis Method (DRBM) by Wang [10],

   c. Indirect Reduced Basis Method (IRBM) developed by the authors.

3. To compare the accuracy and efficiency of the above three methods with the NSM using a 1000-node sample truck model having 46 design parameters. The comparison is based on central processor unit (CPU) time needed by each method. The accuracy was compared with reference to the NSM results.

REFERENCES

1. Fox. R. L. and Kapoor, M.P., "Rate of change of eigenvalue and eigenvectors," AIAA Journal, Vol. 6, pp. 2426-2429, December 1968.
2. Nelson R.B., "Simplified Calculation of Eigenvector Derivative," AIAA Journal, Vol. 14, pp 1201-1205, 1976.
3. MSC/NASTRAN Application Manual, " Application Notes - January 1986," Mac-Neal Schwendler Corporation, Los Angeles, California, January 1986.
4. Ojalvo, I. U., "Gradients for Large Structural Models with Repeated Frequencies,' Society of Automotive Engineers, SAE TP Series 861789, Oct., 1986.
5. Ojalvo, I. U., "Efficient Computation of Mode-Shape Derivatives for Large Dynamic Systems," AIAA Paper 86-0871, May 1986.
6. Wang, B.P., "An Improved Approximate Method for Computing Eigenvector Derivative, "Presented at a Work-in-progress Session, 26th AIAA/ASME/ASCE/AHE Structures, Structural Dynamics and Material Conference, Orlando, Florida, April 1985.
7. Wang, B.P., "Improved Approximate Methods for Computing Eigenvector Derivatives in Structural Dynamics," AIAA Journal, May 1991.
8. Sutter, T.R., Camarda, C.J., Walsh, J. L., and Adelman, W. M, "Comparison of Several Methods for the Calculation of Vibration Mode Shape Sensitivities," AIAA Journal, Vol. 26, pp 1506-1511, 1968.
9. Wang B. P., Caldwell, S. P., and Smith, C. M., " Improved Eigensolution Reanalysis Procedures in Structural Dynamics," Proceedings of MSC/NASTRAN World Users Conference, Los Angeles, CA, March 26-30, 1990.

Structural Identification Using Mathematical Optimization Within a Production Finite Element Analysis Code

Mark S. Ewing, Senior Research Engineer
Wright Laboratory, Wright–Patterson Air Force Base
(WL/FIBRA, WPAFB, OH 45433-6553)

Structural identification in terms of physical variables using mathematical optimization techniques has been established for simple, two-dimensional structures [1]. Consider a structural finite element model with physical parameters, r_i, as well as analytically determined mode shapes, or eigenvectors, $\underset{\sim}{\phi}_j$ (with individual elements,ϕ_j^k), and natural frequencies, ω_j. (For notational convenience, $\lambda = \omega^2$.) The actual structure has physical parameters, $\bar{r}_i$, mode shapes, $\underset{\sim}{\bar{\phi}}_j$, (with individual elements,$\bar{\phi}_j^k$), and natural frequencies, $\bar{\omega}_j$.

To identify the structure in terms of the model's physical parameters, the following optimization problem may be solved. Minimize either:

$$\sum_{j=1}^{J} ||\underset{\sim}{\phi}_j - \underset{\sim}{\bar{\phi}}_j||^2 \tag{1}$$

or:

$$\sum_{j=1}^{J} (\frac{\lambda_j}{\bar{\lambda}_j} - 1)^2 \tag{2}$$

subject to the constraints:

$$| \frac{\lambda_j}{\bar{\lambda}_j} - 1 |< a_j \quad for\ j = 1, 2, ...J' \tag{3}$$

$$||\phi_j - \bar{\phi}_j||^2 < b_j \quad for\ j = 1, 2, ...J" \tag{4}$$

where: J is the number of modes used for the identification task (some inappropriate modes can be neglected); both a_j and b_j are "small" numbers chosen to quantify how close to target is "close enough"; J' is the number of modes which are to have particularly accurate frequency matching; $J"$ is the number of modes which are to have particulary accurate mode shape matching. An objective function based on the frequency response characteristics of a structure is also possible and in some cases actually preferred [2].

The extension of mathematical optimization techniques to a production size finite element code demands allowance for variable linking to reduce the number of variables in the problem [3]. However, it also requires considerable attention to the issue of how to best compare analytical and experimental mode shapes [4].

For large problems, the free vibration problem can only practically be solved through the use of some form of model reduction scheme, for instance, Guyan reduction. This suggests chosing the so-called "master" degrees of freedom from such a reduction to coincide with measurement degrees of freedom. However, this is not always practical or desired [4]. In any event, any updating of mass and stiffness matrices as described above requires mass and stiffness matrix gradients; unfortunately, only the gradients of the non-reduced mass and stiffness matrices are generally known. Gradients in the reduced problem, then must be generated by the reduction rules - or transformations - which define the reduction.

Once the desired sensitivities are known in the reduced problem space, a general updating algorithm must have the facility to either further reduce analytical modeshapes to match measured mode shapes or expand measured modeshapes to match analytical mode shapes. In either case, since the difference between the analytical and measured mode shapes appears in the mode shape constraint (equation 4) and in the objective function, if equation 1 is used, the gradients of these differences must be available to the optimization algorithm. So, gradients known in the full or reduced space must be transformed to the appropriate "comparison space".

The Automated STRuctural Optimization System (ASTROS) [5] is currently being modified to allow the use of equations 1 and 2 as objective functions and equations 3 and 4 as constraints. Examples of the use of ASTROS with the structural identification task option will be presented. Both aircraft and space structures will be addressed.

References

1. Ewing, M. S. and Venkayya, V. B., "Structural Identification Using Mathematical Optimization Techniques", Proceedings, AIAA 32nd Structures, Structural Dynamics and Materials Conference, Baltimore, MD, 8-10 April 1991, pp 840-845. (AIAA paper 91-1135)

2. Hasselman, T. K. and Chrostowski, J. D., "A Recent Case Study in System Identification", Proceedings, AIAA 32nd Structures, Structural Dynamics and Materials Conference, Baltimore, MD, 8-10 April 1991, pp 2154-2168. (AIAA paper 91-1190)

3. Ewing, M. S. and Kolonay, R. M., "Dynamic Structural Model Modification Using Mathematical Optimization Techniques", Proceedings, OPTI '91 – Computer Aided Optimum Design of Structures, Boston, MA, 25-27 June 1991.

4. Imregun, M. and Visser, W. J., "A Review of Model Updating Techniques", (featured Technical Article), Shock and Vibration Digest, vol. 23, no. 1, pp 9-20, Jan 1991.

5. Johnson, E. H. and Venkayya, V. B., "Automated STRuctural Optimization System (ASTROS), AFWAL-TR-88-3028, Dec 1988.

CORRELATION OF FINITE ELEMENT MODELS USING MODE SHAPE DESIGN SENSITIVITY

Christopher C. Flanigan
SDRC Engineering Services Division, Inc.
San Diego, California

Abstract

Procedures for test/analysis correlation based on design sensitivity and optimization methods were enhanced by including mode shape sensitivity to augment modal frequency information. Mode shape sensitivities were calculated and added to the correlation problem as additional state variables. Special techniques were developed to process the mode shape design sensitivity coefficients efficiently. The combination of modal frequency and mode shape state variables increased the amount of data available to the correlation problem, which resulted in improved results.

SENSITIVITY ANALYSIS OF RESPONSES TO DYNAMIC LOADS

Warren C. Gibson
CSA Engineering, Inc.
Palo Alto, CA

Submitted to

ADPA/AIAA/ASME/SPIE Conference on
Active Materials and Adaptive Structures

Nov 5-7, 1991

Sensitivity analysis is a key element of structural optimization, optimization-based methods of parameter identification, and model tuning. Sensitivities of responses to dynamic loads pose special challenges because of the volume and complexity of data in a complete frequency spectrum or time record. This paper presents an efficient method of computing the sensitivity of one or more peaks in a frequency spectrum or time domain. The method accounts for shifts in peak location as well as changes in peak magnitude as design variables are varied. The method has been implemented in NASTRAN's DMAP language and linked to the ADS general-purpose optimizer. Examples are presented that illustrate optimization of both structural mass and viscoelastic damping treatments.

MACROMOLECULAR SMART MATERIALS AND STRUCTURES

Darrell H. Reneker
Wayne L. Mattice
Roderic P. Quirk

Institute of Polymer Science
The University of Akron
Akron, Ohio 44325-3909

Smart molecules are designed to receive a stimulus, transmit or process it, and then to respond by producing a useful effect. Our approach to smart molecules is modular. Modules, consisting of oligomers with particular receptor or functional properties, are synthesized separately and then combined in appropriate ways to serve a variety of needs.

Four oligomers with different functions are now being designed, synthesized, and characterized. The four functions are attachment to a solid surface, attachment to a polymer matrix, a stress sensitive chromophore, and a weak link that will break in a predictable way.

In the future, many other functions can be included in this modular approach. The design of the oligomers and the designs for combinations that constitute smart molecules are being developed with powerful computer modeling facilities and software. These include conformational modeling, molecular dynamics, crystallography, and ab-initio calculations of molecular electronic wave functions.

The solid surface being used in the study of attachment is graphite. Graphite was chosen because it is similar to the carbon or graphite fibers used in composites, and to the carbon black used in the manufacture of rubber. We prepare graphite surfaces with monatomic steps, which are chemically functionalized so that the steps serve as identifiable points of attachment of the oligomer that serves as the attachment module.

The attachment to graphite, of individual oligomers such as octadecyl amine or a polystyrene oligomer with a molecular weight of 8000 which has amine groups at its ends, are directly observed with the scanning tunneling microscope or the atomic force microscope. Octadecyl amine molecules tend to lie flat on the graphite, with the molecule in a planar zig-zag conformation. Clusters of octadecyl amine molecules form near the functionalized monatomic steps. The polystyrene oligomer is observed in a random coil conformation at a monatomic step on the graphite, with several random coils joined together. These observations of oligomers attached to graphite are regarded as preliminary, and it is expected that as the attachment process is varied, other conformations will be produced and observed.

An important ancillary benefit of this part of the work is the development of a predictable way of attaching molecules to graphite for study with the scanning tunneling or atomic force microscopes. The molecule of interest can be chemically provided with a short, flexible chain terminated with an amine group. Then, using the processes for attachment being developed in this work, the molecule can be attached to a monatomic step and observed.

Dye molecules which change shape and color simultaneously are being considered for the stress sensitive chromophore module. Candidates includes stilbenes, azo compounds and spiropyrans. Appropriate oligomers will be used to hold the dye molecule at the point where the stress is to be sensed. The stress sensitive chromophores with appropriate attachment modules are studied experimentally by placing them at the interface between a hard particle and a rubbery matrix, and stretching the matrix. Oligomers with an ionic bond in the backbone to provide a weak link are being investigated in similar ways.

Candidates for the oligomers for attachment to a polymeric matrix have been synthesized and studied as copolymers to stabilize blends of different polymers. Oligomers of polystyrene, polyethylene oxide, and polybutadienes have been prepared with a variety of functional groups at their ends.

The four functions listed above can be combined, for example, to make a smart molecule for use at the interface between a reinforcing fiber and a polymer matrix. This smart molecule minimizes the damage from a broken reinforcing fiber by distributing the local strain from the broken fiber over a larger volume of the matrix, and simultaneously provides a visible indication that the composite structure was damaged.

Such a smart molecule has an oligomer that attaches the smart molecule to the fiber. This attachment molecule is connected to two branches. The first branch contains a stress sensitive chromophore and a weak link in series. The second branch contains a longer, strong chain which carries the load when the weak link is separated. The two branches are rejoined and connected to an oligomer that attaches to the polymer matrix. If the reinforcing fiber breaks, the high local strain causes the weak link to break, changing the color of the chromophore, and contributing to the visual indication of damage. The longer chain maintains the mechanical connection between the fiber and the matrix so that the part does not fail catastrophically.

Time Resolved Photon Echo Measurements of Dynamics in Complex Solids: Organically Doped Inorganic Sol-Gel Glasses

Drew M. L'Esperance, Robert A. Crowell and Eric L. Chronister*
Department of Chemistry, University of California, Riverside, CA 92521

Introduction

We present the first time resolved measurements of homogeneous dephasing of organic dopants in an inorganic sol-gel glass. Organically doped sol-gel glasses have been synthesized and their dynamics investigated by time-resolved optical measurements (photon echo and fluorescence anisotropy) as well as frequency domain optical hole-burning measurements. The chromophore relaxation rate is determined from photon echo measurements at low temperature ($T = 1.35$ K), while thermal homogeneous dephasing mechanisms are investigated by an analysis of activated dephasing at elevated temperatures. Our results are contrasted with recent hole-burning experiments on doped sol-gel glasses. We observe low temperature homogeneous dephasing times, T_2, as short as 800ps for cresylviolet chromophores in a tetra-ethoxy silane (TEOS) and aluminosilicate (ASE) sol-gel glasses. ASE and TEOS glasses have been doped with rhodamine dyes, polyaromatics, cresylviolet, resorufin and a wide range of chromophores with fast nonradiative electronic relaxation rates. A partial list includes naphthalene (170ns), rhodamine 6G (2ns), Rose Bengal (500ps), stilbene (70ps), azulene (2ps), as well as photochemically active dopants such as quinizarin and chlorin.

The dynamics of the sol-gel matrix is analyzed in terms of a distribution of two-level systems which can model the local fluctuations characteristic of the non-equilibrium glassy state. The three dimensional covalent nature of the host polymer structure is different than a van der Waals solid of a frozen liquid (e.g. ethanol glasses) or that produced by the covalent chains of an organic polymer (such as PMMA). The net effect on the local structure is an increase in the barrier height for local structural changes. It is not yet clear whether the sol-gel barrier height distribution has a functional form different from the $V^{-1/2}$ dependence seen in other glassy systems, where V is the barrier height.

Time-Resolved Photon Echo Spectroscopy

Photon echo decays at 1.3 K yield homogeneous dephasing rates. At higher temperatures, thermally populated low frequency modes of the solid or low frequency modes of the dopant molecule induce thermally activated dephasing processes. Temperature dependent photon echo measurements in glasses are complex due to the dynamics of the glass as well as the dynamics of the chromophore within the glassy matrix. At low temperatures we observe an anomalous power law temperature dependence characteristic of glassy systems while at higher temperatures we see activated optical dephasing similar to that observed in doped crystalline systems. This activation energy should correspond to the energy of the low frequency mode that most strongly couples to the optically excited state. Thus, photon echo spectroscopy is a sensitive probe of how the chromophore interacts with the inorganic host matrix.

Time-resolved photon echo results have yielded the homogeneous lifetime of cresylviolet doped into TEOS and ASE glasses. We observe a photon echo decay time of t_{pe}=200ps for a cresylviolet doped ASE glass at 1.3 K, which corresponds to T_2=800ps. Our results for cresylviolet doped TEOS are similar except that the nonresonant intensity spike at zero delay relative to the long time decay is much greater for this system. One possibility for this difference is that vibronic transitions become more dominant in the TEOS matrix. Our photon echo results can be compared with hole-burning experiments performed on cresylviolet adsorbed onto porous silica at a similar temperature. A homogeneous linewidth of 8GHz (corresponding to a dephasing time of ~20ps) was reported for cresylviolet adsorbed onto porous silica at 1.7K yet our measured dephasing time is an order of magnitude longer. We conclude that the doped chromophores are mostly embedded in the bulk sol-gel solid and that they do not reside on the surface of a pore within the undensified xerogel.

Time-Resolved Fluorescence Anisotropy

In addition to relaxation and dephasing processes it is important to know the extent to which energy transfer occurs between chromophores. Energy transfer will be very important in determining the dynamics of mixed chromophore systems. Although optically excited dopant molecules can dispose of excess energy by either reemitting a photon or by thermalizing with lower frequency vibrations and lattice modes, this excitation can also be transfered between chromophores. When the electronic excitation is transfered between molecules with randomly oriented transition dipoles, fluorescence depolarization occurs. Since the energy transfer mechanism is sensitive to the proximity of neighboring molecules, time-resolved fluorescence depolarization is used to measure the average spatial distribution of dopant molecules within the host glass as well as the dynamics of energy transfer.

Our room temperature, time-resolved, fluorescence anisotropy decays have dynamical timescales longer than their dephasing rate from which we conclude that the dopant distribution function does not have a large number of close neighbors (as one would expect if they were confined in solvent pores), but that they appear to be randomly distributed as if in a frozen liquid. In addition, we see no evidence of dispersion at room temperature indicating that $\Delta\nu_{homo}$ is large enough to prevent donor trapping of the excitation to occur.

Spectral Hole Burning

At low temperatures we observe efficient nonphotochemical spectral hole-burning, however, this bleaching does not occur at higher temperatures (e.g. 77K) due to the broadened homogeneous linewidth. We use spectral hole burning measurements as an additional means to obtain the homogeneous dynamics of the chromophore molecules. A photon echo decay and the linewidth of a spectral hole are related by a Fourier transform (in the approximation that the structure of the glass is static), thus these experiments provide complimentary information. The spectral hole-burning process in sol-gel matrices may prove usefull as an alternative to organic polymer host systems in optical (and holographic) memory storage devices.

Acknowledgement

Acknowledgment is made to the Army Research Office (#27185-MS-SM) for support of this research.

PARAMETRIC STUDY OF CHIRAL COMPOSITES

Vasundara V. Varadan, Ruyen Ro, Vijay K. Varadan
Research Center for the Engineering of Electronic and Acoustic Materials
Department of Engineering Science and Mechanics
The Pennsylvania State University
University Park, PA 16802

ABSTRACT

The first complete experimental study of chiral composite material samples at microwave and millimeter wave frequencies is presented. The samples prepared are artificial chiral composites realized by embedding either right- or left-handed helices in an epoxy host material. The electromagnetic field behavior in materials with a handed microstructure is readily described by the use of the constitutive relations $\mathbf{D} = \varepsilon\,(\mathbf{E} + \beta\,\nabla \times \mathbf{E})$ and $\mathbf{B} = \mu\,(\mathbf{H} + \beta\,\nabla \times \mathbf{H})$, where ε and μ are the dielectric permittivity and magnetic permeability and β is a new material parameter called the chirality parameter which results from the non-central symmetric (lack of reflection symmetry) nature of chiral materials. In optics only rotation and dichroism are measured and these two measurements are insufficient to yield three complex material parameters. Another reason perhaps is the weak nature of naturally occurring chirality. The objectives of this study are:

(1) preparation of a family of chiral samples using miniature metallic right- and left-handed helices of various sizes and embedding them in an epoxy host material at different volume concentrations,

(2) measurement of the reflected and elliptically polarized, rotated transmitted field and the evaluation of ε, μ and β from such measurements,

(3) parametric study of the properties resulting from different spring sizes and the estimation of an optimal helix length that results in maximum power absorption.

A free-space measurement system is employed to characterize the chiral composites. The TRL calibration technique and time-domain gating method are used to enhance the experimental accuracy. One reflection and two transmission measurements, at two different polarization directions, are made for a normally incident, linearly polarized wave.

The artificial chiral samples characterized in this research are fashioned by embedding metallic helices into an epoxy material. In order to study the effects of the handedness of chiral inclusions on these properties, the samples are left-handed, right-handed, or racemic (equal amounts of left- and right-handed helices). Also, the samples are made with three different volume concentrations, 0.8%, 1.6%, and 3.2%, to investigate the quantitative effects of chiral inclusions. It is verified that the chirality parameter of the racemic samples has a value nearly equal to zero. For the handed samples, both the real and imaginary parts of the chirality parameter increase in magnitude as the concentration increases for the samples tested and at the frequencies of interest. In addition, the oppositely handed samples have approximately the same values with opposite signs for the real and imaginary parts of the chirality parameter. It is also observed that the usual complex permittivity and permeability have approximately the same values for the same concentration samples regardless of their handedness. The chirality parameter measured here is nearly four orders of magnitude higher than what can be inferred from optical measurements and molecular chirality. The Cotton effect, which combines the unequal absorption and unequal phase velocity of the transmission of left and right circularly polarized waves of chiral samples, has been observed in the region at which the maximum power absorption occurs.

The ratio L/λ_c of one turn length of the chiral inclusion to the wavelength in the chiral medium at frequencies where the maximum power absorption occurs has been calculated. It is shown that the frequencies and the corresponding ratios L/λ_c where the maximum power absorptions are observed are very close for all chiral samples. Hence, it is appropriate to use the averaged frequencies and the corresponding ratios L/λ_c of chiral samples of different handed and different volume concentrations to represent the characteristic frequency and the corresponding ratio L/λ_c of each sample, respectively. The ratio L/λ_c can be used to optimize absorption of chiral samples at different operating frequencies.

A complete experimental characterization of chiral composite samples was attempted in the 8-40 GHz frequency band and has been successfully completed. In addition valuable insight has been obtained on the relationship between microstructural chirality and the resulting macroscopic electromagnetic properties.

SOME MATERIAL ISSUES IN THE ACTIVE MATERIAL SYSTEMS

C. S. CHEN
YALE UNIVERSITY

Many active material systems currently developed, contain external active constituents including sensors, actuators, and avionics. Some active constituents can be fabricated into the composite materials as components, while others are chemically, mechanically, or thermally attached to the host materials. Therefore, performance of these active systems depends on how well the designed functions are transmitted from the sensors and actuators to the host materials. Physical interconnections now become important issues. We will discuss the effect of the induced incompatibility and heterogeneity within the active material systems. We will establish some tailoring guidelines for constructing the active material systems to provide the designed performance.

The sensors, acturators, and avionics are often made of materials different from that of the host. Therefore, many discontinuities in mechanical, electrical, and thermal properties may occur at the interconnections. Not only complicated stress field may occur within the active material systems during and after the fabrication, the systems is also highly heterogeneous.

We have evaluated the effect of the material heterogeneity. Based on the scale of the material inhomogeneity and magnitude of the discontinuities, signal distortion can become very large within the active material systems. By considering an orderly arrangement of the system, we have analyzed the behavior of such heterogeneous systems. Results show that discontinuity in mechanical stiffness alone can lead to large structural damping. The dissipative nature of some active constituents and host materials such as polymers can contribute further to the damping characteristics of the active material systems. On the other hand, we show that within certain frequency range, it is possible to construct a nearly damping-free heterogeneous system from the dissipative elements. The results can be used as a guide to fabricate the more desirable active material systems.

In general, we propose the construction of an interfacial region between the active constituents and host materials for reducing material discontinuities. With explicit examples, we introduce the desirable material variation and promising methods to improve the compatibility within the active material systems. Several practical routes will be presented to improve the transmission of responses from the active constituents to the host materials.

Potential tailoring in molecular scale to produce active material systems will also be briefly discussed. Examples including some liquid crystalline polymers and electrically conductive polymers will be summarized.

Mailing address: C. S. Chen
4305 Oldfield Dr.
Arlington, TX 76016

Tel: (817) 483-4532

Thermal-mechanical Properties of Thin-Film NiTi Deposited on Si

A. P. Jardine, Department of Materials Science
S.U.N.Y. at Stony Brook, Stony Brook, NY 11794-275

ABSTRACT

Thin-film NiTi has been developed by D.C. sputter deposition in a Ultra-high vacuum environment onto Si and Nb thin films predeposited on Si. The thin-film strains associated with the transformation were measured by monitoring the electrical resistance changes associated with the transformation by configuring the material as part of a resistance bridge. Strains were measured by examining the deflection of NiTi deposited on 2 micron thick Si substrates. Significant switching speed increases can be realised on downscaling to thin-film from bulk NiT. The speed of the phase transformation from the high temperature Austenitic B2 phase to the low-temperature martensitic B19 phase were monitored by timing the resistance change induced by applying current to the thin film. The correlation of these properties to processing conditions will be discussed.

Prospects for Electronic Component Distribution in Intelligent Structures

David J. Warkentin* and **Edward F. Crawley†**

Space Engineering Research Center
Department of Aeronautics and Astronautics
Massachusetts Institute of Technology
Cambridge, MA 02139

May 10, 1991

Submitted to the ADPA/AIAA/ASME/SPIE Conference on Active Materials and Adaptive Structures

November 5-7, 1991
Radisson Mark Plaza Hotel, Alexandria, VA

---

* Research Assistant; Student Member, AIAA

† Professor of Aeronautics and Astronautics; Associate Fellow, AIAA; Member, ASME

ABSTRACT

Introduction

An intelligent structure is one which uses a highly distributed, integral control system to achieve a desired control of thermal, shape, dynamic, surface, or other properties. Instead of conventional techniques involving only a few sensors and high authority actuators and a central processor to implement the control algorithm, an intelligent structure would incorporate a large number of distributed sensors, actuators, and processors in order to allow the implementation of continuous, hierarchical, or impedance algorithms. The adaptability of such a structure could offer advantages in the areas of reduced spillover into unmodelled modes and increased robustness in the case of component failure.

The application of intelligent structures depends on the development of suitable actuators, sensors, processors, and algorithms. In the area of strain control, substantial work has been performed to demonstrate the operation of actuators such as piezoelectrics and shape memory alloys (Crawley and de Luis, 1987, Rogers et al, 1989), and sensors such as fiberoptics and other strain sensors (Turner et al, 1990), which would be suitable for embedding in an intelligent structure. Algorithms for the control of structures with large numbers of highly distributed sensors and actuators have also been suggested (Young, 1983, Hall et al, 1989). Recently, work has also been done to demonstrate the feasibility of physically embedding the electronic components used to implement the control algorithms (Warkentin and Crawley, 1991).

The actual application of intelligent structures technology will depend on the realization of tangible benefits and the existence of sufficiently powerful embeddable components. The purpose of this paper is to suggest some of those potential benefits and to demonstrate the implementation of structural control with single chip microcomputer.

Benefits of Distribution and Embedding

A substantial amount of circuitry is required to support the function of the actuators and sensors. Sensors and actuators must be powered, and signals to and from the transducers must be conditioned and perhaps digitized. The control processing, either in analog, digital, or hybrid form, must also be performed by integrated circuits. The components which perform these functions may be centrally located as in traditional control systems, or distributed to a greater or lesser degree.

Two possible options for the placement of distributed components are surface mounting and embedding, both of which have advantages and disadvantages. These issues are identified and briefly discussed, after which the potential advantages of a distributed, embedded control system are investigated in more detail. This is done by considering the distribution of different levels of functionality and examining how this affects the resulting number of communications lines run into the structure, the number of chips required, and the computational load and speed of the control system. A comparison is drawn between a conventional centralized system and a hierarchic one which appears to naturally lend itself to implementation in an intelligent structure.

Some amount of physically distributed analog circuitry would allow signal conditioning near the sensors and actuators. It is found that for a structure equipped with a large number of sensors and actuators, a reduction in the number of communications lines leading out of the structure from O(n) to O($\log_2 n$) could be achieved by implementing a digital bus system, requiring the futher distribution of A/D and D/A conversion functions as well as the digital bus interface.

Increasing the degree of distribution to include some digital control processing, faster control loop speeds could be achieved through the implementation of a hierarchical control algorithm. In an algorithm such as that presented in (Hall et al, 1989), model reduction at the global level and block diagonalization of the local control may be used with parallel processing at the local level to produce control loop periods with an O(n) dependence on system size, as compared to the O(n^2) dependence of a traditional centralized system.

Demonstration of Single Chip Microcomputer Control

In contrast with conventional computer control systems which employ many specialized chips to perform the various functions of signal conversion, computation, storage, and communication, the embedded control system of an intelligent structure would greatly benefit from the use of processors which could combine as many functions as possible on a single chip. This would greatly simplify the interconnections as well as reduce the adverse effect on structural integrity.

A microcomputer is identified which incorporates on a single chip many components (e.g., A/D and D/A conversion, high speed communication, memory) which are traditionally distributed among many separate chips. It is apparent that a wide variety of functions can now be combined to greatly reduce the number of chips required for a complete control system; combined with the ability to embed electronic components in general, this puts the development of true intelligent structures within reach. The presence of communication ports on these single chip microcomputers makes possible their incorporation into a network of such devices. This system could yield an efficient distribution of computational load, perhaps in the implementation of a hierarchic or decentralized controller.

A commercially available single chip microcomputer is shown to be capable of performing a simple control task with relatively little extra circuitry. The tip displacement disturbance response of a cantilever beam with piezoelectric actuators was reduced by 20 dB from the open-loop level, and damping ratios of 31%, 4%, and 11% were achieved in the first three modes, demonstrating a substantial degree of control effectiveness. This achievement, together with the other functions present on the chip, support the conclusion that with a still greater level of functional integration, leading to a further reduction in the number of components and increased suitability for physical embedding, such a device or a near derivative might well be able to perform as a local controller in the hierarchic control architecture of an intelligent structure.

References

Crawley, E. F. and de Luis, J., 1987, "Use of Piezoelectric Actuators as Elements of Intelligent Structures," *AIAA Journal*, Vol. 25, No. 10.

Hall, S. R., Crawley, E.F., and How, J., 1989, "A Hierarchic Control Architecture for Intelligent Structures," submitted to *AIAA Journal of Guidance, Control, and Dynamics*.

Rogers, C. A., Liang, C., and Barker, D. K., 1989, "Dynamic Control Concepts Using Shape Memory Alloy Reinforced Plates," Smart Materials, Structures, and Mathematical Issues, Technomic Publishing Co.

Turner, R. D., Valis, T., Hogg, W. D., and Measures, R. M., 1990, "Fiber-Optic Strain Sensors for Smart Structures," *Journal of Intelligent Material Systems and Structures*, Vol. 1, No. 1.

Warkentin, D. J., and Crawley, E. F., 1991, "Embedded Electronics for Intelligent Structures," *Proceedings, 32nd AIAA/ASME/ASCE/AHS/ASC Structures, Structural Dynamics, and Materials Conference*, Baltimore, MD.

Young, K. D., 1983, "An Application of Decomposition Techniques to Control of Large Structures," *Proceedings, 4th VPI & SU/AIAA Symposium on Dynamics and Control of Large Space Structures*.

Photonic and Electronic Control of Embedded Bragg Reflection Sensors for Smart Structures Applications

Stanley Reich

ABSTRACT

Requirements for the next generation aircraft have pushed the conventional use of aerospace materials to the limit of their performance. The development of composites, fiber optics, photonics, miniaturized antennas, and processing has progressed sufficiently to permit their integration into airframe and space structures. Advances in the development of composite structures has led to the embedment of sensors which can provide the ability to monitor operating conditions or sense external threats. The ability to sense and dynamically control a material's structural characteristics in "real-time" opens the possibility of providing vastly expanded flight envelopes in future aircraft designs, while avoiding the need for fundamental improvements in material characteristics. These systems, commonly referred to as "Smart Structures", can be viewed as the integration of sensor, distribution and support technology into materials to form a technically workable, yet economic system, to perform real time measurement and control of the dynamic events.

When information such as strain, temperature, pressure, etc. is required from multiple sensors which are discretely located, or distributed along the structure severe design constraints are placed on the control and distribution of the sensor system. Additionally, the system becomes more complex when the design is expanded to include the collection of aircraft usage data (presently an expensive, time consuming task) and the detection of battle damage along with feedback to the pilot, or central computer, for remedial action, or flight abortion.

The specifics of a photonic/electronic system to provide the requirements described, along with an approach to the distribution, processing and control of a multiplicity of length-limited Bragg reflection sensors will be presented. The system approach and monitoring components for each application (ie., fatigue monitoring, battle damage, crack detection and data collection) will be discussed.

Length limited Bragg reflection filters acting as sensor elements each have a specific frequency resonance which is modified, or detuned by the application of an external physical effect. The electronic/photonic processing system is designed to sense the resonance shift, convert it to a value proportional to the magnitude of the applied phenomena and distribute or store the data as required. Included in the presentation will be some trade-offs between employing a processing approach with a tuned laser source, or a broadband source and a monochromater receiver. Additionally, network considerations in terms of how they relate to usage data collection, and detection and control of battle damage will be presented.

COMBINING FIBER OPTICS, RADIO FREQUENCY AND TIME DOMAIN REFLECTOMETRY TECHNIQUES FOR SMART STRUCTURE HEALTH MONITORING

J. S. SCHOENWALD, SCIENCE CENTER
R. H. MESSINGER, SPACE SYSTEMS DIVISION
ROCKWELL INTERNATIONAL CORPORATION

We are in the process of developing a fiber optic sensor system for resolving strain levels of 10^{-4} with differences of 10^{-6} between closely spaced regions within composite samples. The phase amplitude response of swept ultra-high frequency (UHF) laser amplitude modulated light transmitted through composite embedded optical fibers is being used to measure strain in test structures subject to tensile loading. The objective is to identify differences in resulting strain fields due to the presence or absence of defects.

The technique borrows from earlier work [1] in that the change in phase delay of the transmitted UHF modulation signal is a measure of strain in the fiber embedded in the structure. Now, however, we seek to measure small differences in strain along a single direction between adjacent regions of the test structure. This is accomplished by embedding several optical fibers of different penetration lengths, which are simultaneously illuminated by a single modulated laser source through a 3x3 fiber optic coupler. Each fiber is terminated with a cleaved, metalized end, causing the optical signal to reflect back efficiently through the optical splitter to a detector. The phase-magnitude characteristics of the reflected signals from the three fibers are determined by a UHF network analyzer in the transmission measurement mode. A fast Fourier transform (FFT) performed automatically by the network analyzer yields a transmission-mode time domain image of the reflection from the fiber ends, which appear as distinct reflections. Time and frequency domain information are simultaneously available. Using a signal subtractions process, small strains are detected using differencing techniques, and information about the nature of the strain field should make identification of defect generation possible. Figure 1 is an illustration of the basic elements of the system.

We are testing the sensor system on composite specimens with and without embedded teflon inserts to simulate defects. A description of the computer controlled load testing and data acquisition system will be presented, together with results obtained so far.

This work is supported by the Rockwell International Independent Research and Development Program.

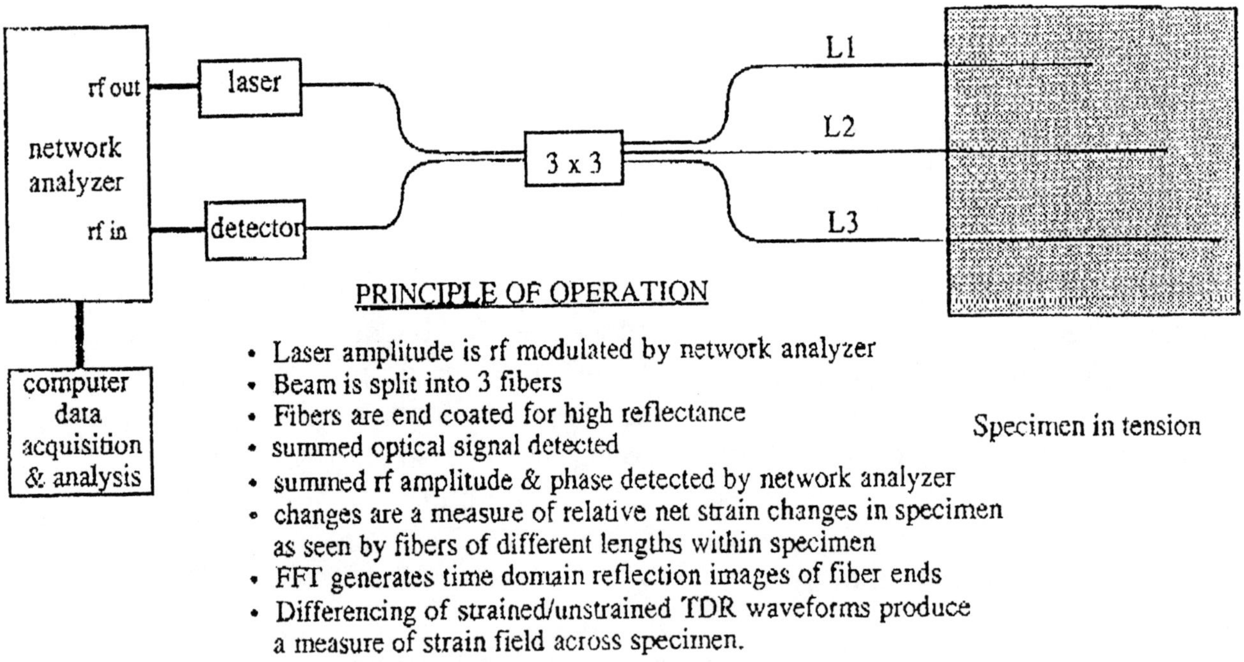

Figure 1. Fiber optic rf interferometric/time domain reflectometry differential strain sensor.

MULTIPLEXED STRAIN SENSORS AND ACTUATORS FOR EMBEDMENT IN ACTIVELY DEFORMED STRUCTURES

S.C. Jacobsen, M.G. Mladejovsky, M. Rafaelof, D.K. Backman,
Center for Engineering Design
University of Utah, Salt Lake City Utah 84112

Abstract

Smart materials (SM) are currently being considered for many applications in which the dynamic control of structural shape, or surface properties, can favorably alter boundary interactions between material surfaces and fluid or electromagnetic fields. Clearly, if SM's were available for use, their application could improve the performance of many systems.

In a number of ways the building blocks of SM's are like robots and other machines capable of producing controlled movement. SM's consist of combinations of the five machine-element subgroups: (1) structures, (2) actuators, (3) sensors, (4) controllers and (5) conduits. In SM's, however, problems are more difficicult in three ways: (a) It is planned the movement be achieved through material distortion rather than relative displacement at joints. (b) Due to the distributed nature of movements, many more sensors and actuators will be necessary, and (c) It is planned that sensors and actuators be embedded to become a part of the material.

These three requirements lead to substantial, and numerous, problems which must be solved if SM's are to become practical components in real systems. Example problems include: (1) Achieving strength and life in materials undergoing large and repeated strains and which include embedded, non load bearing foreign objects. (2) Designing and producing sensor systems capable of reliable operation, embedded in hostile, flexing environments and interconnected by high bandwidth nets capable of monitoring distributed material distortion. (3) Developing high power density actuator systems capable of generating high force levels and bandwidth and also operating close to or within deforming structures. (4) Defining control systems capable of blending command and sensor information in order to supervise many actuators, non-colocated with their sensors. (5) Designing conduits power transfer and communication between sensors, actuators and controllers .

Efforts at the CED are currently focused on solving the sensor problem which can be stated as: How to embed a large number of small interconnected sensors throughout a flexing material in which large strains are being generated and simultaneously maintain acceptable performance, reliability and economy. The approach at the CED is to develop a system of intelligent, multiplexed, mechanical transducers based on Micro Electro Mechanical Systems (MEMS) technology.

The paper will discuss an approach to transducer design based on the idea that the relative movement of very small, appropriately patterned conductive field emitting substrate, can be precisely measured by a silicon chip which contains an array of FET-based field detectors in close proximity to that emitter layer. The FET-based detectors coexist in the silicon chip with other circuit elements which can amplify, digitize and multiplex (onto a common three wire bus) signals produced by detector/emitter interactions. By appropriate arrangement of emitter and detector arrays, transducers are being designed to achieve desired tradeoffs between resolution, dynamic range, bandwidth, sensor population size, and other features. With detector and associated circuitry on the same chip, transducers are being designed to implement local intelligence necessary for functions such as: self calibration, correction for subsystems failure, interaction with adjacent sensors.

Two figures below show electrostatic field emitter and detector chips which function as the basis for digital, multiplexed strain and displacement transducers which are powered and monitored via a common three-wire bus.

Two figures below show digital, multiplexed strain and rotation transducers which are powered and monitored via a common three-wire bus.

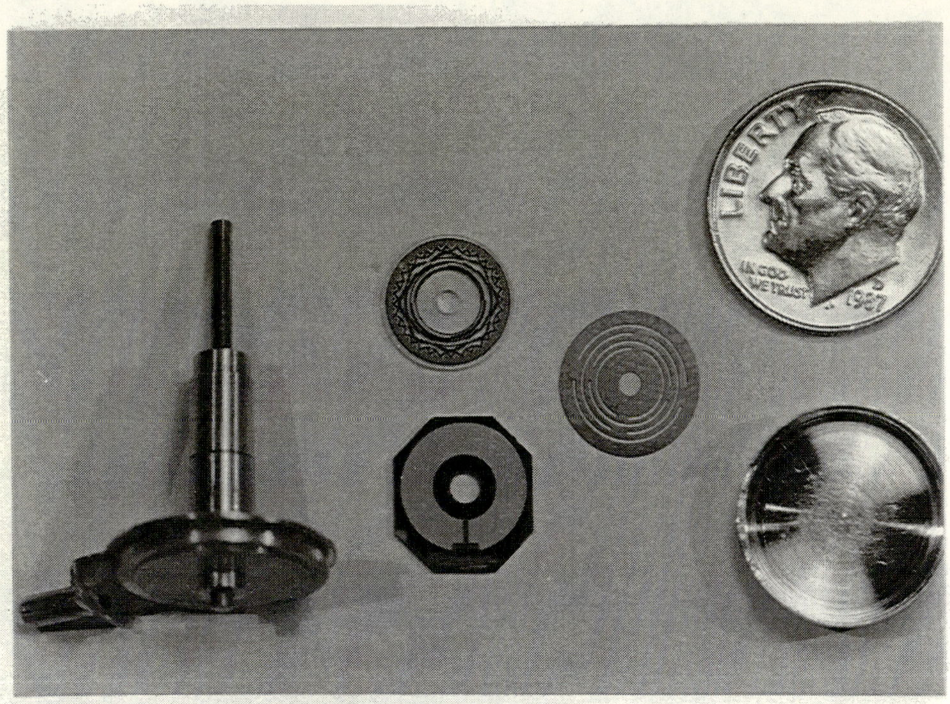

Figure 1 - Elements of a rotation sensor which uses: (a) A CMOS chip, complete with on-board supporting circuitry, (b) a patterned electrostatic field emitter and (c) supporting bearings and structures as the basis for a thirteen bit, digital, absolute and multiplexed measurement system.

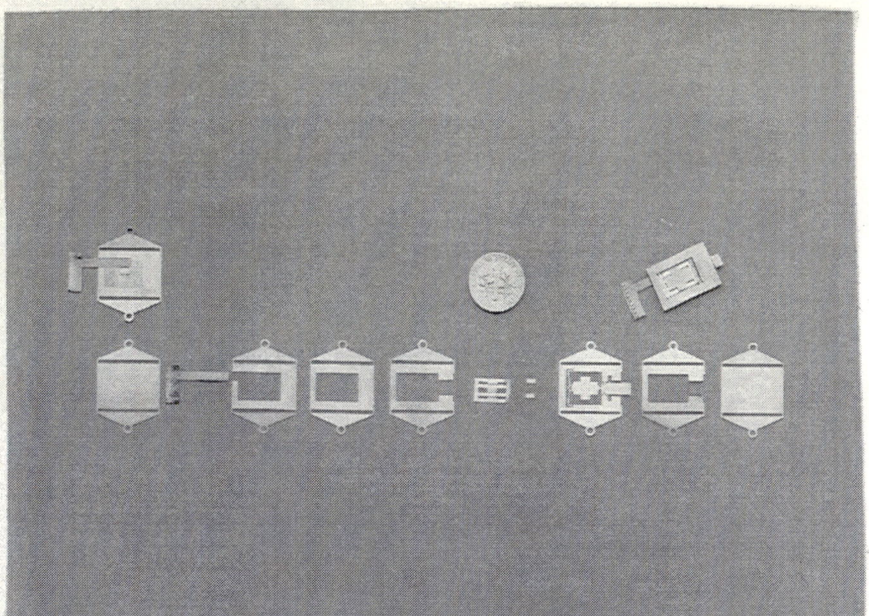

Figure 2 -Elements of a strain sensor which uses: (a) A CMOS chip, complete with on-board supporting circuitry, (b) a patterned electrostatic field emitter and (c) supporting laminated suspension structures as the basis for a digital, absolute and multiplexed strain measurement system.

CONTROL OF SMART TRAVERSING BEAMS

A.BAZ, S.POH, J.RO and J.GILHEANY

Mechanical Engineering Department
The Catholic University of America

ABSTRACT

Shape memory Nickel-Titanium alloy (NITINOL) is utilized in developing smart traversing beams for use in critical applications where weight and deflection are of utmost importance such as the launching and crossing operations of long support bridges. NITINOL fibers are placed at critical locations inside the traversing beams in order to sense and control the static deflection characteristics of the beams. With such capabilities, it would possible to manufacture light weight and long span support bridges which can be easily handled in short emplacement time.

The smart traversing beam replaces the conventional traversing beam , traditionally manufactured from heavy metallic cross sections, which is launched from a truck provided with a massive counterbalance weight to resist the tipping moment generated by the dead weight of the extended beam. The excessive weight of the metallic traversing beam imposes severe stress and deflection constraints on the maximum length of its span which, in turn, limits its use to narrow gap support bridges. To avoid these problems, the smart beam concept offers a viable alternative as the metallic beam is replaced by an actively controlled light weight composite beam. The smart beam will operate with smaller counterbalance weights over wider gaps.

The smart beam relies in its operation on two sets of NITINOL wires embedded in the composite fabric of its thin-walled hollow cross section. The first set consists of large diameter wires to serve as actuators whereas the second set of wires is of much smaller diameter to act as sensors. When the sensors' set detects deflections, during launching or cross over operations, a signal is generated which is proportional to the associated load acting on the beam. This signal will activate a proportionate number of actuator wires past their transition temperature. The resulting phase transformation forces times the distance to the neutral axis of the beam generates a moment which counterbalances the gravitational moment acting on the beam. In this manner, the beam deflection and stresses will be reduced considerably.

A finite element model is developed to describe the static and dynamic behavior of a thin-walled squared-cross section cantilevered beam when subjected to a concentrated load at its free end. The model accounts for the load generated by the NITINOL wires as they undergo their phase transformation.

A closed-loop computer-controlled system is built to demonstrate the feasibility of the concept and validate the developed finite element model. The system is used to control the deflection of a 30 cm long hollow beam that has a 5cmx5cm cross section. The sides of the beam are made of photoelastic plate (PS-4 from Measurements Group, Inc., Raleigh, NC) which is 0.625 cm thick. The top and bottom faces are made of fiberglass-polyester composite. In the top face, a set of 0.55 mm diameter NITINOL actuators are embedded where as in the bottom face a single loop of 0.15 mm diameter NITINOL sensor wire is embedded. The actuators are fixed from one end to the beam free end and their other end is connected to a load cell to monitor the load generated by the wires as they undergo their phase transformation upon activation. The sensor wire is connected to a Wheatstone bridge and acts as a strain gage. When a load is applied to the beam, the sensor will detect the developed strains. The strain signal is amplified and sent to the computer to compute a control action based on an ON-OFF controller with a dead band. The control action is sent to a power amplifier to heat the NITINOL actuator set in order to bring the beam back to its undeflected position. The stress distribution inside the beam are monitored continuously by placing the beam inside a polariscope and impact it with diffused light. The generated fringes are recorded by the computer and by a video camera during the different phases of operation. The fringes, which represent the contours of constant shear stress, are compared with the results of the finite element model.

The results obtained suggest the potential of the smart traversing beam and demonstrate successfully the effectiveness of the NITINOL wires in sensing and controlling the deflection of that class of beams. [Work supported by Fort Belvoir R D & E center]

Shape Control of a Three-Dimensional Composite Elastica with Embedded Shape Memory Wires

by

Iradj G. Tadjbakhsh and **Dimitris C. Lagoudas**
Department of Civil Engineering
Rensselaer Polytechnic Institute, Troy, N.Y. 12180

Abstract

A consistent theory of a one-dimensional elastic continuum undergoing a three dimensional motion (elastica) has recently been developed by I.G. Tadjbakhsh primarily for engineering applications. The rod is assumed to be extensible, and naturally curved and twisted in its unstrained state with measures of bending and torsion defined as the difference in the curvature vector. For a hyperelastic material a frame invariant strain energy function is defined in terms of the curvature strain and the center line extension. Compared to classical theories extensibility is an important ingredient for SMA reinforced elastic rods.

A number of thin shape memory wires is assumed to be embedded in the cylindrical rod parallel to the reference center line or in a helical configuration. Assuming some initial prestraining of the SMA wires before they are perfectly bonded to the rod, raising the temperature beyond A_f will induce an eigenstrain in the surrounding matrix due to the martensitic-austenitic phase transformation in the wires and the constraint of the shape recovery imposed by the elastica. The net force on any cross section will cause resultant bending moments and for helically imbedded SMA wires it will also cause a resultant twisting moment.

The shape recovery and the stiffness change due to the phase transformation will enter into the constitutive modeling of the elastica by the replacement of the total curvature strain by the elastic curvature strain and of the total center line extension by the elastic extension, subtracting therefrom the contribution of the shape memory effects from the elastic energy. It is important to mention that for the nonlinear case the effect of the imposed elastic strain on the shape memory induced strain should be included. The magnitude of this interaction depends on the placement configuration of the SMA wires with respect to the cross section of the rod and additionally on the magnitude of the superimposed elastic response. It is stressed that this is merely the effect of nonlinearity and is not related to stress induced martensitic transformation or superelasticity effects.

The proposed theory of shape memory composite rods in three dimensions is tested by predicting the positioning of the free end of a cylindrical composite tube in certain prescribed spatial positions. The quasi-static problem is considered for various reference configurations as well as different boundary conditions. Optimum design parameters such as the placement of the SMA wires with respect to the rod cross section, the geometry of the cross section and the initial prestraining of the SMA wires are obtained for given movability constraints of the rod as well as positioning requirements for certain points or slopes of the center line of the rod.